# Fundamentals of Astronomy

# Fundamentals of Astronomy

## Second Edition

Cesare Barbieri and Ivano Bertini

CRC Press
Taylor & Francis Group
Boca Raton  London  New York

CRC Press is an imprint of the
Taylor & Francis Group, an **informa** business

Second edition published 2021
by CRC Press
6000 Broken Sound Parkway NW, Suite 300, Boca Raton, FL 33487-2742

and by CRC Press
2 Park Square, Milton Park, Abingdon, Oxon, OX14 4RN

© 2021 Taylor & Francis Group, LLC

First edition published by CRC Press 2006

CRC Press is an imprint of Taylor & Francis Group, LLC

ISBN: 978-0-367-25349-3 (hbk)
ISBN: 978-0-367-25320-2 (pbk)
ISBN: 978-0-429-28730-5 (ebk)

Typeset in Times
by codeMantra

*To Our Families*

# Contents

Preface..........................................................................................................................xiii
Acknowledgments........................................................................................................xv
Authors.......................................................................................................................xvii

**Chapter 1**   Spherical Astronomy.............................................................................1

    1.1    Elements of Plane Trigonometry.................................................1
    1.2    Some Properties of Plane Triangles............................................4
    1.3    Elements of Spherical Trigonometry...........................................5
    1.4    Cartesian and Polar Coordinates.................................................7
    1.5    Terrestrial Latitude and Longitude on the Spherical Earth...........9
    1.6    Elements of Vector Calculus.....................................................11
    Notes..............................................................................................13
    Exercises........................................................................................14

**Chapter 2**   First Notions on Astronomical Reference Systems.......................17

    2.1    The Alt-Azimuth System..........................................................18
    2.2    The Hour Angle and Declination System..................................25
    2.3    The Equatorial System..............................................................26
    2.4    Telescope Mounts.....................................................................29
    2.5    The Ecliptic System..................................................................29
    2.6    The Galactic System.................................................................31
    2.7    Other Systems...........................................................................32
    Notes..............................................................................................32

**Chapter 3**   Transformations of Coordinates....................................................35

    3.1    Transformations by Matrix Rotation.........................................35
    3.2    Transformations by Spherical Trigonometry.............................39
    3.3    Some Examples and Applications..............................................41
    Exercises........................................................................................44

**Chapter 4**   First Notions on the Movements of the Earth and the Astronomical Times.............47

    4.1    The Movements of the Earth.....................................................47
    4.2    The Sidereal Time $ST$..............................................................48
    4.3    The Solar Time $T_\odot$ and the Equation of Time $E$..................48
    4.4    The Universal Time UT.............................................................52
    4.5    The Tropical Year and the Rates of $ST$ and UT.......................52
    4.6    The Year and the Julian Calendar..............................................53
    4.7    The Besselian Year or Annus Fictus.........................................53
    4.8    The Seasons...............................................................................54
    4.9    The Julian Date.........................................................................55
    Notes..............................................................................................55
    Exercises........................................................................................56

**Chapter 5**   The Movements of the Fundamental Planes .............................................................. 59

    5.1   First Dynamical Considerations .................................................................. 59
    5.2   The Precession of the Equinox ................................................................... 60
    5.3   The Movements of the Fundamental Planes ................................................ 62
    5.4   First-Order Effects of the Precession on the Stellar Coordinates .................. 63
    5.5   The Nutation ........................................................................................... 68
    5.6   Approximate Formulae for General Precession and Nutation ...................... 70
    5.7   Newcomb's Rotation Formulae for Precession ............................................ 71
    5.8   Precession and Position Angles .................................................................. 73
    5.9   Solar System Objects ................................................................................ 73
    Notes  ...................................................................................................... 74
    Exercises ................................................................................................... 75

**Chapter 6**   Dynamics of Earth's Rotation ........................................................................... 77

    6.1   Newton's Lunisolar Precession .................................................................. 77
    6.2   The Lunisolar Torque ............................................................................... 79
    6.3   The Precessional Potential ......................................................................... 81
    6.4   The Earth's Free Rotation ......................................................................... 84
    Notes  ...................................................................................................... 91
    Exercise ................................................................................................... 91

**Chapter 7**   Aberration of Light ........................................................................................... 93

    7.1   The Solar Aberration ................................................................................ 94
    7.2   The Annual Aberration ............................................................................. 96
    7.3   The Lorentz Transformations .................................................................... 98
    7.4   Effects of Annual Aberration on the Stellar Coordinates ............................ 99
    7.5   The Diurnal Aberration ............................................................................. 102
    7.6   Planetary Aberration and Planetary Perturbations ..................................... 103
    7.7   The Gravitational Deflection of Light ........................................................ 104
    Notes  ...................................................................................................... 106

**Chapter 8**   The Parallax .................................................................................................... 109

    8.1   The Trigonometric Parallax ...................................................................... 109
    8.2   The Diurnal Parallax ................................................................................ 110
    8.3   Solar and Lunar Parallaxes ....................................................................... 115
    8.4   The Annual Parallax ................................................................................. 116
    8.5   Secular and Dynamical Parallaxes ............................................................. 120
    Notes  ...................................................................................................... 121
    Exercises ................................................................................................... 122

**Chapter 9**   Radial Velocities and Proper Motions ............................................................... 125

    9.1   Radial Velocities ...................................................................................... 126
    9.2   Proper Motions ........................................................................................ 129
    9.3   Variation of the Equatorial Coordinates .................................................... 132
    9.4   Interplay between Proper Motions and Precession Constants ...................... 133
    9.5   Astrometric Radial Velocities ................................................................... 134
    9.6   Apex of Stellar Motions and Group Parallaxes .......................................... 135

9.7   The Peculiar Motion of the Sun and the Local Standard of Rest.................. 137
9.8   Secular and Statistical Parallaxes.......................................................... 139
9.9   Differential Rotation of the Galaxy and Oort's Constants ........................... 140
Notes .......................................................................................................... 141
Exercises..................................................................................................... 142

**Chapter 10**  The Astronomical Times, the Atomic Time and the Earth Rotation Angle ........... 145

10.1   The Sidereal Time *ST*......................................................................... 145
10.2   The Solar Time $T_\odot$.............................................................................. 146
10.3   The Year ............................................................................................ 147
       10.3.1   Tropical Year ........................................................................ 147
       10.3.2   Besselian Year B or Annus Fictus ........................................... 148
       10.3.3   Sidereal Year ........................................................................ 148
       10.3.4   Anomalistic Year.................................................................... 149
       10.3.5   Draconitic (or Eclipse) and Gaussian Years ............................. 149
10.4   The Dynamical Ephemeris Time ET ....................................................... 149
10.5   The Atomic Time ................................................................................ 152
10.6   The Earth Rotation Angle (ERA)............................................................ 153
Notes .......................................................................................................... 155
Exercise ...................................................................................................... 155

**Chapter 11**  The Terrestrial Atmosphere ............................................................................ 157

11.1   The Vertical Structure of the Atmosphere ............................................... 157
11.2   The Refraction .................................................................................... 160
11.3   Effects of Refraction on the Apparent Coordinates ................................. 164
11.4   The Chromatic Refraction of the Atmosphere ......................................... 165
11.5   Relationships between Refraction Index, Pressure and Temperature .......... 166
11.6   Scintillation and Seeing....................................................................... 168
Notes .......................................................................................................... 170

**Chapter 12**  The Two-Body Problem ................................................................................. 173

12.1   The Barycentric Treatment................................................................... 173
12.2   The Gravitational Attraction ................................................................ 177
12.3   The Relative Movement........................................................................ 179
12.4   Planetary Masses from Kepler's Third Law............................................. 183
12.5   Escape Velocity .................................................................................. 184
12.6   Some Considerations on Artificial Satellites........................................... 185
Notes .......................................................................................................... 186
Exercise ...................................................................................................... 187

**Chapter 13**  Orbital Elements and Ephemerides................................................................. 189

13.1   Kepler's Equation ............................................................................... 189
13.2   Ephemerides from the Orbital Elements ................................................ 192
13.3   Planetary Configurations and Titius–Bode Law ...................................... 194
13.4   Orbital Elements from the Observations................................................. 196
13.5   Application to Visual Binary Stars........................................................ 198
Notes ..........................................................................................................204

**Chapter 14**  Elements of Perturbation Theories................................................205

    14.1   Perturbations of the Planetary Movements .................................205
    14.2   Planet Plus Small Moon ..............................................................209
    14.3   Case Earth–Moon........................................................................210
    14.4   The Lunar Month and the Librations .........................................212
    14.5   The Case Planet Plus Planet.......................................................214
    14.6   The Restricted Circular Three-Body Problem ............................216
    14.7   A Non-Spherical Body Plus a Small Nearby Satellite ...............219
    14.8   Other Interesting Cases ..............................................................222
    Notes  .....................................................................................................223
    Exercise..................................................................................................224

**Chapter 15**  Eclipses, Occultations and Transits.............................................225

    15.1   Moon's Phases ............................................................................225
    15.2   Conditions for the Occurrence of an Eclipse .............................226
    15.3   Solar Eclipses .............................................................................227
    15.4   Lunar Eclipses ............................................................................229
    15.5   Besselian Elements and Magnitude of the Eclipse.....................230
    15.6   Number and Repetitions of Eclipses ..........................................231
    15.7   Stellar Occultations ....................................................................233
    15.8   Transits of Exoplanets ................................................................235
    Notes  .....................................................................................................238
    Exercises................................................................................................239

**Chapter 16**  Elements of Astronomical Photometry ........................................241

    16.1   Visual Magnitudes......................................................................241
    16.2   Extension of the Definition of Magnitude .................................243
         16.2.1   The Reflectivity of the Optics and Transmissivity of Filters ...........243
         16.2.2   The Efficiency of the Detectors...............................246
    16.3   Extension by the Earth's Atmosphere.........................................249
    16.4   The Black Body ..........................................................................252
    16.5   Color Indices and Two-Color Diagrams.....................................255
    16.6   Calibration of the Apparent Magnitudes in Physical Units.........257
    16.7   Apparent Diameters and Absolute Magnitudes of the Stars .......258
    16.8   The Hertzsprung–Russell Diagram.............................................261
    16.9   Interstellar Absorption and Polarization .....................................263
    16.10 Extension to the Bodies of the Solar System..............................265
    16.11 Radiation Quantities....................................................................268
    Notes  .....................................................................................................269
    Exercises................................................................................................270

**Chapter 17**  Elements of Astronomical Spectroscopy .....................................273

    17.1   Spectroscopic Techniques ..........................................................274
    17.2   The Analysis of the Spectral Lines ............................................277
    17.3   Detailed Balance and the Boltzmann Equation ..........................281
    17.4   The Saha Equation ......................................................................284
    17.5   Spectral Classification of Stars and the Abundance of the Elements ...........288
    17.6   The Harvard and the MK Classification Schemes ......................291

17.7 Very Low-Temperature Stars ..................................................................294
17.8 Relationship between the MK Classification and Photometric Parameters ....294
17.9 Spectra of Peculiar Stars ....................................................................295
17.10 Spectra of Solar System Objects ........................................................296
Notes ..........................................................................................................300

**Bibliography** ....................................................................................................303
References ....................................................................................................306
Web Sites ....................................................................................................314
**Index** ..........................................................................................................319

# Preface

The first edition of *Fundamentals of Astronomy* in 2006 originated from the introductory courses to astronomy and astrophysics given at the University of Padova, Italy. The present edition, written more than 12 years after the first, largely maintains the original structure, providing a broad overview of several classical topics, from spherical astronomy to celestial mechanics, closing with two chapters on the elements of astronomical photometry and spectroscopy. Several concepts have been clarified; some imprecision has been corrected; and the numerical values, explanatory notes and exercises inserted at the end of many chapters have all been updated to contemporary values.

The topics have been selected based on the belief that classical concepts deserve a basic role in the education of students. Spherical astronomy, for instance, is presented in detail, with a recollection of millennia old ideas. To some students, this material can appear unduly formal and perhaps obsolete. However, the possibility of performing exact measurements from modern ground optical and radio telescopes and from space satellites has revived a wide interest in astrometry. The precise data that were provided by meridian circles, such as the Carlsberg, those of the Very Long Baseline Interferometry (VLBI) at radio wavelengths and of the Hipparcos and GAIA satellites, require a careful treatment of the very fundamentals of coordinates, motions and time measurements. Moreover, the flourishing multi-wavelength astrophysics requires the best knowledge of the positions of the emitting regions, in order to identify the different regions where radio waves or optical photons or X and gamma rays originate.

This book, while broad in coverage, does not aim to reach the level of advanced theoretical knowledge or the highest numerical accuracy sometimes required by these contemporary applications. In particular, general relativity will be mentioned only in specific occasions. However, we hope that this introductory text, with the indication of its limits when appropriate, will stimulate young researchers to go beyond the present treatment. Several important topics had to be left out. Such decision was facilitated by the availability of excellent textbooks at all levels. The Bibliography section contains a mixture of old and new books, selected by personal taste, again with the intention of stimulating further reading.

During the years between the first and second editions, a great revolution has occurred in the way science is made available; the Internet gives easy access to both well-established and developing knowledge. Many research and educational institutions provide excellent websites where the reader can follow the unfolding of experiments and observations, almost in real time. Navigation of the Internet takes the reader through an incredible number of sites, some of which are very good, but others are very bad and even deceiving. Therefore, we have indicated in the Web Sites section, those sites we consider reliable and valuable to science and education; again, this section does not attempt to present a complete list. The student will quickly develop the capability to judge what is most useful. The References section contains a selection of papers published in the principal scientific journals.

It must be emphasized that after the first edition of this book was published, important revisions to astronomical concepts and quantities took effect. Some of these changes have been implemented in the present edition, using the *Astronomical Almanac* (see the Web Sites section) and the third edition of its *Explanatory Supplement* (Urban and Seidelmann in the Bibliography section) as guidelines. Published in 2012, this third edition contains major changes to basic concepts, but it is still pre-GAIA.

Furthermore, important revisions to the basic constants of the International System (SI) have been enforced since May 20, 2019 (World Metrology Day), according to decisions taken by the member states of the Bureau International des Poids et Measures (BIPM) to change the definitions of the kilogram, ampere, kelvin and mole. See the Web Sites section for the BIPM resolutions.

# Acknowledgments

Several colleagues kindly provided comments, unpublished material and updated information. We wish to thank in particular Dainis Dravins, Craig Mackay, Roberto Mignani, Giampiero Naletto, Roberto Nesci, Paolo Ochner, Roberto Ragazzoni, Andrea Richichi, Costantino Sigismondi, Alessandro Siviero, Gert Weigelt and Luca Zampieri. Thanks are due to Tommaso and Silvia Occhipinti for the kind help with the figures.

Moreover, we wish to thank all the students who used the first edition, discovered errors and asked for better clarity. As educators, we warmly hope that some of the readers will be able to bring personal contributions to the fascinating field of astronomy and astrophysics.

# Authors

**Cesare Barbieri** is a Professor Emeritus of astronomy at the University of Padova, Italy. He directed the Astronomical Observatory of Padova and the construction of the 3.5 m Telescopio Nazionale Galileo (Canary Islands). He held several positions in the European Space Agency and European Southern Observatory advisory committees, with an active role in several space missions (the Halley Multicolor Camera on board GIOTTO to comet Halley, the Faint Object Camera on the Hubble Space Telescope, the imaging system OSIRIS on board ROSETTA to comet Churyumov–Gerasimenko).

His main research themes are astrometry, photometry and discovery of quasi-stellar objects; telescopes and their instrumentation for ground and space; spectrophotometry and discovery of asteroids; and the exosphere of Mercury. More recently, he became interested in quantum optics applied to astronomy. He has organized several international conferences and exhibits. He has received the NASA Group Award for the FOC/HST, the ESA recognition awards for GIOTTO and ROSETTA, and the Gold Medal of Italian Ministry for Public Education.

The Main Belt asteroid 13992 Cesarebarbieri is dedicated to him.

**Ivano Bertini** is a senior researcher of astronomy and astrophysics at the Parthenope University of Naples, Italy, and an associate scientist at the Institute for Space Astrophysics and Planetology-INAF, Rome, Italy. He also worked at the Physikalisches Institut of the University of Bern (Switzerland), at the Instituto de Astrofísica de Andalucía in Granada (Spain), at the European Space Astronomy Centre of the European Space Agency in Madrid (Spain), and at the University of Padua (Italy). He has played active roles in several space missions (the imaging system OSIRIS and the dust-collector instrument GIADA on board ROSETTA to comet Churyumov–Gerasimenko, several instruments on board ESA Comet Interceptor and ASI LICIACube as part of NASA DART, scientific support of the proposed ESA missions CASTALIA and HERA) and in ground-based European projects to discover and monitor hazardous Near-Earth Asteroids (NEO-Shield 2 and EURONEAR).

His main research themes are astrometry, photometry and spectroscopy of small Solar System objects from ground and space; space instrumentation; discovery and follow-up of hazardous asteroids; and dust in interplanetary space and comets. He has received the ESA recognition awards for ROSETTA.

The Main Belt asteroid 95008 Ivanobertini is dedicated to him.

# 1 Spherical Astronomy

The name *spherical astronomy* indicates the systematic and formal representation of the positions and angular movements (*motions*) projected on the celestial sphere of the heavenly bodies. The name *astrometry* indicates the several methods of measurement and data reduction. When the distances of these bodies are known, their position in the Cartesian space, velocities and accelerations can be determined. Celestial mechanics, which one may also call dynamical astronomy when extended from Solar System objects to stars and galaxies, utilizes positions, motions, acceleration and masses to determine the forces responsible for the observed orbits and calculate the past and future ones. At a higher level of investigation, the shape, inner structure and rotation of the celestial bodies will be taken into account, but at the moment, we proceed as if the planets or comets or stars are geometrical points.

In this chapter, we begin the study of spherical astronomy, using essentially geometric concepts, without considering the physical bases of the phenomena. Historically, this is the way astronomy began several millennia ago, leading to the geocentric conception of the world that found its maximum expression in the Ptolemaic theory. Such vision prevailed until Nicolaus Copernicus put forward his heliocentric model of the Solar System (*De revolutionibus orbium coelestium*, 1543). The geocentric vision was finally abandoned after the discoveries made by Galileo Galilei in the early 17th century. At any rate, this geocentric vision is fully adequate for the present purposes. In subsequent chapters, when investigating the dynamics of planets, asteroids, comets or finally the movements of the stars with respect to the Sun and to the entire Milky Way, we will be led to the operational definition of reference frames which are as inertial as possible, in order to correctly apply the Newtonian dynamical laws. Occasionally, we will discuss the inadequacies of the Newtonian theory and introduce some concepts of general relativity.

The following sections will provide notions of plane and spherical trigonometry needed in the rest of the book.

## 1.1 ELEMENTS OF PLANE TRIGONOMETRY

In theoretical expressions, such as the trigonometric functions, the angles on a plane must be expressed in radians. However, the practical units of sexagesimal degrees (360 degrees in the circle, symbol °; minutes of arc, symbol ′, $1° = 60′$; seconds of arc, symbol ″, $1′ = 60″$) are employed in several applications. Geodetic instrumentation often employs centesimal degrees (400 in the circle), but we will avoid this system.

Given that $\pi/2$ radians correspond to 90°, one radian is

$$1\,\text{radian} \approx 57°.2957795 \approx 3437′.74677 \approx 20624″.806$$

Inversely,

$$1″ \approx 0.000004848\,\text{radians}, \quad 1′ \approx 0.000290888\,\text{radians},$$

$$1° \approx 0.017453292\,\text{radians}$$

$$\theta(\text{radians}) \approx \theta″/206264.806, \quad \theta″ \approx 206264.806\,\theta(\text{radians})$$

With repeated applications,

$$\theta = \frac{\theta''}{206264.806} \text{radians}, \quad \theta^n \frac{(\theta'')^n}{(206264.806)^n},$$

$$(\theta'')^n = \theta'' \left( \frac{\theta''}{206264.806} \right)^{n-1} = (\theta'')\theta^{n-1}$$

The constant $R'' \approx 206{,}264.806$, often written as $R'' = 1/\sin 1''$, will appear frequently throughout the book.

The reason for the difference between theoretical and practical units is the following: by definition, the radian is the arc equal to the radius, but the length of the circumference is not a rational multiple of the arc. However, in order to perform measurements with graduated circles or digital encoders, rational units are needed.

In addition to radians and sexagesimal units, another system is used in astronomy, that of hours ($^h$), minutes ($^m$) and seconds ($^s$); although the names are equal, those units represent angles, not times! Given that the circumference is divided in $24^h$, the conversion factors are

$$24^h = 360° = 2\pi \,\text{radians}, \quad 1^h = 15° \approx 0.26179935 \,\text{radians}$$

$$4^m = 1° \approx 0.017453292 \,\text{radians}, \quad 1^h = 15° \approx 0.26179935 \,\text{radians}$$

Some useful formulae of plane trigonometry are the following:

$$\sin(\alpha \pm \beta) = \sin\alpha \cos\beta \pm \cos\alpha \sin\beta \qquad (1.1)$$

$$\cos(\alpha \pm \beta) = \cos\alpha \cos\beta \mp \sin\alpha \sin\beta \qquad (1.2)$$

$$\tan(\alpha \pm \beta) = \frac{\tan\alpha \pm \tan\beta}{1 \pm \tan\alpha \tan\beta} \qquad (1.3)$$

which give the duplication formulae when $\alpha = \beta$.

The inverse formulae are

$$\sin\alpha = x, \quad \arcsin x = n\pi + (-1)^n \alpha$$

$$\cos\alpha = y, \quad \arccos x = 2n\pi \pm \alpha$$

$$\tan\alpha = z, \quad \arctan z = n\pi + \alpha$$

The expressions of the trigonometric functions in terms of the imaginary unit $i$ are

$$\sin\theta = \frac{e^{i\theta} - e^{-i\theta}}{2i}, \quad \cos\theta = \frac{e^{i\theta} + e^{-i\theta}}{2}, \quad \tan\theta = \frac{1}{i}\frac{e^{2i\theta} - 1}{e^{2i\theta} + 1}$$

Frequent use will be made of the series expansions:

$$\sin\theta = \theta - \frac{1}{3!}\theta^3 + \frac{1}{5!}\theta^5 - \cdots + (-1)^k \frac{\theta^{2k+1}}{(2k+1)!} + \cdots \quad \left(\text{all } \theta's\right) \qquad (1.4)$$

$$\cos\theta = 1 - \frac{1}{2!}\theta^2 + \frac{1}{4!}\theta^4 - \cdots + (-1)^k \frac{\theta^{2k}}{(2k)!} + \cdots \quad \left(\text{all } \theta's\right) \quad (1.5)$$

$$\tan\theta = \theta + \frac{1}{3}\theta^3 + \frac{2}{15}\theta^5 + \cdots (-1)^{n-1} \frac{2^{2n}\left(2^{2n}-1\right)B_{2n}\theta^{2n-1}}{(2n)!} + \cdots \quad (-\pi/2 < \theta < \pi/2) \quad (1.6)$$

$$e^{\sin\theta} = 1 + \theta\frac{1}{2}\theta^2 - \frac{1}{6}\theta^4 + \cdots$$

where the $B_{2n}$ are Bernoulli's numbers ($B_0 = 1$, $B_2 = 1/6$, $B_4 = -1/30$, etc.). The inverse expressions are

$$\theta = \sin\theta + \frac{1}{6}\sin^3\theta + \frac{3}{40}\sin^5\theta + \frac{5}{112}\sin^7\theta + \cdots \quad (1.7)$$

$$\theta = \tan\theta - \frac{1}{3}\tan^3\theta + \frac{1}{5}\tan^5\theta + \frac{1}{7}\tan^7\theta + \cdots \quad (1.8)$$

These formulae show that the first differences between the functions $\sin\theta$ and $\tan\theta$ and the arc $\theta$ are quantities of third degree (and of opposite sign), while the function $\cos\theta$ differs from 1 by a term of second degree. Therefore, when the angles are very small, it is legitimate to put arc $\theta$ equal to $\sin\theta$ or $\tan\theta$ and with less precision $\cos\theta$ equal to 1 (it is very inefficient to calculate the angle from its cosine for small angles).

As already said, when approximating the arc with its sin or tan, the function restitutes the value of the arc in radians. To obtain it in seconds of arc, one must multiply the value in radians by $R''$. For instance, if $\sin a = 0.00141$, then $a'' \approx 0.00141 \times R'' \approx 290''.83$.

Let us see an example where practical and theoretical units are mixed together: suppose we wish to determine the maximum value of $\theta$ for which $\tan\theta - \theta \leq 1''$. We obtain

$$\tan\theta - \theta \approx \frac{\theta^3}{3} \leq \frac{1}{206264.804}, \quad \theta \leq \sqrt[3]{3/206264.804} \approx 0.02441\,\text{radians}$$

$$\theta'' = R''\theta \leq 5034''.9 \approx 1°24'$$

Another example is the calculation of the Taylor expansion of a function $y = f(x)$ using small finite increments $\Delta\theta$:

$$\Delta f = \frac{df}{d\theta}\Delta\theta + \frac{1}{2!}\frac{d^2}{d\theta^2}\Delta^2\theta + \cdots$$

Suppose we have chosen $\Delta\theta = 50'' = 0.00024$ radians. The square of the increment is $(\Delta\theta)^2 = 0.00024 \times 50'' = 0''.012$ (notice, arcsec, *not* arcsec square!). As seen previously, the rule to calculate the $n$th power of an arc expressed in arcsec is to multiply the value in arcsec for the value in radians raised to the $(n-1)$ power:

$$(\theta'')^n = (\theta'')\theta^{n-1}$$

Thus, ignoring the second and higher terms, an increment of $\Delta\theta = 50''$ produces an error of $\pm 0''.01$ on the function, multiplied by the amplitude of the first derivative. Should a precision of $0''.1$ on the function be sufficient, the increment of the variable could be raised to $150''$; to $500''$ if an error of

$\pm 1''.0$ is tolerated. These considerations are also applicable when deriving the value of the angles from expansions of trigonometric formulae; for instance, near $0°$ the error on the angle $\theta$ can be much greater than the error on $\cos\theta$, while near $90°$ the error on the angle can be much greater than the error on $\sin\theta$. As shown in the "Notes" section, it is more precise to derive $\theta$ from its tangent rather than from its sine or cosine.

Let us briefly consider how to invert series. Any converging series of the type

$$\theta = \varphi + a\varphi^2 + b\varphi^3 + c\varphi^4 + d\varphi^5 + \cdots$$

can be inverted as

$$\varphi = \theta + \alpha\theta^2 + \beta\theta^3 + \gamma\theta^4 + \delta\theta^5 + \cdots$$

where

$$a = \alpha, \quad \beta = b - 2a^2, \quad \gamma = c + 5a^3 - 5ab,$$

$$\delta = d - 14a^2 - 3b^2 + 18a^2 b - 4ac$$

## 1.2  SOME PROPERTIES OF PLANE TRIANGLES

Let us now consider a plane triangle, having vertices A, B, C, sides $a$, $b$, $c$ and angles $\alpha$, $\beta$, $\gamma$ (the general convention is to name $a$ the side opposite to angle $\alpha$ and so on). It is well known that $\alpha + \beta + \gamma = \pi$, so that the knowledge of two angles is sufficient to determine the third one. In general, the knowledge of three elements is sufficient to find the other three, with a notable exception: given the three angles, the three sides are not unequivocally determined, while given the three sides, the three angles are unequivocally determined. Useful relations are

$$a^2 = b^2 + c^2 - 2bc\cos\alpha \tag{1.9}$$

$$\frac{\sin\alpha}{a} = \frac{\sin\beta}{b} = \frac{\sin\gamma}{c} = \frac{1}{2R}, \quad \frac{a-b}{a+b} = \frac{\tan\frac{1}{2}(\alpha-\beta)}{\tan\frac{1}{2}(\alpha+\beta)},$$

$$\sin\frac{\alpha}{2} = \sqrt{\frac{(s-b)(s-c)}{bc}}, \quad \cos\frac{\alpha}{2} = \sqrt{\frac{s(s-a)}{bc}}, \quad \tan\frac{\alpha}{2} = \sqrt{\frac{(s-b)(s-c)}{s(s-a)}} \tag{1.10}$$

(known as Brigg's formulae),

$$K = s \cdot r = \frac{1}{2}bc\sin\alpha = c^2\frac{\sin\alpha\,\sin\beta}{2\sin\gamma} = \frac{abc}{2R}$$

where $a + b + c = 2s$ is the perimeter, $R$ the radius of the circumscribed circle, $r$ the radius of the inscribed one and $K$ the area.

In the context of plane triangles, it is useful to recall the binomial series expansion:

$$(1-y)^{-1/2} = 1 + \frac{1}{2}y + \frac{1\times3}{2\times4}y^2 + \frac{1\times3\times5}{2\times4\times6}y^3 + \cdots \quad (|y| < 1) \tag{1.11}$$

Consider then Equation 1.9 in the case $b/c \ll 1$, and write

$$a^2 = c^2 \left[ 1 + \frac{b^2 - 2bc\cos\alpha}{c^2} \right], \quad b\cos\alpha = x, \quad (b/c)^2 \approx 0$$

After simple passages, taking into account that $a$, $b$ and $c$ are all positive quantities, from Equation 1.11 we derive

$$\frac{1}{a} \approx \frac{1}{c}\frac{1}{\sqrt{1 - \frac{2x}{c}}} = \frac{1}{c}\left[ 1 + \frac{x}{c} + \frac{(3x^2 - b^2)}{2c^2} + \frac{(5x^3 - 3xb^2)}{2c^3} + \cdots \right] \approx \frac{1}{c}\left[ 1 + \frac{x}{c} + \frac{3x^2}{2c^2} + \frac{5x^3}{2c^3} + \cdots \right] \quad (1.12)$$

a relation which will be used in several occasions.

## 1.3   ELEMENTS OF SPHERICAL TRIGONOMETRY

Spherical trigonometry finds its origins in the Chaldean and Greek cultures, notably thanks to Hipparchus of Nicea (circa 180 B.C.) and then Claudius Ptolemaeus (in Latin, in English Ptolemy, 2nd century A.D.). The great German mathematician and astronomer Carl F. Gauss (1777–1855) put in a systematic form many of the relations used in the following paragraphs. In the three-dimensional Cartesian space, the sphere is the locus of points having the same distance $R$ from a given center O. Its surface is finite but unlimited. Each plane passing through the center of the sphere defines on it a great circle with radius $R$; one of these planes is assumed as the equatorial plane. A straight line passing through O and perpendicular to this plane intersects the sphere in two points, say P and Q, which are the poles of the plane. Let the sphere have a unit radius $R = 1$, so that the area on the surface can be expressed in steradians (abbreviated to sr) or for practical applications in square degrees (sq deg). The area of the entire sphere is then

$$4\pi \text{ sr} = 4\pi(360/2\pi)^2 = 129600/\pi \approx 41,252.96125 \text{ sq deg}$$

The angle from the center O between two points on the sphere, say C and D, namely, the angle COD, is equal to the length of the arc of the great circle passing through C and D. A plane parallel to the equator intersects the sphere is a small circle, called a *parallel*. Draw two great circles (called *meridians*) passing through the poles P and Q and making an angle $\lambda$ among them (see Figures 1.1 and 1.3). Let H, H′, K and K′ be the intersections of the two meridians with the parallel and the equator; $z$ be the angle POH = POK. Given that the length of arc H′K′ is equal to the angle $\lambda$, the length of arc HK is $\lambda \cdot \sin z$. Later, we shall show that this length is larger than the distance measured along the great circle passing for H and K.

From Figure 1.1, we see that three great circles divide the sphere in eight portions: in rigorous terms, a spherical triangle is that part whose sides are all smaller than $\pi$, namely, that part that is all contained in a hemisphere.

However, there are cases where it is convenient to relax this strict definition. For instance, in Chapter 2 we will examine the triangle determined by the Sun moving along the ecliptic (vertex A), its projection on the celestial equator (vertex B) and the intersection between the two planes (vertex C). With the passage of days, the two sides AC and BC will exceed $\pi$. The triangle ABC will cease to be a spherical triangle as defined here. One could then reason in terms of the supplementary triangle CAB. At any rate, there are formulae valid for triangles with sides exceeding $\pi$. See, for instance, Woolard and Clemence or Chapter 14 of the *Explanatory Supplement to the Astronomical Almanac* (Urban and Seidelmann), both in the Bibliography section.

The sum of the angles at the vertices of the spherical triangle is always larger than $\pi$; indeed, a spherical triangle can have all three angles = $\pi/2$. Let the capital italics letters $A$, $B$ and $C$ indicate

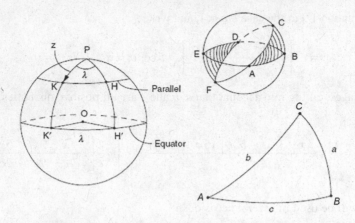

**FIGURE 1.1** A sphere with great circles. Three intersecting great circles define spherical triangles, such as PH'K', ABC and DEF. Instead, PHK is not a spherical triangle.

the angles at the vertices A, B and C, while letters $a$, $b$ and $c$ be the length of the opposing arcs. All six quantities, expressed in circular units, satisfy the limitations:

$$0 < a + b + c < 2\pi, \quad \pi < A + B + C < 3\pi$$

The difference $\sigma = A + B + C - \pi$ is called *spherical excess*.

The principal relations among the six elements $A$, $B$, $C$, $a$, $b$ and $c$ are usually indicated as Gauss's groups. For a complete list, refer, for instance, to the books of Smart (1965, *op. cit.*) and Danjon (1980, *op. cit.*). The three relations of the first such group are

$$\cos a = \cos b \cdot \cos c + \sin b \cdot \sin c \cdot \cos A \tag{1.13}$$

$$\sin a \cdot \cos B = \cos b \cdot \sin c - \sin b \cdot \cos c \cdot \cos A \tag{1.14}$$

$$\frac{\sin A}{\sin a} = \frac{\sin B}{\sin b} = \frac{\sin C}{\sin c} \tag{1.15}$$

Another relation and a variant of the first formula we shall apply are

$$\cos a \cdot \cos C = \sin a / \tan b - \sin C \cdot \cot B,$$

$$\cos A = -\cos B \cdot \cos C + \sin B \cdot \sin C \cdot \cos a$$

Analogous to Brigg's formulae are the following ($2s = a + b + c$):

$$\sin \frac{A}{2} = \sqrt{\frac{\sin(s-b)\sin(s-c)}{\sin b \sin c}}, \quad \cos \frac{A}{2} = \sqrt{\frac{\sin s \sin(s-a)}{\sin b \sin c}}, \quad \tan \frac{A}{2} = \sqrt{\frac{\sin(s-b)\sin(s-c)}{\sin s \sin(s-a)}}, \tag{1.16}$$

Therefore, in a spherical triangle the three angles unequivocally determine the three sides, a property not present in plane trigonometry.

When the sides of the spherical triangle are very small in comparison with the radius $R$ of the sphere, one can resort to the approximation of plane trigonometry, by considering the plane triangle on the plane tangent to the sphere. As shown by the French mathematician A.M. Legendre, a better approximation can be obtained by using the spherical excess $\sigma$:

$$\tan\frac{a}{2} = \sqrt{\frac{\sin\dfrac{\sigma}{2}\sin\left(A-\dfrac{\sigma}{2}\right)}{\sin\left(B-\dfrac{\sigma}{2}\right)\sin\left(C-\dfrac{\sigma}{2}\right)}} \tag{1.17}$$

and similarly for $b$ and $c$. Thus, a very small spherical triangle is equivalent to a plane triangle having the same sides $a$, $b$, $c$, angles

$$\alpha = A - \sigma/3, \quad \beta = B - \sigma/3, \quad \gamma = C - \sigma/3$$

and area $K = \sigma$ sr.

## 1.4  CARTESIAN AND POLAR COORDINATES

Let us now consider a Cartesian rectangular reference frame $O(x, y, z)$ passing through O, as in Figure 1.2. Operatively, there will be cases where it is convenient to define the vertical $Oz$ with appropriate devices (e.g., a plumb line) and then derive the horizontal plane $Oxy$ as the perpendicular to $Oz$. In other cases, the horizontal plane can be better measured (e.g., with a mercury bath) and then derive the vertical. Given a point P, its position is specified by the three Cartesian coordinates $(x, y, z)$, and alternatively by the distance $r$ to O and the two angles $(\lambda, \beta)$, namely, by the vector $\mathbf{r}(r, \lambda, \beta)$. The transformation laws between the two systems are

$$\begin{cases} x = r\cos\beta\,\cos\lambda \\ y = r\cos\beta\sin\lambda, \\ z = r\sin\beta \end{cases} \qquad \begin{cases} \lambda = \arctan\dfrac{y}{x} \\ \beta = \arcsin\dfrac{z}{r} \\ r = \sqrt{x^2 + y^2 + z^2} \end{cases} \tag{1.18}$$

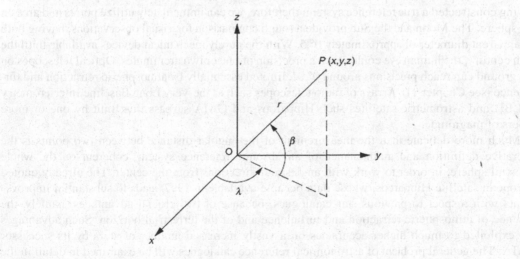

**FIGURE 1.2**  Cartesian and polar spherical coordinates.

The differentials, useful to evaluate both small variations and the influence of the measurement errors, are

$$
\begin{cases}
dx = \dfrac{x}{r}dr - z\,\cos\lambda d\beta - y d\lambda \\[2mm]
dy = \dfrac{y}{r}dr - z\,\sin\lambda d\beta - x d\lambda, \\[2mm]
dz = \dfrac{z}{r}dr + r\cos\beta d\beta
\end{cases}
$$

$$
\begin{cases}
dr = \cos\beta\cos\lambda dx + \cos\beta\sin\lambda dy + \sin\beta dz \\
r d\beta = -\sin\beta\cos\lambda dx - \sin\beta\sin\lambda dy + \cos\beta dz \\
r\cos\beta d\lambda = -\sin\lambda dx + \cos\lambda dy
\end{cases}
$$

Usually, the latitude $\beta$ is given in degrees, minutes and seconds of arc ($^\circ{}'\,''$) between $(0^\circ, \pm 90^\circ)$, positive to the north of the fundamental plane and negative to the south. Instead of $\beta$, it is often useful to give its complement to $90^\circ$, namely, the polar distance $z = 90^\circ - \beta$, $0^\circ \le z \le 180^\circ$. There is more freedom in the choice of units for the longitude $\lambda$: degrees, or radians, but often hours, minutes and seconds ($^{h\,m\,s}$) are employed. The direction of the longitude can also vary according to the application; sometimes, it increases in the conventional way of trigonometry (counterclockwise, or direct or prograde), and in other instances, the opposite verse is adopted (clockwise or retrograde).

In astronomy, the direct measurement of the distance $r$ to the generic celestial body P is usually impossible, except for particular cases (Solar System bodies, nearby stars). For the vast majority of stars and galaxies, one must resort to indirect distance indicators, which are subject to a variety of systematic errors. Therefore, in general the astronomer cannot measure the Cartesian coordinates $(x, y, z)$, while he can measure with utmost accuracy the two angles $(\lambda, \beta)$. The consequence of the immense distance of stars and galaxies is that we see all celestial bodies as sources of light (point-like in the case of stars) projected on a two-dimensional celestial sphere with the observer at its center. The radius of the sphere is arbitrary, we can consider it of infinite extent (when we look at it from the inside) or of unit radius (when we look at it from the outside).

On such a sphere, we can measure the relative angular distance between two points even before having constructed a true reference system; therefore, we can immediately utilize points and arcs on the sphere. The Moon and the Sun provide a rough comparison for visual observations, having both an apparent diameter of approximately $0^\circ.5$. With the purely mechanical devices available until the 16th century, the human eye could reach a precision of one or two arcminutes. Optical telescopes on the ground can reach precisions around $0''.01$, limited essentially by atmospheric refraction and turbulence (see Chapter 11). Arrays of radio telescopes such as the Very Long Baseline Interferometry (VLBI) and astrometric satellites such Hipparcos and GAIA surpass this limit by one or more orders of magnitude.

Much more delicate than the measurement of the angular distance between two points is the operative definition and maintenance of an absolute reference system, coherent on the whole celestial sphere, in order to work with angles and directions from the center. The already quoted European satellite Hipparcos, whose data became available in 1997, leads to substantial improvements with respect to previous star catalogues, because of two crucial advantages, namely, the absence of atmospheric refraction and turbulence and of the terrestrial horizon. Such advantages are exploited to much higher accuracies on a vastly increased number of stars by its successor GAIA. The general problem of astronomical reference catalogues will be examined in detail in the following chapters.

## 1.5   TERRESTRIAL LATITUDE AND LONGITUDE ON THE SPHERICAL EARTH

As an example of the previous considerations, let us assume that the Earth is a sphere of unit radius; let us draw from its center O a line coinciding with the rotation axis, which intersects the terrestrial poles in the points P and Q. The plane perpendicular to axis PQ through O will define on the surface a circle named terrestrial *equator*. By extending the rotation axis and the equatorial plane to intersect the celestial sphere, we define the celestial poles and the celestial equator. The real Earth's figure is certainly more complex than a sphere, but irrespective of its true shape, the celestial poles and equator are univocally defined by the diurnal rotation axis, which passes through the barycenter but does not necessarily coincide with the axis of figure, as we'll see in detail in later chapters. Let us ignore for the moment these complications and imagine regarding the unitary terrestrial sphere from the outside. Consider à particular place J on the surface (see Figure 1.3) and draw the great semi-circle through J passing through P and Q. This semicircle will be the terrestrial *meridian* of J, intersecting the equator in J′. Now, draw the small circle parallel to the equator through J, namely, the terrestrial *parallel* passing through J. Notice that through any given two points on the sphere, there is only one great circle joining them (unless the points are diametrically opposed), but infinite small circles. Therefore, the angular distance between two points is univocally defined by the arc of the great circle joining them. We show in the exercises that this is the minimum distance. In other words, great circle arcs are the *geodesics* on the sphere, equivalent to straight lines in Euclidean geometry.

Let us now arbitrarily assume a second point G on a different meridian. The intersection of this meridian with the equator, namely G′, is defined by convention as the zero point of the *longitudes*. At present, the point G coincides with the point where the meridian circle of the Royal Astronomical Observatory in Greenwich (England) was located in the second half of the 19th century. The decision to give Greenwich this special status was reached only in 1887, after considerable debate. For instance, in the 16th century, the great geographer Mercator had assumed the island of Hierro in the Canaries as zero point.

Now, the arc J′J or the angle J′OJ is the terrestrial *latitude* $\phi$, and the arc G′J′ or the angle G′OJ′ is the terrestrial *longitude* $\Lambda$ of J. In this manner, any point on the Earth has associated the ordered pair $(\Lambda, \phi)$, except the two poles P and Q for which $\Lambda$ is undetermined. Instead of $\phi$, its complementary angle $90° - \phi$ (named *co-latitude*) is often used.

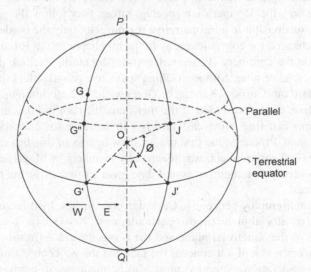

**FIGURE 1.3**   The terrestrial unit sphere. The generic point J is determined by the ordered pair of angles $\Lambda = G'OJ'$, $\phi = J'OJ$.

Traditionally, $\Lambda$ is given in ($^h$ $^m$ $^s$) east or west of Greenwich, while $\phi$ is expressed in ($^\circ$ ' ")
north or south of the equator, $-90^\circ \leq \phi \leq +90^\circ$. As already remarked, care must be exercised for
$\Lambda$, which can be found expressed in circular units and increasing toward east ($-180^\circ \leq \Lambda \leq 180^\circ$,
west longitudes being regarded as negative), but also toward west. In The Astronomical Almanac,
the longitudes west of Greenwich are negative, and east longitudes are positive. For instance, the
Observatory of La Palma in the Canaries has $\Lambda \approx -17^\circ$, and the Observatory of Asiago in North Italy
has $\Lambda \approx +11^\circ$. The decision of reversing the sign of the terrestrial longitude $\Lambda$, taken by International
Astronomical Union (IAU) in 1982, has been questioned by several authors, because it goes not
only against the previous century-long practice, but also against the longitude system adopted for
all other planets of the Solar System. Different space missions have adopted different sign conven-
tions for their longitude systems. For instance, the NASA missions to Mercury, Messenger, have
adopted the opposite sign of Mariner 10. The same is true for previous and contemporary missions
to Mars. Therefore, one has to be aware of this change before comparing the respective cartogra-
phies. More details on the coordinate systems on the surfaces of planets, asteroids and comets will
be given in Chapter 2.

It is clear that ($\Lambda$, $\phi$) are angular coordinates on the unitary sphere. If linear distances are needed,
one must take into account that the radius of the Earth is approximately $a_\oplus = 6380$ km. For instance,
the length of the arc J'J $= a_\oplus \phi$ ($\phi$ in radians) gives the distance in kilometer from the equator to J
along the meridian, while the length of the arc G"J $= a_\oplus \Lambda \cos \phi$ ($\Lambda$ in radians) gives the distance in
kilometer from the fundamental meridian to J along the parallel of latitude. Another way of giving
distances is by using the *nautical mile*, namely, that arc on the sphere corresponding to 1'; according to the definition of 1929, a nautical mile equals 1852 m (1.15 statute miles). Because the Earth is
not really a sphere, this length of the nautical mile corresponds to that of 1' on the meridian at $45^\circ$
latitude. Other useful numbers to remember are the following: $1^\circ$ in latitude corresponds approximately to 111.13 km (68 miles), and the circumference of a meridian is approximately 40 million m
long. The original definition of the meter was indeed in terms of the 10-millionth part of the length
of the arc from pole to equator. This definition was the basis for relating the measurements of distance made in the terrestrial laboratory to those made on heavenly bodies by means of the diurnal
parallax.

See Chapter 2 for a more precise discussion and values on the ellipsoidal Earth.

For everyday life, particular importance is attached to the 23 meridians at $15^\circ$ ($1^h$) and multiples
from Greenwich, which ought to define the *time zones*. However, the actual borders of the time
zones do not coincide with the $15^\circ$ meridian spacing; rather, they follow the geographic extension
of the states and their subdivision in administrative regions. Not only the borders, even the number
of time zones can be changed for convenience, as frequently happened in Russia. In the USA, there
are four time zones on the continent (Eastern, Central, Mountain, Pacific), plus Alaska, Hawaii
and other islands for a total of nine. Although China spans five geographical time zones, civil time
follows a single standard time, offset $8^h$ ahead of Greenwich. The meridian opposite to Greenwich
is the so-called date line: on the west side of the meridian, the date is 1 day more than on the east
side. A traveler crossing that line going eastward has to decrease his calendar date by 1 day and
*vice versa* going westward. Probably, the first practical awareness of this fact (already supposed by
medieval philosophers and geographers) was reached by members of Magellan's expedition in the
16th century, by discovering that counting actual days gave a result differing by one full day with
respect to calendar dates.

It is often convenient to employ geocentric Cartesian coordinates, by directing the $x$-axis toward
a given point X in the equatorial plane (in this particular case, X = G'), the $y$-axis toward a point Y
$90^\circ$ counterclockwise on the equatorial plane and the $z$-axis toward Z (in this case, the north pole
P). Mathematical convenience will not conceal the fact that the geocentric Cartesian system is not
directly accessible to observation. Attention must also be made not to confuse the Cartesian distance between two points (necessary, e.g.,, in interferometric applications) with the distance along
the great circle joining them.

## 1.6   ELEMENTS OF VECTOR CALCULUS

Let $\mathbf{i}$, $\mathbf{j}$, $\mathbf{k}$ be the unit vectors along $(x, y, z)$; the position of a point J on the unitary sphere can be represented by the unit vector $\mathbf{r}_J$:

$$\mathbf{r}_J(x, y, z) = \mathbf{r}_J(\Lambda, \phi) = x\mathbf{i} + y\mathbf{j} + z\mathbf{k}$$

where $x = \cos(xJ)$, $y = \cos(yJ)$, $z = \cos(zJ)$ are the direction cosines of the line OJ. The transformations between Cartesian and polar spherical coordinates are

$$x = \cos\Lambda\cos\phi, \quad y = \sin\Lambda\cos\phi, \quad z = \sin\phi$$

We might also write

$$\hat{\mathbf{r}}_J = (x, y, z) = \begin{pmatrix} \cos\Lambda\cos\phi \\ \sin\Lambda\cos\phi \\ \sin\phi \end{pmatrix} \tag{1.19}$$

The general problem of coordinate transformations will be examined in Chapter 3.

It is useful to recall some basic notions of vector algebra. The sum of two vectors $\mathbf{a}$, $\mathbf{b}$ is a third vector $\mathbf{c}$ lying in the plane defined by the first two:

$$\mathbf{c} = \mathbf{a} + \mathbf{b} = (a_x + b_x)\mathbf{i} + (a_y + b_y)\mathbf{j} + (a_z + b_z)\mathbf{k}, \quad \mathbf{a} - \mathbf{b} = \mathbf{a} + (-\mathbf{b})$$

The sum has the commutative and associative properties:

$$\mathbf{c} = \mathbf{a} + \mathbf{b} = \mathbf{b} + \mathbf{a}, \quad \mathbf{a} + (\mathbf{b} + \mathbf{c}) = (\mathbf{a} + \mathbf{b}) + \mathbf{c}$$

The scalar (or inner) product $c$ between two vectors whose directions make an angle $\theta$ ($0° \leq \theta \leq 180°$) is the number:

$$c = \mathbf{a} \cdot \mathbf{b} = \mathbf{b} \cdot \mathbf{a} = ab\cos\theta = a_x b_x + a_y b_y + a_z b_z$$

The vector product between $\mathbf{a}$ and $\mathbf{b}$ is a third vector $\mathbf{c}$ perpendicular to the plane defined by the first two:

$$\mathbf{c} = \mathbf{a} \times \mathbf{b} = (a_y b_z - a_z b_y, a_z b_x - a_x b_z, a_x b_y - a_y b_x) = ab\sin\theta\,\mathbf{u}$$

being $\mathbf{u}$ the unit vector parallel to $\mathbf{c}$. *The alternative notation* of the vector product, $\mathbf{a}\char`^\mathbf{b}$ instead of $\mathbf{a} \times \mathbf{b}$, will be used when confusion might arise. Notice that the vector product has the distributive property $\mathbf{a} \times (\mathbf{b} + \mathbf{c}) = \mathbf{a} \times \mathbf{b} + \mathbf{a} \times \mathbf{c}$, but not the commutative property, because $\mathbf{c} = \mathbf{a} \times \mathbf{b} = -\mathbf{b} \times \mathbf{a}$. In particular,

$$\mathbf{i} \times \mathbf{j} = \mathbf{k}, \quad \mathbf{j} \times \mathbf{k} = \mathbf{i}, \quad \mathbf{k} \times \mathbf{i} = \mathbf{j}$$

Therefore, the vector product can be written in matrix form:

$$\mathbf{c} = \mathbf{a} \times \mathbf{b} = \begin{vmatrix} \mathbf{i} & \mathbf{j} & \mathbf{k} \\ a_x & a_y & a_z \\ b_x & b_y & b_z \end{vmatrix}$$

The double mixed product

$$[\mathbf{a},\mathbf{b},\mathbf{c}] = \mathbf{a} \cdot (\mathbf{b} \times \mathbf{c}) = \mathbf{b} \cdot (\mathbf{c} \times \mathbf{a}) = \mathbf{c} \cdot (\mathbf{a} \times \mathbf{b}) = \begin{vmatrix} a_x & a_y & a_z \\ b_x & b_y & b_z \\ c_x & c_y & c_z \end{vmatrix}$$

is a number that can be interpreted as the volume of the parallelepiped of sides $\mathbf{a}$, $\mathbf{b}$, $\mathbf{c}$.

The triple vector product is defined by

$$\mathbf{a} \times (\mathbf{b} \times \mathbf{c}) = (\mathbf{a} \cdot \mathbf{c})\mathbf{b} - (\mathbf{a} \cdot \mathbf{b})\mathbf{c}, \quad (\mathbf{a} \times \mathbf{b}) \times \mathbf{c} = (\mathbf{a} \cdot \mathbf{c})\mathbf{b} - (\mathbf{b} \cdot \mathbf{c})\mathbf{a} \neq \mathbf{a} \times (\mathbf{b} \times \mathbf{c})$$

The vector product can be applied in particular to define the moment of a force $\mathbf{F}$: consider a point O in space, and a force $\mathbf{F}$ whose direction does not intercept O. Draw from O a vector r to any point along the direction of $\mathbf{F}$. The moment of $\mathbf{F}$ with respect to O is the vector $\mathbf{K}$:

$$\mathbf{K} = \mathbf{r} \times \mathbf{F} = (rF\sin\theta)\mathbf{k} \tag{1.20}$$

$\theta$ being the angle between the two vectors, perpendicular therefore to the plane defined by $\mathbf{r}$ and $\mathbf{F}$; notice that $\mathbf{K}$ is independent of where $\mathbf{r}$ terminates on the direction of $\mathbf{F}$.

The derivative and the integral of a vector are, respectively,

$$\frac{d\mathbf{a}}{dt} = \frac{da_x}{dt}\mathbf{i} + \frac{da_y}{dt}\mathbf{j} + \frac{da_z}{dt}\mathbf{k}, \quad \frac{d}{dt}(\mathbf{a} \cdot \mathbf{b}) = \mathbf{b} \cdot \frac{d\mathbf{a}}{dt} + \mathbf{a} \cdot \frac{d\mathbf{b}}{dt}$$

$$\frac{d}{dt}(\mathbf{a} \times \mathbf{b}) = \frac{d\mathbf{a}}{dt} \times \mathbf{b} + \mathbf{a} \times \frac{d\mathbf{b}}{dt}$$

$$\int \mathbf{a}(t)dt = \mathbf{i}\int a_x(t)dt + \mathbf{j}\int a_y(t)dt + \mathbf{k}\int a_z(t)dt$$

If a vector $\mathbf{b}$ exists such that $\mathbf{a} = d\mathbf{b}/dt$, then

$$\int \mathbf{a}(t)dt = \mathbf{b} + \mathbf{c}\big(\mathbf{c} = \text{constant vector}\big),$$

$$\int_{t_1}^{t_2} \mathbf{a}(t)dt = \mathbf{b}(t_2) - \mathbf{b}(t_1)$$

In dynamics and in celestial mechanics, it is usual to denote the first and second derivatives of a vector with the notations:

$$\frac{d\mathbf{a}}{dt} = \dot{\mathbf{a}}, \quad \frac{d^2\mathbf{a}}{dt^2} = \frac{d\dot{\mathbf{a}}}{dt} = \ddot{\mathbf{a}}$$

It is important to remember that finite angular rotations of a rigid body do not obey the commutative property. However, infinitesimal rotations and therefore angular velocities do obey all vector laws. In the following chapters, we will encounter vectors $\mathbf{r}$ referred to a coordinate system itself rotating with angular velocity $\boldsymbol{\omega}$ with respect to an inertial system. The total derivative of $\mathbf{r}$ relative to inertial axes will be

$$\left(\frac{d\mathbf{r}}{dt}\right)_{\text{inertial}} = \left(\frac{d\mathbf{r}}{dt}\right)_{\text{relative}} + \boldsymbol{\omega} \times \mathbf{r} \tag{1.21}$$

## NOTES

- Precision of trigonometric functions. Suppose your calculator gives the value of the trigonometric functions $f$ with $n$ decimal digits. The difference between the true value of the function and that given by the calculator can be at most 1 on the last digit, namely, $\Delta f \leq 1 \times 10^{-n}$, which can be regarded small enough to be considered as the differential of $f$ in such a point. This is because

$$d\operatorname{sen}\theta = \cos\theta \, d\theta, \quad d\cos\theta = -\operatorname{sen}\vartheta d\theta, \quad d\tan\theta = \frac{1}{\cos^2\vartheta} d\theta$$

The errors $\varepsilon_i = d\theta_i$ on the angles due to the errors on the functions will be

$$\varepsilon_1 = \left|\frac{1}{\cos\theta}\right|10^{-n}, \quad \varepsilon_2 = \left|\frac{1}{\operatorname{sen}\theta}\right|10^{-n}, \quad \varepsilon_3 = \left|\cos^2\theta\right|10^{-n}$$

While the third expression deriving from the tangent never exceeds $10^{-n}$ for any angle, the first two can become very large if the angle approaches $\pi/2$ or 0, respectively. For instance, suppose that your calculator gives eight decimal digits (perhaps it is prudential to assume as exact the one before the last visualized digit) and you wish $\varepsilon_3$ in arcsec. We have to multiply the previous expression by $R'' = 206,264.8$:

$$\varepsilon_3 \leq 2.06 \times 10^5 \times 10^{-8} \approx 2 \times 10^{-3} \operatorname{arcsec}$$

for all $\theta$'s. Should we have used $\varepsilon_1$ in the proximity of 90°, e.g., at 89°.5, the precision would have deteriorated by $1/\cos\theta \approx 115$ times.

As a rule, in order to have an error less than $30''$, four digits are needed, six for a precision better than $0''.5$ and eight for $0''.005$. When astrometry must reach precisions better than a thousandth or even few millionths of an arcsec (as was the case of the Hipparcos and now GAIA), the task of the computer becomes truly formidable.

- Precision of series developments. Let us recall the expressions:

$$\sin a = a - \frac{1}{3!}a^3 + \frac{1}{5!}a^5 + \cdots \approx a\left(1 - \frac{1}{6}a^2\right),$$

$$\tan a = a + \frac{1}{3}a^3 - \frac{1}{15}a^3 + \cdots \approx a\left(1 - \frac{1}{3}a^2\right)$$

When the arc $a$ is small, the error made by confusing sin with arc is smaller than the first ignored term, namely, the term in $a^3$; in the case of the tangent, the error is slightly larger than twice that made with the sin and with opposite sign (the term in $a^5$ exceeds $0''.01$ only above 5°). Let us calculate the maximum value of $a$ before the error exceeds a wanted amount, as given by Table 1.1. For instance, the Moon has an average diurnal parallax of $3422''.70$, so that if we retain only the first two terms:

$$\sin 3422''.70 \approx 3422''.70\left[1 - \frac{1}{6}\left(\frac{3422.70}{206,264.8}\right)^2\right]$$

$$\approx 3422''.70[1 - 0.000046] = 3422''.54$$

Namely, by considering only the second terms the sin differs from the parallax by $0''.16$, a quantity that can or cannot be considered negligible, according to the application.

**TABLE 1.1**
**Admitted Error and Maximum Angles**

| $a - \sin a$ | $a_{max}$ | $a - \tan a$ | $a_{max}$ | $1 - \cos a$ | $a_{max}$ |
|---|---|---|---|---|---|
| 0″.001 | 0°10′34″.3 | 0″.001 | 0°08′23″.5 | $10^{-6}$ | 0° 04′51″.7 |
| 0″.010 | 0°22′46″.7 | 0″.010 | 0°18′04″.7 | $10^{-5}$ | 0°15′22″.5 |
| 0″.100 | 0°49′04″.4 | 0″.100 | 0°38′57″.0 | $10^{-4}$ | 0°48′37″.0 |
| 1″.000 | 1° 45′43″.7 | 1″.000 | 1°23′54″.5 | $10^{-3}$ | 2°33′43″.2 |

## EXERCISES

Experience tells that all too often the students forget to check the quadrant of the angle. Remember, between 0 and $\pi$, the inverse cosine is single valued, but the inverse sine is not!

1. Find the elements and the area of the spherical triangles having the following vertices:

$$(\Lambda, \phi) = (0^h, 0°), \quad (\Lambda, \phi) = (4^m, 0°), \quad (\Lambda, \phi) = (4^m, 1°)$$

$$(\Lambda, \phi) = (0^h, 0°), \quad (\Lambda, \phi) = (1^h, 0°), \quad (\Lambda, \phi) = (1^h, 45°)$$

$$(\Lambda, \phi) = (0^h, 0°), \quad (\Lambda, \phi) = (6^h, 0°), \quad \text{the North pole}$$

$$(\Lambda, \phi) = (0^h, 0°), \quad (\Lambda, \phi) = (18^h, 0°), \quad \text{the North pole}$$

$$(\Lambda, \phi) = (0^h, 0°), \quad (\Lambda, \phi) = (21^h, 0°), \quad \text{the North pole}$$

2. Calculate the distance between two Earthly sites $A$ and $B$ both north of the equator, and the coordinates of northernmost point on the great circle joining them, in the hypothesis of spherical Earth of radius $a_\oplus = 6380\,\text{km}$.

   Let the two places in the northern hemisphere have the geographic coordinates as shown in Table 1.2. We wish to determine the following:
   • the length of the great circle arc passing through A, B and of the joining chord.
   • P being the north pole, and PAB the corresponding spherical triangle, the amplitude of the angles $A = \text{PAB}$, $B = \text{PBA}$.
   • that $0 < \Sigma\,(\text{arcs}) < 2\pi$, $\pi < (\Sigma\,\text{angles}) < 3/2\pi$.

   Recall the first formula of the first Gauss's group (Equation 1.13), which we write here as

$$\cos p = \cos a \cos b + \sin a \sin b \cos P$$

$$= \sin \phi_A \sin \phi_B + \cos \phi_A \cos \phi_B \cos \Delta\Lambda \tag{1.22}$$

   where arc AB = $p$ unknown, arc PA = $b$ = $90° - \phi_A$ = 65°.7, arc PB = $a$ = $90° - \phi_B$ = 53°.2, angle P = $\Lambda_A - \Lambda_B$ = 100°.95 (A is east, and B is west of Greenwich). From Formula 1.22, we derive

$$p = 83°.80676 = 83°48′24″ = 5028′.4$$

**TABLE 1.2**
**Geographic Coordinates of the Two Sites**

|   | Longitude $A$ | Latitude $\phi$ |
|---|---|---|
| A | 133°.65 E | +24°.3 |
| B | 125°.40 W | +36°.8 |

By definition, the nautical mile is the arc corresponding to 1′, so that 5028.4 is also the distance in nautical miles, equivalent to 9328 km, between A and B. The length of the chord, useful in measurement of the diurnal parallax or in interferometric applications, can be derived from

$$L = 2a_\oplus \sin\frac{p}{2} = 8521.5\,\text{km}$$

The amplitude of the other two angles can be derived from the tangent formula 1.16, where

$$2s = a + b + p = 202°.2, \quad s = 101°.1.$$

Whence $A = 52°.26$; by the same procedure, $B = 64°.17$, so that $b + c + p = 202°.70$, $A + B + P = 217°.38$.

To determine C, the northernmost point on circle AB, we notice that the meridian for c will be perpendicular in C to the circle. Therefore, in the spherical triangle PCA (or PCB), the angle in C will be 90°. All other elements being known, we immediately obtain arc PC (the co-latitude) and the angle APC, namely, the longitude:

$$\Lambda_C = 164°.38\,\text{W}, \quad \phi_C = 43°.88\,\text{N}$$

3. Calculate the distance between the astronomical observatories of Padova (Italy) and La Silla (Chile), and the longitude where the great circle between the two sites cuts the equator (ignore the elevation above the sea level).

   The coordinates of the two observatories are

$$\Lambda_{\text{Pd}} = 11°52'.3\,\text{E}, \quad \phi_{\text{Pd}} = +45°24',$$
$$\Lambda_{\text{LS}} = 70°43'.8\,\text{W}, \quad \phi_{\text{LS}} = -20°14'$$

For the first question, consider the spherical triangle between Padova, the north pole P and La Silla, and let $p$ be the angle at the pole between the meridians of the two places. This angle coincides with the difference in longitude between Padova and La Silla, with due regard to the first being east and the second west of Greenwich. Two sides of the triangles are given by the co-latitudes. Then, as in the previous exercise, Equation 1.13 easily provides the third side, namely, the wanted angular distance of 105°.5, corresponding to 6336 nautical miles, or to 11,733 km. Try on Google Earth.

For the second question, if E is the intersection point between the arc through Padova, La Silla and the equator, notice that the latitude of E is zero; therefore, the spherical triangle drawn through P, E and La Silla has one side of 90°. The angle opposite to this side is easily found, and after a few calculations, we finally obtain $\Lambda_E = 44°$ W.

4. Two observers on the Earth surface, A and B, separated by a distance of $50 \pm 0.1$ km, observe a satellite C. The value of the angle CAB, as measured by A, is $(87.5 \pm 0.01)$ deg; the value of the angle CBA, as measured by B, is $(88.5 \pm 0.01)$ deg. Determine the other elements of the triangle ABC, and discuss the relative influence of the errors of the result.

# 2 First Notions on Astronomical Reference Systems

The measurement of the positions on the celestial sphere was among the earliest operations performed on heavenly bodies. On several aspects, the procedure is similar to that performed on the Earth by the geographer. However, there are peculiarities that have led to the definition of reference systems unique to astronomy. These systems originated from observations carried out over several millennia from the surface of the Earth; an astronomer in the extra-terrestrial space, or living on another planet, might have found solutions more appropriate to his particular location.

The solid Earth prevents seeing the entire celestial sphere; the line of sight is limited above a great circle called *horizon*. Such circle is more readily visible from a high mountain or on the open sea. Due to the diurnal direct rotation of the Earth (direct, from west to east, retrograde from east to west), the celestial sphere appears to rotate from east to west around an ideal axis, which cuts the sphere at two points called the *celestial poles*, north P and south Q (a generic observer only sees one of them). Due to the great distances of all celestial bodies, the axis passing for any observer on the Earth and parallel to the rotation axis defines exactly the same two points P and Q. Therefore, in the course of the day, each star will appear to describe exactly the same minor circle, irrespective of the position of the observer. In the same manner, the plane passing through any observer and perpendicular to the rotation axis defines the same *celestial equator*. Only for nearby objects and with difficult measurements, we are able to detect the daily rotation of the observer itself.

The enormous distances to the stars have another important consequence, namely, that their relative movements are small and remained undetected until the 17th century. For this reason, the term "fixed stars" was used and is encountered even today. On the contrary, the Sun, the Moon, the planets, the comets and asteroids display an appreciable motion with respect to this fixed canopy of stars; the name *planet* itself has the meaning of "wandering body" in Greek. The movements of the Sun and the Moon were used to define units of time, as we will see in detail in later chapters. While the diurnal rotation is the basis of the concept of the *day*, the movement of the Sun through the constellations defines the *solar year* and that of the Moon defines the *lunar month*. The yearly motion of the Sun takes place along a great circle inclined by approximately 23°27′ to the celestial equator. This circle is named the *ecliptic*; the angle $\varepsilon$ between equator and ecliptic is called *obliquity* of the ecliptic. Horizon, equator and ecliptic are three great circles on the celestial sphere used to define different astronomical reference systems. The common principles at the basis of each system are

- a plane through the observer, selected as the fundamental plane, defines a fundamental great circle on the sphere. A particular point G′ on this fundamental circle is chosen as origin of angular abscissae;
- a great circle is drawn through the poles of the fundamental plane and any star X; this circle intersects the fundamental plane in X′;
- arc G′X′ is the angular abscissa $\lambda$ and arc X′X the angular ordinate $\beta$ of X, so that any star is univocally determined by the ordered pair $(\lambda, \beta)$, with the obvious exceptions of a star exactly at one of the two poles, for which $\lambda$ is undefined.

## 2.1 THE ALT-AZIMUTH SYSTEM

The horizontal plane and its perpendicular, namely, the *vertical*, define the Alt-Azimuth (or horizon) system. This system can be realized, at least in principle, by very simple devices sensitive to gravity, such as a plumb line or the free surface of a liquid. The points where the vertical cuts the celestial sphere are called, respectively, Zenith Z (above) and Nadir Z′ (below, unobservable from the Earth's surface). The plane passing through the observer and perpendicular to the vertical encounters the celestial sphere in the astronomical *horizon*.

Consider now an observer in the northern hemisphere, and draw the great circle through the visible celestial north pole P and his Zenith Z; this circle is called the *meridian* of the observer; it contains also the Nadir Z′ and the other pole Q. The meridian cuts the horizon on two points, the so-called true north N (on the same side of P with respect to Z) and true south S. Any other great circle passing through Z and Z′ is called a *vertical* circle. In particular, the vertical circle at 90° to the meridian is named *first vertical*; on the horizon, it defines two points called true east E and true west W, respectively. Those four points on the horizon are referred to as *cardinal points*.

In order to determine the two angular coordinates of a star X, let us draw the vertical circle through Z and X (see Figure 2.1, drawn for an observer in the northern hemisphere) that intersects the horizon in X′, thus defining the two arcs $A$ and $h$:

1. the *azimuth A*, namely, the arc SX′ counted clockwise from the south through the west, usually measured in (° ′ ″), $0° \leq A \leq 360°$. This convention of origin and direction has been adopted here in order to coincide with that of the hour angle *HA*, described in the following paragraph. Several authors, the *Astronomical Almanac* and many telescope pointing algorithms prefer to count $A$ from the north through the east; others use a counterclockwise direction, and still others have $A$ between −180° and +180°;

2. the *altitude* (or *elevation*) *h*, namely, the arc X′X counted from the horizon toward the star and measured in (° ′ ″), $-90° \leq h \leq +90°$. Alternatively, the Zenith distance $z = 90° - h$ might be preferred, with $0° \leq z \leq 180°$. The terrestrial observer cannot see stars below the horizon ($h < 0°$, $z > 90°$). With an Arabic term, the parallels of elevation are called *almucantar* (or *almucantarat*). The height of the pole P above the horizon, namely, the arc NP, defines the astronomical latitude of a northerly site. For an observer in the southern hemisphere, the height of the visible pole Q taken with negative sign is his astronomical latitude.

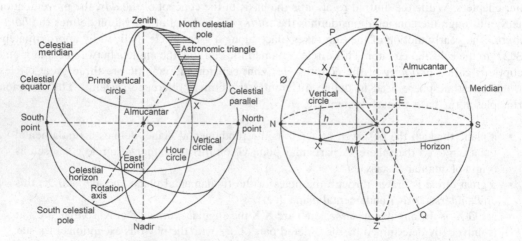

**FIGURE 2.1** (drawn for an observer in the northern hemisphere). (a) The local celestial sphere and the astronomic triangle as seen from the east. (b) The Alt-Az coordinates of star X with Azimuth origin from the south (as seen from the west). Angle SOX′ is the azimuth $A$ and angle X′OX the altitude $h$ (its complement ZOX is the Zenith distance $z$ of X).

Let us see the modifications for an observer in the southern hemisphere. As before, the meridian is the great circle through the Zenith and the visible pole Q. The diurnal rotation of the celestial sphere takes always place from east to west. The height in meridian of the pole Q taken with negative sign is the astronomical latitude; alternatively, the arc PZ $= 90° - \phi$ is the co-latitude of the observer. Be aware of the different aspect of the celestial sphere in the two hemispheres. Assume that you are in the northern hemisphere, standing with your shoulders to the visible pole, namely, with the south cardinal point in front, east to the left and west to the right. All stars will describe their parallel of declination in the clockwise direction. If you are in the southern hemisphere with the same attitude, namely, with your shoulders to the visible pole Q, the celestial north pole will be in front, and the stars will move in an anticlockwise direction, but of course always from east to west. Moreover, the figures of the Moon and of the constellations will appear upside down.

Therefore, each site has associated with it a system of fixed cardinal points and great circles, with respect to which the celestial sphere is in continuous rotation. In other words, the pair $(A, h)$ of star X depends on the location of the observer and changes with the passage of time. Therefore, it is necessary to discuss more precisely the relationship between the geographic and astronomic coordinates on one hand and the influence of time on the other. We stated in Chapter 1 that the Earth is only approximately a sphere; indeed, already in the 18th century, the accuracy of the measurements conclusively showed that the shape is better approximated by an oblate *ellipsoid of revolution* (called also a *spheroid*) having the polar axis slightly shorter than the equatorial one.

Dynamical considerations led to the introduction of the *geoid*, to indicate the equilibrium surface the free water of the oceans would have under the sole influence of gravity and rotation. In other words, all points on the geoid have the same potential. Such surface is extended by convention through the continental lands. The mathematical ellipsoid of revolution is usually below the geoid on the land and above it on the oceans, but the difference in general does not exceed 100 m. Actually, the Earth is remarkably smooth, and the difference in elevation between the highest and the lowest points is less than 20 km, and on the average much smaller. Therefore, we are justified to ignore, at least for the moment, the slight deviation of the real Earth from an oblate ellipsoid of revolution.

The slight ellipticity of the Earth has far-reaching consequences, from both the geometrical and physical points of view:

- regarding geometry, with reference to Figure 2.2, the geodetic vertical is mathematically defined as the normal to the ellipsoid. It does not encounter the center O of the Earth; it intersects the rotation axis in a latitude-dependent point O'. The convenience of using geo-centric Cartesian coordinates requires the consideration of the direction from the center of the Earth to any particular place on the surface, namely, of the geocentric vertical. It will be shown in the following that the latitude-dependent difference between geocentric and geo-detic vertical reaches 12' at mid-latitudes. Actually, the direction of the astronomic vertical is defined by the overall gravitational field in any given place. This "true" vertical does not necessarily intersect the rotation axis in O'. The small deviations of the astronomic from the geodetic vertical are caused by the presence of nearby masses or voids. These so-called *anomalies of the gravity* rarely exceed 10″, although in peculiar places, they can reach 50″. Accurate measurements clearly reveal these two slightly different Zeniths;
- regarding physical considerations, a most important astronomical effect is due to the influ-ence of the Moon and the Sun on the direction of the rotation axis, namely, the lunisolar (or equator) precession. Moreover, even in the absence of external forcing bodies, the rotation axis would not necessarily coincide with the polar axis of the ellipsoid (free or Eulerian nutation). These effects will be discussed in particular in Chapters 4–6.

Limiting the following considerations to the geometrical aspects, thus ignoring the small dif-ferences between geodetic and astronomic vertical and the lunisolar precession, let us use the

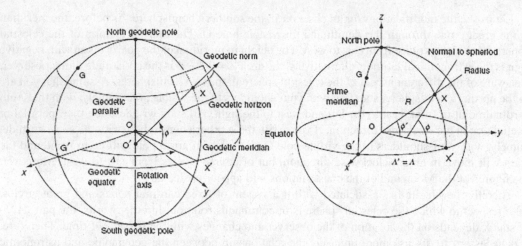

**FIGURE 2.2** Geodetic, astronomic and geocentric coordinates of a point X on the surface of the ellipsoid.

ellipsoidal model of the Earth with two equal equatorial axes and a slightly shorter polar axis, namely, an oblate spheroid with rotational symmetry and meridians represented as ellipses. Several mathematical models have been adopted in the course of the centuries to describe the ellipsoid; a long list is given in Table 5.1 of the *Explanatory Supplement* from the time of Airy, *circa* 1830, to the present. Today, the model named WGS84, having equatorial radius $a_{\oplus} = 6378.137$ km and polar axis $c = 6356.752$ km, finds a widespread utilization for navigation. The original WGS84 oblate reference ellipsoid had associated a geoid with a resolution of about 200 km modeled with spherical harmonic series of degree 180 to define the nominal sea level surface. Later, the Earth Gravitational Model 1996 (EGM96) produced by the National Geospatial-Intelligence Service (NGA) used the same WGS84 ellipsoid but with resolution increased to 100 km and spherical harmonics series of degree 360. In 2008, the EGM96 was refined using spherical harmonics of degree 2159 with a resolution approaching 10 km, producing the EGM2008. The geoid models are constantly improved using satellite data, in particular those from the European Space Agency (ESA) Gravity Field and Steady-State Ocean Circulation Explorer (GOCE) satellite. See the Web Sites section for the link to the NGA products.

Important parameters of each spheroidal model are the flattening $f$ and the eccentricity $e$, namely, the quantities:

$$f = \frac{a_{\oplus} - c}{a_{\oplus}} \approx 0.00335 \approx \frac{1}{298.25}, \quad 1 - f = \frac{c}{a_{\oplus}},$$

$$e^2 = \frac{a_{\oplus}^2 - c^2}{a_{\oplus}^2} \approx 0.006694, \quad e \approx 0.0818. \tag{2.1}$$

More precisely, $1/f = 298.25642$ (according to IERS resolution, 2010).

The two parameters $e$ and $f$ are connected by the following relationships:

$$e^2 = f(2 - f), \quad (1 - f)^2 = 1 - e^2, \quad c^2 = a_{\oplus}^2\left(1 - e^2\right) = a_{\oplus}^2\left(1 - f\right)^2 \tag{2.2}$$

With reference to Figure 2.2, let $(r, z)$ be the coordinates of a point X on the ellipse (namely, on the mathematical surface), distant $R = a_{\oplus}\rho$ from the center O:

$$r = R\cos\phi', \quad z = R\sin\phi', \quad \tan\phi' = \frac{z}{r}, \qquad 0 \le r \le a_\oplus, \quad 0 \le z \le c,$$

$$\left(\frac{r}{a_\oplus}\right)^2 + \left(\frac{z}{c}\right)^2 = 1$$

or equivalently

$$(1-f)^2 r^2 + z^2 = a_\oplus^2 (1-f)^2, \quad (1-e^2) r^2 + z^2 = a_\oplus^2 (1-e^2)$$

The geodetic latitude $\phi$ is defined by the angle between the direction of the semi-major axis and the normal to the tangent to the ellipse for X. The angular coefficient $m$ of the tangent is given by

$$m = -\frac{\mathrm{d}z}{\mathrm{d}r} = \pm\left(\frac{c}{a_\oplus}\right)^2 \frac{r}{z}$$

The angular coefficient $m'$ of the normal is immediately found by $mm' = -1$, so that

$$m' = \tan\phi = \mp\left(\frac{a_\oplus}{c}\right)^2 \frac{z}{r}$$

On its turn, the geocentric latitude is simply

$$\tan\phi' = \frac{z}{r} = \left(\frac{c}{a_\oplus}\right)^2 \tan\phi = (1-f)^2 \tan\phi = (1-e^2)\tan\phi \tag{2.3}$$

Numerically,

$$\tan\phi' \approx 0.99330552 \tan\phi, \quad \tan\phi \approx 1.00673960 \tan\phi' \tag{2.4}$$

Given that the latitude is constrained between $\pm 90°$, no quadrant ambiguity can arise. Therefore, the module of the geodetic latitude is always greater or equal to the geocentric one.

From the trigonometric formulae seen in Chapter 1, we also derive that

$$\tan(\phi' - \phi) = \frac{\tan\phi' - \tan\phi}{1 + \tan\phi'\tan\phi} = \frac{q\sin 2\phi}{1 - q\cos 2\phi}, \quad q = -\frac{e^2}{2-e^2}$$

In our case, the difference between the two angles is always a small quantity, so we can apply the following series development, demonstrated by Lagrange using the complex expressions of the trigonometric functions:

$$\phi' = \phi - \left(f + \frac{1}{2}f^2\right)\sin 2\phi + \frac{1}{2}f^2(1+f)\sin 4\phi + \cdots$$

$$\approx \phi - 692''.737\sin 2\phi + 1''.163\sin 4\phi \tag{2.5}$$

where we have made use of the relations (2.2). The difference $|\phi' - \phi|$ reaches its maximum value of $11'.54$ at latitude $45°$. The results given by Equation 2.5 do not differ by those given by Equation 2.4 by more than $0''.1$.

It is useful to remember that this result helps to solve the transcendent equation of frequent utilization:

$$\tan x = m \tan y \quad (m > 0) \tag{2.6}$$

Putting

$$\frac{1-m}{1+m} = q(|q| < 1)$$

the solution is

$$x = y - q \sin 2y + \frac{q^2}{2} \sin 4y - \frac{q^3}{3} \sin 6y + \frac{q^4}{4} \sin 8y + \cdots \tag{2.7}$$

In the same manner, $R$ is found as function of $\phi$ or $\phi'$:

$$r = \frac{a_\oplus \cos\phi}{\sqrt{1 - e^2 \sin^2\phi}} = R\cos\phi', \quad z = \frac{a_\oplus (1 - e^2)\sin\phi}{\sqrt{1 - e^2 \sin^2\phi}} = R\sin\phi'$$

$$R = \sqrt{r^2 + z^2} = a_\oplus \sqrt{\frac{1 - e^2(2 - e^2)\sin^2\phi}{1 - e^2 \sin^2\phi}} \tag{2.8}$$

Or else, by dropping the terms in $e^4$ and higher and using (2.2):

$$R = a_\oplus \left[ \left( 1 - \frac{1}{2}f + \frac{5}{16}f^2 \right) + \frac{1}{2}f\cos 2\phi - \frac{5}{16}f^2 \cos 4\phi + \cdots \right] \tag{2.9}$$

$$\approx 6367.45 + 10.69\cos 2\phi - 0.02\cos 4\phi + \cdots \quad (\text{km})$$

At $28°$ latitude, $R$ is approximately $4.9\,\text{km}$ shorter than $a_\oplus$ and $10.6\,\text{km}$ shorter at $45°$.

We can also express the radius of curvature $k$ (in the same units of $a_\oplus$, say in km) at each location $X(r, z)$:

$$\frac{dz}{dr} = -\frac{1}{\tan\phi}, \quad \frac{d^2z}{dr^2} = \frac{d\tan\phi}{dr} \frac{1}{\tan^2\phi}$$

$$k = \frac{\left( 1 + \left( \dfrac{dz}{dr} \right)^2 \right)^{3/2}}{\left| \dfrac{d^2z}{dr^2} \right|} = a_\oplus \frac{1 - e^2}{\left( 1 - e^2 \sin^2\phi \right)^{3/2}}$$

Therefore, the length $K$ (say in km) of the $1°$ arc of meridian passing for X is

$$K = 2\pi \frac{k}{360} = a_\oplus \frac{\pi}{180} \frac{1 - e^2}{\left(1 - e^2 \sin^2 \phi\right)^{3/2}}$$

By measuring $K$ at several places along the meridian (ideally from pole to pole), the overall curvature of the surface is found. From the comparison of its values for the different places, the quantities $a_\oplus$, $c$, $e$, $f$ can be derived. At the equator, $K \approx 110.6$ km, and at the pole, $K \approx 111.7$ km.

Notice that we do not need a third dimension to determine the radius of curvature of the surface. Of course, observations of the Earth from the outer space are of fundamental value to determine its true shape.

Let us now consider a point X at an altitude $H$ (measured along the geodetic vertical) above the surface of the Earth, such as the top of a mountain, or a low-flying satellite such as the Hubble Space Telescope (HST), which is orbiting at approximately $H = 600$ km above the ground. Because

$$dr = H \cos\phi, \quad dz = H \sin\phi$$

we also have

$$dR = H \cos(\phi - \phi'), \quad d\phi' = \frac{H}{R} \sin(\phi - \phi')$$

From the geodetic coordinates longitude $\Lambda$ and latitude $\phi$, we can derive the geocentric Cartesian coordinates $(x, y, z)$ by taking into account $H$ and the deviation of the vertical. An approximate procedure is the following: calculate the geocentric latitude $\phi'$ by means of Equation 2.5 and the radius $R$ by means of Equation 2.9. Add $H$ to $R$ (no sensible error here from ignoring the deviation of the vertical), and then calculate

$$x = (R(\phi) + H)\cos\phi' \cos\Lambda, \quad y = (R(\phi) + H)\cos\phi' \sin\Lambda, \quad z = (R(\phi) + H)\sin\phi'$$

with due account to the quadrant of $\Lambda$.

Chapter 5 of the *Explanatory Supplement* provides a more accurate formalism to calculate the geocentric Cartesian coordinates $(x, y, z)$ of the point X, by means of two auxiliary functions $C$, $S$:

$$C = \frac{1}{\sqrt{\cos^2\phi + (1 - f)^2 \sin^2\phi}}, \quad S = (1 - f)^2 C$$

which give

$$\begin{cases} x = a_\oplus \rho \cos\phi' \cos\Lambda = (a_\oplus C + H)\cos\phi \cos\Lambda \\ y = a_\oplus \rho \cos\phi' \sin\Lambda = (a_\oplus C + H)\cos\phi \sin\Lambda \\ z = a_\oplus \rho \sin\phi' = (a_\oplus S + H)\sin\phi \end{cases} \tag{2.10}$$

The inverse transformation to derive $(\Lambda, \phi, H)$ from $(x, y, z)$ is usually performed by successive iterations.

As an example of the previous considerations, Table 2.1 shows the geocentric, geodetic and astronomical positions of the intersections of the principle axes of the Copernicus Telescope at Cima Ekar (Asiago, Italy) and of the Italian National Telescope (Telescopio Nazionale Galileo, TNG) in La Palma (Canary Islands, Spain). Notice that the geodetic altitude $H$ does not coincide with the so-called height above sea level (a.s.l., given in parentheses), being determined by a best fitting procedure with a reference ellipsoidal surface that, e.g., in the Canaries, is approximately 50 m below the average sea level. Table 2.1 also shows the appreciable gravity anomaly at the two sites.

To derive these coordinates, the Global Positioning System (GPS), namely, a network of NASA satellites whose positions have been referred to that of WGS84 and a set of stars from the fundamental catalog FK5 were used.

Some practical considerations are worth mentioning. First, the astronomic horizon is not the visible one, even in open sea: the eye of the observer is usually located at a certain height above the surface of the water, so that the visible horizon is a minor circle depressed with respect to the astronomical one according to this height and to the horizontal refraction. See, e.g., a discussion of the effect in Smart (1965, *op. cit.*). Second, the celestial north pole at the present epoch is near the bright star α Ursae Minoris (Polar Star, Polaris), a second magnitude star; the south pole, in the constellation of Octans and approximately 20° distant from the bright galaxy Large Magellanic Cloud, has no bright nearby star. The nearest is σ Octanctis, some 5° away, which is used in the almanacs as south pole reference. Therefore, the celestial poles are not directly materialized. However, they can be identified with great accuracy by using a number of circumpolar stars that permit the accurate determination of the position of the meridian and of the astronomic latitude, as we will discuss in the next paragraph and in Chapter 3, provided the atmospheric refraction (see Chapter 11) is taken into consideration. Much more intriguing was in the past the determination of the longitude of a given location. Many ways were devised (for instance, the ephemerides of the Moon or of the moons of Jupiter, as proposed by Galileo in the 17th century), but finally the superb mechanical clocks produced during the 19th century allowed precise time to be kept, even on board the great ships (see, for instance, Sobel, in the Bibliography section). Since then, continuous improvements in clock technology (quartz oscillators, atomic clocks, optical clocks), the broadcasting of radio signals by networks of ground stations and satellites, have eliminated the problem. This matter is further discussed in Chapter 10.

---

## TABLE 2.1
## Coordinates of the Copernicus and Galileo Telescopes in the WGS84 Reference Ellipsoid

| Telescope | 182 cm Copernicus | 352 cm Galileo (TNG) |
|---|---|---|
| | **Cartesian Geocentric** | |
| $X$ | +4,360,893.8 | +5,327,423.3 |
| $Y$ | +892,690.4 | −1,719,592.5 |
| $Z$ | +4,554,619.0 | +3,051,176.2 |
| | **Geodetic** | |
| $\Lambda$ | (E) +11°34′07″.92 | (W) −17°53′20″.6 |
| $\Phi$ | +45°50′54″.92 | +28°45′14″.4 |
| $H$ (m) | 1435 (1380 a.s.l.) | 2427.6 (2370 a.s.l.) |
| | **Astronomic** | |
| $\Lambda$ | +11°34′22″.14±0″.45 | −17°53′37″.9±0″.45 |
| $\Phi$ | +45°50′36″.99±0″.41 | +28°45′28″.3±0″.60 |

## 2.2   THE HOUR ANGLE AND DECLINATION SYSTEM

We consider now the time variation of the coordinates due to the diurnal rotation of the Earth. On the celestial sphere, the meridian cuts the celestial equator in a point M (see Figure 2.3).

For any given star X, let us draw the great circle passing through X and the visible pole P. This circle, which also passes through the other pole Q, is called the *hour circle* of X. Let X′ be the intersection of the hour circle on the equator, and call *hour angle* of X, $HA$(X), the arc MX′, usually counted westward from M in ($^{h\,m\,s}$) between $0^h$ and $24^h$. Other choices are possible. For example, it is very intuitive to measure $HA$ positive westward from $0^h$ to $+12^h$ and negative eastward from $0^h$ to $-12^h$. At upper culmination on the meridian, $HA = 0^h$, and at lower culmination, $HA = 12^h$. The second coordinate of X is the arc X′X, counted in (° ′ ″) from 0° to 90°, positive toward the north pole and negative toward the south pole. This second coordinate is the *declination* of X, indicated with $\delta$(X): $\delta$(X) = arc X′X. With the passage of time during the day, each star X will describe its parallel of declination, namely, the small circle $\delta$(X) = *const*, by continuously increasing its $HA$. The Sun behaves in the same way, but, in addition, it has an easterly motion of approximately 1°/day along a great circle called the *ecliptic* (see later) with respect to the fixed stars. In other words, the ecliptic is the great circle described by the Sun in 1 year. Being the ecliptic inclined by ≈23°.5 to the celestial equator, the daily movement of the Sun from dawn to dusk is not along a great circle, and not even along a parallel of declination, with quite interesting consequences on the illumination of walls and solar panels, a topic outside the scope of this book. Similar but more complex considerations must be applied to the Moon, the planets and other moving bodies such as comets and asteroids.

For a site at latitude $\phi$ in the northern hemisphere, if $\delta$(X) > 90° − $\phi$, the star will never rise or set, being always above the horizon; these stars are called *circumpolar* stars. If instead $\delta$(X) < −(90° − $\phi$), the star will never be visible above the local horizon. Stars with intermediate declinations between circumpolar and invisible will rise and set. The same considerations apply to sites in the southern hemisphere, with due account to the signs of $\phi$ and $\delta$. By means of circumpolar stars, we can determine $\phi$ even without knowing the declination of the star. Consider indeed a circumpolar star: the semi-sum of its altitudes $h$ above the horizon in upper and lower culmination will immediately give $\phi$; at the same time, the semi-difference will provide $\delta$(X).

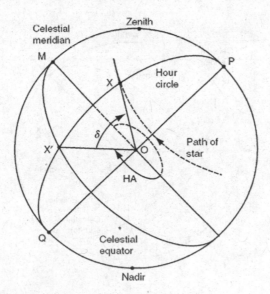

**FIGURE 2.3**   The hour angle and declination system.

## 2.3   THE EQUATORIAL SYSTEM

The $(HA, \delta)$ system introduced in the previous paragraph has the virtue of fixing in time one of the two coordinates, namely, the declination. To make constant also the second coordinate, a point on the equator is needed which remains as fixed as possible with respect to the stars. To determine this point, let us consider the locus occupied by the Sun during its yearly motion, namely, the great circle called *ecliptic* on which the Sun (indicated with ⊙) moves eastward by approximately 1°/day. The ecliptic is inclined to the equator by an angle $\varepsilon$ (obliquity of the ecliptic) of approximately 23°27′. The equator and ecliptic intersect each other in two opposite points called *equinoxes*: the vernal equinox is that point where the Sun transits at the beginning of the spring, around March 21. The autumn equinox occurs 6 months later, around September 21. On both points $\delta_\odot = 0°$, but the time derivative $(d\delta_\odot/dt)$ is positive in the first date and negative in the second. The spring vernal point is usually indicated with the astrological sign of Aries ♈, graphically approximated in the present text with the Greek letter $\gamma$ (gamma), the autumn point with the astrological sign of Libra ♎, approximated with the Greek letter $\Omega$ (Omega).

The points on the ecliptic at 90° from the equinoxes are called *solstices*: summer solstice around June 21 and winter solstice around December 22; the declination of the Sun in these points is $\delta_\odot = \pm\varepsilon$, while its derivative is zero (hence the name, the solar declination is stationary). The great circles passing through the poles and the equinoxes, or the solstices, are called *colures* of the equinoxes or solstices, respectively. The poles of the ecliptic, indicated with E and E′, are points belonging to the colure of the solstices.

With reference to Figure 2.4, given a star X draw the great circle through P and X intersecting the equator in X′. Choose the vernal equinox $\gamma$ as the origin of the first angular coordinate, and measure the angle $\gamma X'$ in direct sense: this angle is the *right ascension* of star X. The second angular coordinate of the equatorial system remains the *declination* $\delta(X)$ defined in the previous paragraph.

The right ascension of the star X, indicated with $\alpha$, $\alpha(X) = $ arc $\gamma X'$, is usually measured in $(^{h\,m\,s})$ from $0^h$ to $24^h$. Notice the sense of $\alpha$, opposite to that of $HA$. The right ascension can also be defined as the angle at the pole between the hour circle passing through X and the vernal colure.

For the Sun, at the equinoxes $\alpha_\odot = 0^h$ and $12^h$, respectively, and at the solstices, $\alpha_\odot = 6^h$ and $18^h$. The north pole of the ecliptic E has $\alpha(E) = 18^h$, $\delta(E) = 90° - \varepsilon$. Point E is in the constellation of

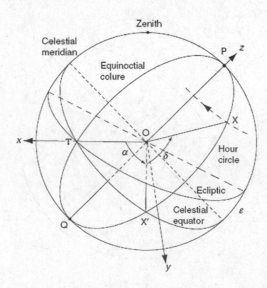

**FIGURE 2.4**   The equatorial system. The figure also shows a Cartesian $(x, y, z)$ system having the celestial equator as $xy$ plane; the axis $z$ is directed to the celestial north pole P.

Draco, near the gaseous nebula NGC 6543; the nearest bright star in the vicinity of E is $\omega$ Draconis, of the fourth visual magnitude, 3° away from it. Point E′ is in the constellation of Dorado.

The right ascension will remain constant in time as much as $\gamma$ remains fixed with respect to the stars. We will see in Chapter 5 that $\gamma$ is subject to secular and periodic motions (general precession, nutation). However, during short time spans (say 1 year), the pair of coordinates $(\alpha, \delta)$ will be almost constant. For longer periods, suitable and accurate correction formulae will be derived. The equatorial system is the fundamental one for all precise descriptions of the celestial sphere; it is the system employed by all major star catalogues.

Let us define *sidereal time*, *ST*, the hour angle of point $\gamma$:

$$ST = HA(\gamma)$$

Because of the rotation of the Earth, the sidereal time is a continuously varying angle between $0^h$ and $24^h$. In the limits of the present simplified assumptions, the *ST* measured in the fundamental meridian of Greenwich can be identified with the universal time, UT. Subsequent chapters will provide more accurate definitions of UT.

Taking into account the opposite sense of $\alpha$ and *HA*, for any star X we have the fundamental relationship:

$$\alpha(X) = ST - HA(X) \tag{2.11}$$

The practical application of Equation 2.11 requires due consideration to the adopted convention for *HA*, because by definition $0^h \leq \alpha \leq 24^h$. In any case, when a star transits through the meridian in upper culmination, its right ascension coincides with the sidereal time. This coincidence can be read in both ways: either we measure *ST* by observing the upper transits of fundamental stars having accurate right ascensions, or we derive the right ascension of a transiting star from the precise knowledge of *ST*. Let us examine in more depth this second possibility. *ST* is a quantity that varies with time (meant as the fundamental variable of all mechanical laws) in a fairly regular way; better yet, suppose that all deviations from uniformity are so small that we can disregard them. We can therefore build a clock whose reading coincides with *ST*. For all practical applications, we could then legitimately identify *ST* with a time, although the rigorous definition is the instantaneous angle on the celestial equator between the meridian and point $\gamma$. This approach was used by all astronomers of the past to build fundamental catalogues of stars. They did know since Hipparchus how to correct for the secular motion of the polar axis due to lunisolar precession of the equinox; since 1736, they also knew how to correct for the shorter period nutation discovered by Bradley (see Chapters 4 and 5). These corrections provided an almost uniform march of their clocks. To be sure, modern data show that both the position of the observer with respect to the direction of the rotation axis in an inertial frame and the daily angular velocity show small and unpredictable but well-measurable fluctuations. This argument will be expounded in detail in Chapters 6 and 10; in the following paragraphs, we shall continue our discussion as if *ST* were by all purposes a uniform time.

Therefore, let us build a telescope with a mechanical mount having only one degree of freedom and that of elevation, while the optical axis is constrained as accurately as feasible in the meridian plane. The focal plane of the telescope is equipped with a high precision device (e.g., grid of wires), in order to measure accurately the instant of transit of the star. This telescope is known as a *meridian circle* or *transit instrument*, according to the several possible practical realizations. If we can identify the instant of transit of $\gamma$ and have our *ST* clock start from zero at that precise moment, then by measuring the Zenith distance and the *ST* of passage of any star, we will derive its right ascension and declination at the date of observation. This procedure is easier said than done, all sort of systematic errors being possible, such as errors of the latitude of the observatory, of the alignment of the optical axis in meridian, of the zero point and march of the clock, of the correction formulae for precession and nutation, and of the correction for atmospheric refraction. As an example, let us discuss one of these systematic errors, namely, the dependence of the precision in right ascension from the declination. Indeed, a hypothetic polar star would never cross the meridian, while an equatorial star has the maximum linear speed in crossing the wires on the focal plane. Let $s$ be the thickness of the wire; the time $\Delta T$ employed by the star to cross the wire will be

$$\Delta T = \frac{ks}{\cos\delta}$$

where $K$ is an instrumental constant. Therefore, the error in $\Delta T$ is

$$d\Delta T = Ks\left(\frac{\sin\delta}{\cos^2\delta}\right)d\delta$$

namely with a systematic dependence from the declination.

The construction of a fundamental catalogue is indeed one of the most complex operations of all astronomy. We will see in Chapter 3 how the $ST$ can be directly linked to the position of the Sun on the ecliptic (in a somewhat pedantic sense, sidereal is a deceiving adjective: the Sun, not the stars, actually defines $ST$). Many stellar catalogues have simply a differential nature, giving positions relative to a set of fundamental stars. After the works carried out by many astronomers in the 17th, 18th and 19th centuries (we recall the names of Flamsteed, Maskelyne, Bessel, Santini, Auwers, etc.), in more recent times the so-called Fundamental Katalog FK was adopted, containing about 1500 bright stars. From 1964 to 1988, the standard catalogue was its fourth edition, FK4. The fifth revision, FK5, was published in 1988 (see Fricke et al., 1988). Positions and proper motions of 1535 bright stars were derived after a new determination of the origin of right ascension, with the adoption of the precession constants recommended by the IAU in 1976 and with the elimination of the elliptical aberration from the mean coordinates (as explained in Chapter 7). An extension of the FK5 contained information on an additional 3117 secondary stars (most of them fainter than the primary stars, down to magnitude 9.5).

A distinction must be made at this point between reference systems and reference frames (see Kovalevsky, in the Bibliography section). The final step to materialize a reference *system* is done by assigning coordinates to a set of fiducial objects. The result is a reference *frame* presented as a catalogue of positions and proper motions.

In 1991, the IAU decided that the future standard celestial reference frame would be centered in the barycenter of the Solar System. The frame had to be realized by an ensemble of very distant point-like extra-galactic radio sources, in such a way to be as consistent as possible with the FK5 and with a full account of the relativistic metrics. The extragalactic distances would insure for centuries to come an essentially rotation-free system. This fundamental catalog, named ICRS (International Celestial Reference System, see the Web Sites section for updated information), replaced the FK5 since 1997. It is based on the equatorial coordinates at J2000.0 of 608 extra-galactic radio sources selected by the International Earth Rotation Service (IERS; see the *IERS Technical Note* July 21, 1996 quoted in the Web Sites section) for astrometric observations with the Very Long Baseline Interferometry (VLBI). The VLBI is a worldwide network of radio telescopes feeding common reduction stations; for references to its description, see the Web Sites section. These sources constitute the International Celestial Reference Frame (ICRF). This reference frame does not depend on the accuracy of the models of precession and nutation, because the movements of the pole are derived from the same ICRF sources. The origin of the Right Ascension is implicitly determined, establishing that the quasi-stellar object (Quasar) 3C 273 has $\alpha$(J2000.0) = $12^h29^m06^s.6997$ (Hazard et al., 1971). This quasar, at cosmological distance, is also the only one bright enough in visible wavelengths (visual magnitude $\approx$ 13) to have been accessible to the satellite Hipparcos. Therefore, the positional system of Hipparcos was referred to the ICRS, and so were the ephemerides of the Solar System bodies calculated by the Jet Propulsion Laboratory (designated JPL DE 400 to 403, see Standish 1990, 1998a,b). Owing to the fact that Hipparcos contains all 1535 FK5 stars, the position of the FK5 pole and equinox are known in the ICRF to better than a few milliarcseconds; the systematic difference of the equinox FK5-ICRS is $87 \pm 10$ milliarcsec (see *Astrometry of Fundamental Catalogues* by Walter and Sovers in the Bibliography section and Kovalevsky et al., 1997 in the References section). Not only VLBI measurements contributed to the definition of accurate reference frames. Folkner et al. (1994) combined lunar laser ranging data (see, e.g., the Apollo project in the Web Sites section) with VLBI data to tie the non-rotating extra-galactic frame to an inertial planetary ephemeris frame. The data release nr. 2 of the astrometric satellite GAIA provides a dense system of faint stars providing the optical connection with VLBI positions (Mignard et al., 2018).

Given the continuously growing precision of time and Earth rotation angle (ERA) measurements, in 2000 the IAU adopted several resolutions (including a revision of nomenclature), which were enforced since 2006. In addition to new precession and nutation theories, a more precise connection between the UT and ERA was adopted (see *The Explanatory Supplement*). A fuller treatment will be given in Chapters 4–10; see, in particular, the "Notes" section of Chapter 5.

## 2.4 TELESCOPE MOUNTS

The meridian circle, in its several realizations, was the main instrument to build catalogs of fundamental stars. For general astronomical applications, the previous considerations on the coordinate systems on the celestial sphere have shown the advantages of building telescopes having an equatorial mount. Indeed, only the hour angle axis needs an accurate tracking rate, the declination axis being used for pointing, fast slewing and corrections to the pointing, due, e.g., to atmospheric refraction or flexures of the structure. However, for all large telescopes, modern engineering has favored the Alt-Azimuth mount, because the structural flexures are much more controllable with vertical and horizontal axes. Additional advantages are lighter structures and smaller domes, which favor a better thermalization, of great importance for image quality (see Chapter 11). The disadvantages are the need to control, with great precision and continuously variable speed, not two but three axes including the field rotation and the small dead zone at the Zenith, as discussed in Chapter 3. Digital electronics has made it possible to solve this problem in an accurate and economic way. The first large Alt-Az telescope was the 6-m Special Astrophysical Observatory of the Russian Academy of Science, built in the Caucasus mountains already in 1966. Another example is the already quoted 3.5 m TNG, shown in Figure 2.5; the light is brought by a third diagonal mirror to the Nasmyth focus visible on the left of the telescope, where a filed-counter-rotator is mounted. Notice that a slanted reflection will introduce spurious polarization, variable on the sky with the field rotation.

## 2.5 THE ECLIPTIC SYSTEM

The ecliptic, inclined to the equator by the obliquity $\varepsilon$, is the fundamental plane of the system of ecliptic coordinates (see Figure 2.6). The origin of the ecliptic longitudes is the same of the equatorial system, namely, the point $\gamma$, which belongs both to the ecliptic and to the equator. Ecliptic longitudes $\lambda$ are given in ($° ' ''$) between 0° and 360°, or in ($^{h\,m\,s}$) between $0^h$ and $24^h$, increasing in the same sense of the right ascensions. Ecliptic latitudes $\beta$ are given in ($° ' ''$) between 0° and ±90°, as the declinations. All planets revolve around the Sun in the same direct sense, with orbital planes having small inclinations $i$ to the ecliptic. The notable exception is Mercury ($i = 7°$); the formerly planet Pluto has $i = 17°$. However, to the terrestrial observer the planets appear to have complex motions with respect to the fixed stars, for instance, reversals of movement, and even stationary points. Such observational complexity was at the basis of the Ptolemaic system.

**FIGURE 2.5**   The 3.5 m Telescopio Nazionale Galileo (TNG). The vertical and horizontal axes are clearly seen. The field rotation is compensated by counter-rotation at the Nasmyth foci on the horizontal axis. (Photo by CB.)

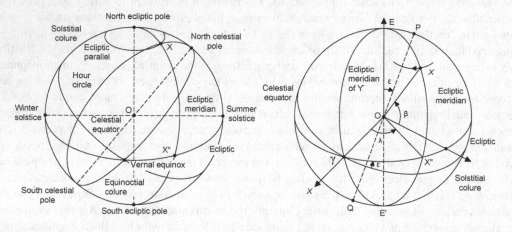

**FIGURE 2.6**   The ecliptic system. (a) The ecliptic sphere, (b) the ecliptic coordinates longitude and latitude and the Cartesian ecliptic axes.

The ecliptic passes through an ensemble of stars organized by the ancient Chaldean and Greek astronomers in 12 constellations, most of which bear the names of animals. These constellations define a band in the sky that was named Zodiac, from the Greek name of animals. The names given by the Latin poet Ausonius (4th century A.D.) are Aries, Taurus, Gemini, Cancer, Leo, Virgo, Libra, Scorpio, Sagittarius, Capricornus, Aquarius and Pisces. Those are also the names of the astrological signs. Those signs are meant to be fixed with respect to the calendar but not with respect to the stars because of the precession of the equinox (see Chapters 4 through 6). These constellations are distributed in a band of ±8° around the ecliptic. The Moon and the visible planets are also confined in such band. Actually, there is a 13th constellation that perks through the Zodiac, namely, Ophiuchus, between Scorpio and Sagittarius, which is ignored by astrologers. Additional information about star names and constellations is given in the "Notes" section.

For objects of the Solar System, an ecliptic Cartesian coordinate system ($x$, $y$, $z$) can be established, having its center at the Earth or at the Earth–Moon barycenter, or at the Sun's center or at the barycenter of the Solar System, according to the convenience. The x-axis is directed toward point $\gamma$, the y-axis at 90° in direct sense in the ecliptic plane and the z-axis directed toward the ecliptic north pole E.

Since centuries, it is known that the obliquity of the ecliptic is not absolutely fixed; the planetary perturbations cause a very slight decrease (at the present epoch) amounting to $0''.47$ per year, as explained in Chapter 5.

## 2.6 THE GALACTIC SYSTEM

The galactic system is a reference system of a different nature than those illustrated in the previous paragraphs, which are based on the Earth, the Sun and geometrical considerations. Its origin coincides by definition with the observer; however, the fundamental plane is determined by the distribution in space of cosmic matter. Therefore, the system does not rely on measurements of directions and time, but on counts of stars in its initial construction and later on the determination of hydrogen gas surface brightness in the various areas of the sky.

The first ("old") system of galactic coordinates was based on counts of stars, which confirmed the visual impression that the Solar System is in the symmetry plane of a flattened system of stars. The pioneering work of William Herschel at the end of the 18th century, and later that of his son John in the southern hemisphere, played a fundamental role to build such model. The Herschels set forth a coherent plan of *stellar gauges* over the entire celestial sphere, which was continued by many other astronomers and finally led to the first galactic system $(l^I, b^I)$. The longitudes $l^I$ are counted in degrees between 0° and 360° in the direct sense. The latitudes $b^I$ are counted in degrees between 0° and ±90°. The north pole G of $(l^I, b^I)$ is in the direction of the cluster of galaxies in Coma ($\alpha_G = 12^h.8$, $\delta_G = +27°.4$), almost in the plane of the ecliptic. This orthogonality between the ecliptic and the galactic plane has been the subject of many theoretical studies on the tidal effect of the Milky Way on the present structure of the Solar System.

The procedure of counting stars according to their position and apparent luminosity is by no means an easy task, even applying refined statistical methods (see, e.g., the book by Trumpler and Weaver in the Bibliography section). The determination of the fundamental plane as the one passing through the areas of maximum density of stars is hampered by the presence of a strong interstellar absorption, whose amount increases precisely going into the plane (we shall examine the interstellar absorption again in Chapter 16). Immune to this absorption is instead the surface brightness of the interstellar gaseous hydrogen clouds, which are concentrated toward the galactic plane. Thanks to their low density and low temperature, they emit a strong spectral line in the radio-frequency domain, at $\lambda = 21$ cm (corresponding to 1420 MHz). It has been found that the galactic plane determined by the areas of maximum 21-cm brightness is inclined by approximately 3° to that of $(l^I, b^I)$. Furthermore, a strong and point-like radio source in Sagittarius, in the very direction of the Milky Way center, having equatorial coordinates ($\alpha_{GC} = 17^h42^m$, $\delta_{GC} = -28°55'$ at 1950.0), can be used as zero point for the longitudes. This radio-frequency galactic system, which became available around 1960, was initially named second galactic system $(l^{II}, b^{II})$. Its superiority over the old system was soon evident, so that by a resolution of IAU in 1976, the system based on star counts was discontinued, and the radio system was simply named $(l, b)$. The conversion between equatorial coordinates (at 1950.0) and the old and new galactic coordinates was published in the *Annals of the Observatory of Lund* numbers 15, 16 and 17, 1962.

With more precision, referring to the equinox and equator at B1950.0 (see Chapter 4 for the definition of the Besselian year), the equatorial coordinates of the galactic north pole G, and the position angle $\theta_{GC}$ of the galactic center from such a pole are

$$\alpha_G = 12^h49^m \qquad \delta_G = +27°.4, \qquad \theta_{GC} = 123° \quad (\text{B1950.0})$$

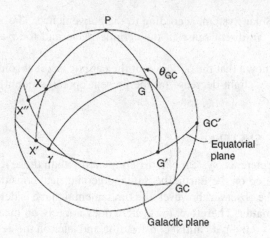

**FIGURE 2.7** The system of galactic coordinates.

The position of the galactic system with reference to the equatorial one is shown in Figure 2.7.

The transformations between equatorial and galactic coordinates are given in Section 3.2.

The galactic coordinates are never used for high precision positional work. Actually, using the most precise determination of the position of the radio source, named, today SgrA* ($17^h45^m40^s.0409$, $-29°00'28''.118$ at J2000), we could notice a slight difference with the defined origin of the coordinates of the center, about 4' in longitude. Moreover, while the galactic system for definition passes through the Sun, the Sun itself is slightly above the hydrogen plane, approximately 17 pc; therefore, the latitude of SgrA* is $\approx -7'$. Finally, pay attention that the galactic rotation system does not rotate with time, while the galactic rotation takes the Sun in slow rotation about the center (see Chapter 9); the effect is of the order of 6 milliarcsec per year.

Do not confuse the galactic coordinates, whose center is the observer (but in practice the Sun), with the *galacto-centric* coordinates (X, Y, Z) having the galactic center as the origin. The transformation between the two requires a model of the Milky Way, in particular the determination of the distance of the Sun from the center of the Galaxy, presently estimated at around 8 kiloparsecs. The extraordinary properties of the galactic center will be illustrated in Chapter 9 when speaking of the gravitational redshift and gravitational waves.

## 2.7 OTHER SYSTEMS

There are other systems of coordinates used by astronomers and not covered in this book, for instance, those on the surface of the Sun or those associated with the magnetic fields of the Earth or Jupiter or Saturn. See, for instance, Hapgood (1992, 1995) and Franz (2002) in the References section. Regarding cartography and the rotation of the Sun (see the "Notes" section), the Earth, the Moon and the other planets, satellites, minor planets and comets, the IAU has instituted a Working Group on Cartographic Coordinates and Rotational Elements. This Group revises its recommendations every 3 years or so. For the latest report, see Archinal et al. (2018). We mention that the distribution of the nearest galaxies, in particular those seen in Virgo, has been used to determine a *super-galactic* coordinate system (L, B); see de Vaucouleurs and Peters (1984).

## NOTES

- Names of stars and constellations. The stars visible to the naked eye (numbering about 3000 in each hemisphere) are designated by Greek letters followed by the abbreviated Latin name of the constellation, letter $\alpha$ indicating the brightest star of that constellation,

$\beta$ the second brightest and so on until $\zeta$, the last visible class (e.g., $\alpha$ UMa, $\beta$ CMi, $\gamma$ Dra). Since telescopic observations, the lettering system was extended to cover the entire Greek alphabet. The visual brightness indicated by the Greek letter is usually valid only inside a particular constellation, because keeping a uniform magnitude scale over the visible celestial hemisphere is a difficult task, impossible to achieve with the naked eye (see Chapter 16). The brightest naked-eye stars often bear a proper name, deriving either from Greek or Arabian designations (e.g., the proper name of $\alpha$ Lyr is Vega, that of $\alpha$ Ori is Betelgeuse, that of $\beta$ Ori is Rigel, that of $\beta$ Per is Algol and so on). These bright stars have long since been arranged in *constellations*, namely, in groups of stars defining broad areas over the celestial sphere, to be used as a first rough indication of direction. The names and borders of these groupings (named also *asterisms*) changed according to epoch and civilization, the constellations of the Chinese astronomers, for instance, or of the Maya, being different from those of the Arabs. According to the present international convention (since 1922), there are 88 constellations in the sky with precisely defined borders, bearing the names of animals (Leo, Draco, Ursa Major, etc.), of mythological heroes (Cassiopeia, Andromeda, etc.) and of particular figures (Libra, Corona Borealis, etc.). Useful books for the names of stars and constellations are the one by Allen and the one by Bakich quoted in the Bibliography section.

- Sun's rotation. Using the photospheric sunspots as fiducial marks, the Sun displays a differential rotation, being faster at the equator (sidereal period of 24.47 days, namely $14°.713$/day, synodic one of 26.24 days) and lower at higher latitudes. The system of heliographic coordinates goes back to the classical papers by Carrington (1863, see the Bibliography section), whose average sidereal rotation period is 27.2753 days. The solar equator is inclined to the ecliptic by $7°.25$. The rotation is counterclockwise if seen by an observer in the northern hemisphere. The internal rotational field is studied via the helioseismology.

# 3 Transformations of Coordinates

In this chapter, we consider several rules for transforming coordinates from one system to another. Two techniques will be used: matrix rotation and spherical trigonometry. In the majority of cases, the transformations will be rigid rotations around the origin. In other cases, a translation of origin must be added, e.g., in passing from the geocentric to the heliocentric coordinates of a comet. Later on, we will encounter phenomena that give rise to slight distortions of the celestial sphere, such as the aberration and the gravitational deflection of light.

## 3.1 TRANSFORMATIONS BY MATRIX ROTATION

Consider two right-handed Cartesian orthogonal systems, say $(x, y, z)$ and $(X, Y, Z)$, having the same origin O. In order to transform the coordinates $(x, y, z)$ of a point P in one system to those in the other, $(X, Y, Z)$, the following relationships can be used:

$$\begin{cases} X = x\cos xX + y\cos yX + z\cos zX \\ Y = x\cos xY + y\cos yY + z\cos zY \\ Z = x\cos xZ + y\cos yZ + z\cos zZ \end{cases} \tag{3.1}$$

or else with matrix notation

$$\begin{pmatrix} X \\ Y \\ Z \end{pmatrix} = \mathbf{R} \begin{pmatrix} x \\ y \\ z \end{pmatrix}, \quad \mathbf{R} = \begin{pmatrix} \cos xX & \cos yX & \cos zX \\ \cos xY & \cos yY & \cos zY \\ \cos xZ & \cos yZ & \cos zZ \end{pmatrix}, \tag{3.2}$$

where $\mathbf{R}$ is the *rotation matrix* which must be specified for each case. The distance $r$ of P from O is clearly invariant under this rotation:

$$r^2 = x^2 + y^2 + z^2 = X^2 + Y^2 + Z^2$$

Let us consider the polar system $(r, \lambda, \beta)$ defined by $O(x, y, z)$ and a rotated one $(r, \Lambda, B)$ defined by $O(X, Y, Z)$. The transformations are

$$\begin{cases} x = r\cos\beta\cos\lambda \\ y = r\cos\beta\sin\lambda \\ z = r\sin\beta \end{cases}, \quad \begin{cases} X = r\cos B\cos\Lambda \\ Y = r\cos B\sin\Lambda \\ Z = r\sin B \end{cases} \tag{3.3}$$

As pointed out in Chapter 1, several authors prefer to use the complement of $\beta$ as polar angle. By substituting Equation 3.3 in the previous matrix notation (3.2), we obtain

$$
\begin{pmatrix} \cos B \cos \Lambda \\ \cos B \sin \Lambda \\ \sin B \end{pmatrix} = \mathbf{R} \begin{pmatrix} \cos \beta \cos \lambda \\ \cos \beta \sin \lambda \\ \sin \beta \end{pmatrix} \tag{3.4}
$$

Notice that all dependence on $r$ has disappeared, so that those relations also apply to the sphere with unit radius. The inverse transformation is obtained by interchanging the role of $(x, y, z)$ with $(X, Y, Z)$, paying attention to maintain the positive sense on the angles. This implies that the matrix of the inverse rotation is the *transpose* of $\mathbf{R}$:

$$
\mathbf{R}^{-1} = {}^{T}\mathbf{R}, \qquad \left( {}^{T}R_{ij} = R_{ji} \right), \qquad \mathbf{R}^{-1}(\theta) = \mathbf{R}(-\theta), \qquad \left( \mathbf{R}_i \mathbf{R}_j \right)^{-1} = \mathbf{R}_j^{-1} \mathbf{R}_i^{-1}
$$

Furthermore, a generic rotation can always be represented as the result of three different successive rotations, $\mathbf{R}_1$ around the $x$-axis, $\mathbf{R}_2$ around the $y$-axis and $\mathbf{R}_3$ around the $z$-axis, $\mathbf{R} = \mathbf{R}_1 \mathbf{R}_2 \mathbf{R}_3$, with

$$
\mathbf{R}_1(\phi_1) = \begin{pmatrix} 1 & 0 & 0 \\ 0 & \cos\phi_1 & \sin\phi_1 \\ 0 & -\sin\phi_1 & \cos\phi_1 \end{pmatrix},
$$

$$
\mathbf{R}_2(\phi_2) = \begin{pmatrix} \cos\phi_2 & 0 & -\sin\phi_2 \\ 0 & 1 & 0 \\ \sin\phi_2 & 0 & \cos\phi_2 \end{pmatrix},
$$

$$
\mathbf{R}_3(\phi_3) = \begin{pmatrix} \cos\phi_3 & \sin\phi_3 & 0 \\ -\sin\phi_2 & \cos\phi_3 & 0 \\ 0 & 0 & 1 \end{pmatrix}
$$

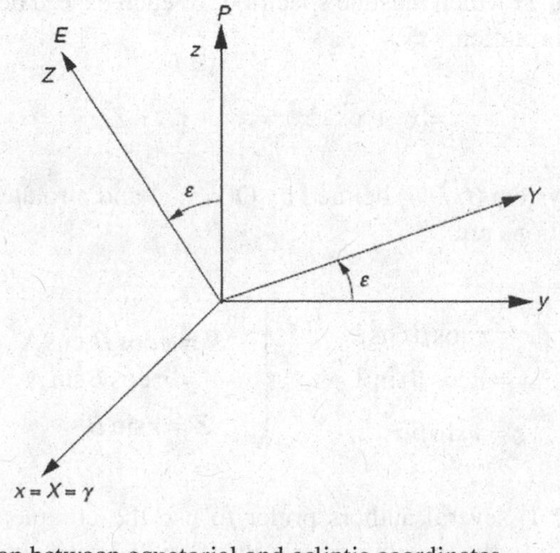

**FIGURE 3.1**   Transformation between equatorial and ecliptic coordinates.

As a first example, consider the transformation from equatorial coordinates $(\alpha, \delta)$ to ecliptic coordinates $(\lambda, \beta)$, by orienting the axes $x$ and $X$ from O toward the vernal point $\gamma$, the $z$-axis to the celestial north pole P and the $Z$-axis to the ecliptic north pole E. The values of the angles are

$$xX = 0, \quad xY = \frac{\pi}{2}, \quad xZ = \frac{\pi}{2}, \quad yY = \varepsilon, \quad zZ = \varepsilon, \quad zY = \frac{3}{2}\pi + \varepsilon, \quad \text{etc.}$$

where $\varepsilon \approx 23°27'$ is the obliquity of the ecliptic. The two systems are therefore connected by a rotation of $\varepsilon$ around the $x$-axis, $\mathbf{R}_1(\varepsilon)$ or inversely of $-\varepsilon$ around the $X$-axis (see Figure 3.1).

$$\mathbf{R}_1(\varepsilon) = \begin{pmatrix} 1 & 0 & 0 \\ 0 & \cos\varepsilon & \sin\varepsilon \\ 0 & -\sin\varepsilon & \cos\varepsilon \end{pmatrix} \approx \begin{pmatrix} 1 & 0 & 0 \\ 0 & 0.9171 & 0.3987 \\ 0 & -0.3987 & 0.9171 \end{pmatrix}$$

The sought-for transformation is therefore

$$\begin{cases} \cos\beta\cos\lambda = \cos\delta\cos\alpha \\ \cos\beta\sin\lambda = \cos\delta\sin\alpha\cos\varepsilon + \sin\delta\sin\varepsilon \\ \sin\beta = -\cos\delta\sin\alpha\sin\varepsilon + \sin\delta\cos\varepsilon \end{cases} \tag{3.5}$$

The inverse is (pay attention to the signs)

$$\begin{cases} \cos\delta\cos\alpha = \cos\beta\cos\lambda \\ \cos\delta\sin\alpha = \cos\beta\sin\lambda\cos\varepsilon - \sin\beta\sin\varepsilon \\ \sin\delta = \cos\beta\sin\lambda\sin\varepsilon + \sin\beta\cos\varepsilon \end{cases} \tag{3.6}$$

Notice that three equations are necessary to determine two angles and their signs (quadrants). We have already remarked that great care is needed in numerical calculations, especially in the proximity of the poles of the systems.

With the same technique, we can transform Alt-Az $(A, h)$ coordinates (with our adopted origin from the south; several authors prefer the Zenith distance $z$ instead of the altitude $h$) in hour angle and declination $(HA, \delta)$ and then to equatorial $(\alpha, \delta)$ by the knowledge of the sidereal time $ST$. In this case, axes $x$ and $X$ will both point to W, axis $y$ to S, axis $z$ to Z, axis $Y$ to M on the celestial equator and axis $Z$ to the celestial north pole P (see Figure 3.2). Clearly, the astronomical latitude $\phi$ of the site is needed. In this case, the rotation matrix will be

$$\mathbf{R} = \begin{pmatrix} 1 & 0 & 0 \\ 0 & \sin\phi & \cos\phi \\ 0 & -\cos\phi & \sin\phi \end{pmatrix} \tag{3.7}$$

However, by convention, the sense of the Cartesian angles is opposite to that of $HA$ and $A$, both increasing in the retrograde sense, and therefore

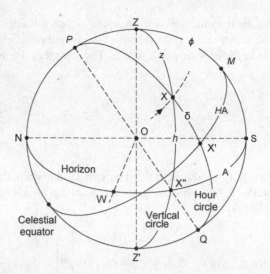

**FIGURE 3.2** Transformation between Alt-Az and hour angle and declination of star X. Arc X″X is the elevation $h$, arc X′X the declination $\delta$, arc MX′ the hour angle $HA$, arc SX″ the Azimuth $A$ and arc NP = arc MZ the latitude $\phi$.

$$\begin{cases} \sin HA \cos\delta = \sin A \cos h \\ \cos HA \cos\delta = \cos A \cos h \sin\phi + \sin h \cos\phi \\ \sin\delta = -\cos A \cos h \cos\phi + \sin h \sin\phi \end{cases} \quad (3.8)$$

The inverse transformation is

$$\begin{cases} \cos h \sin A = \cos\delta \sin HA \\ \cos h \cos A = \cos\delta \cos HA \sin\phi - \sin\delta \cos\phi \\ \sin h = \cos\delta \cos HA \cos\phi + \sin\delta \sin\phi \end{cases} \quad (3.9)$$

Suppose now that we know the equatorial coordinates $(\alpha, \delta)$ of a given star X and the sidereal time $ST$, so that $HA$ is also immediately known. In order to point a telescope having an Alt-Az mount toward X, we need to calculate $(A, h)$ from Equation 3.9 (here, we ignore the atmospheric refraction, see Chapter 11). The third equation will tell if the star is visible above the horizon; the visibility limit $h = 0°$ is reached when

$$\cos HA = -\tan\delta \tan\phi \quad (3.10)$$

(hour angle of rising or setting). In the same way, the third equation in 3.8 gives the Azimuth of rising or setting:

$$\cos A = -\sin\delta \sec\phi$$

Notice that only the equatorial stars ($\delta = 0°$) raise and set exactly on the E and W points. For the circumpolar stars ($|\delta| > |\phi|$), it is always true that $|\tan\delta \cdot \tan\phi| > 1$, so that they are visible above the horizon at all time. The two following relations also apply:

$$A = \arctan \frac{\sin HA \cos \delta}{\cos HA \cos \delta \sin \phi - \sin \delta \cos \phi} \qquad (3.11)$$

$$h = \arcsin \left( \cos \delta \cos HA \cos \phi + \sin \delta \sin \phi \right) \qquad (3.12)$$

Therefore, the Azimuth remains constrained between a maximum and a minimum value (occidental or oriental digression, or elongation), for which the following relations hold:

$$\cos HA = \tan \phi \cot \delta, \qquad \sin h = \sin \phi / \sin \delta, \qquad \sin A = -\cos \delta / \cos \phi$$

Notice that there are two solutions for $HA$ and $A$: one is for the occidental and the other for the oriental digression, that can be easily discriminated.

It is useful also to derive the angular velocities $\dot{A}, \dot{h}$ of a generic star by using the sidereal time as the time variable $t = ST$. In a first approximation, neglecting the effects of atmospheric refraction:

$$\frac{dHA}{dt} = 1, \quad \dot{\delta} = 0, \quad \dot{h} \cos h = -\cos \delta \sin HA \cos \phi$$

From this and from the first equation in 3.8, we obtain

$$\dot{h} = -\cos \phi \sin A, \qquad \dot{A} = \sin \phi + \cos A \tan h \cos \phi \qquad (3.13)$$

The velocity in altitude $\dot{h}$ is always restricted between $\pm 1$ (namely, $\pm 15°$/sidereal hour); it is nil for a telescope at the geographic poles and maximum for an equatorial telescope. More complex is the behavior of the Azimuthal velocity. At the horizon, $\dot{A} = \sin \phi$, and therefore, it is positive in the northern hemisphere, and negative in the southern one both at rising and setting and obviously stationary at the terrestrial equator. This result can be understood by remembering that we have defined the apparent sense of rotation of the celestial sphere by turning our shoulders to the visible pole.

In the particular case of a star transiting at the Zenith ($\delta = \phi$), the Azimuthal velocity becomes infinitely large when the star approaches the meridian; for this reason, a telescope with an Alt-Az mount has a blind spot around the Zenith, a cone whose aperture can be made smaller than $1°$ with careful selection of the motors and associated controls.

For a circumpolar star, at the maximum digressions the velocity is all in altitude; this fact can be advantageously utilized for better determination of the meridian and of the latitude.

In modern telescopes, sophisticated pointing algorithms take into account refraction and several other factors here neglected, including flexures of the structure and misalignments of the optics (see the Web Sites section for Active and Adaptive Optics).

## 3.2  TRANSFORMATIONS BY SPHERICAL TRIGONOMETRY

Spherical trigonometry is the second method of transformation. As an example, Figure 3.3 gives the elements necessary to carry out the transformation between equatorial and ecliptic coordinates. We would easily find again Equations 3.5, 3.8 and their inverse.

More complex is the transformation from equatorial to galactic, so it is better to see it in detail (remember though that low precision is usually needed for galactic coordinates). With reference to Figure 2.4, let P be the celestial north pole, G the galactic center and CG the galactic plane. Given a star X, from the spherical triangles we obtain

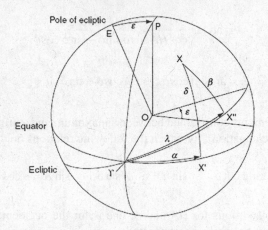

**FIGURE 3.3**    Transformation from equatorial to ecliptic coordinates.

$$
\left\{
\begin{aligned}
\cos b \sin(\theta_{GC} - l) &= \cos \delta \sin(\alpha - \alpha_{GC}) \\
\cos b \cos(\theta_{GC} - l) &= \cos \delta_G \sin \delta - \sin \delta_G \cos \delta \cos(\alpha - \alpha_G) \\
\sin b &= \sin \delta_G \sin \delta + \cos \delta_G \cos \delta \cos(\alpha - \alpha_G)
\end{aligned}
\right.
\tag{3.14}
$$

Inversely,

$$
\left\{
\begin{aligned}
\cos \delta \sin(\alpha - \alpha_G) &= \cos b \sin(\theta_{GC} - l) \\
\cos \delta \cos(\alpha - \alpha_G) &= \sin b \cos \delta_G - \cos b \sin \delta_G \cos (\theta_{GC} - l) \\
\sin \delta &= \sin b \sin \delta_G + \cos b \sin \delta_G \cos(\theta_{GC} - l)
\end{aligned}
\right.
\tag{3.15}
$$

Should one prefer the technique of matrix rotation, recall the equatorial coordinates at epoch B1950.0, as defined by the IAU, of the three points:

$$
\gamma(0, 0), \qquad G(192°.3, +27°.4), \qquad GC(265°.6, -28°.9)
$$

Then, calculate the angular distances:

$$
\cos \gamma G = \cos xZ = -0.86760, \qquad \gamma G = xZ = 150°.2,
$$

$$
\cos \gamma GC = \cos xX = -0.06690, \qquad \gamma GC = xX = 93°.9
$$

and so on. Finally, the complete direct rotation matrix at epoch B1950.0 is

$$
\mathbf{R}_G =
\begin{pmatrix}
-0.06690 & +0.49273 & -0.86760 \\
-0.87276 & -0.45035 & -0.18838 \\
-0.48354 & +0.74459 & +0.46020
\end{pmatrix}
\tag{3.16}
$$

Notice that the equatorial coordinates of the star have to be precessed to B1950.0 before carrying out the transformation. Although not formally defined the IAU, we can assume the following values for J2000:

$$G(192°.84, +27°.13) = \left(12^h51^m, +27°07'.7\right),$$

$$GC(266°.41, -28°.94) = \left(17^h45^m.6, -28°56'.2\right)$$

## 3.3 SOME EXAMPLES AND APPLICATIONS

1. Let us apply the previous relationships to calculate the angular distance between two stars $X_1$ and $X_2$, a number that is independent of the particular system of coordinates. To be specific, consider equatorial coordinates and draw the great circle passing through the two stars (see Figure 3.4). Irrespective of the position of $\gamma$, we have

$$\cos X_1X_2 = \sin \delta_1 \sin \delta_2 + \cos \delta_1 \cos \delta_2 \cos \Delta\alpha, \quad \Delta\alpha = |\alpha_1 - \alpha_2| \tag{3.17}$$

Imagine viewing the celestial sphere from O. The great circle $X_1P$ gives the direction of the celestial north pole through $X_1$. The angle $PX_1X_2$, counted from the north toward east, is the *position angle p* of star $X_2$ with respect to star $X_1$, which is also the angle at the vertex $X_1$ of the spherical triangle $X_1PX_2$. Therefore,

$$\sin X_1X_2 \sin p = \cos \delta_2 \sin \Delta\alpha$$

$$\sin X_1X_2 \cos p = \cos \delta_1 \sin \delta_2 - \sin \delta_1 \cos \delta_2 \cos \Delta\alpha$$

The same result is found using the spherical triangle $X_1QX_2$, where Q is the pole opposite to P, as in Figure 3.1.

In many applications, e.g., binary stars and pairs of galaxies, the angular distance between the two objects is so small that no error is committed by allowing

$$\cos X_1X_2 = 1, \quad \sin X_1X_2 = X_1X_2 = s, \quad s \sin p = \cos \delta_1 \Delta\alpha,$$

$$s \cos p = \Delta\delta, \quad s = \sqrt{\left(\Delta\alpha \cos \delta_1\right)^2 + \Delta\delta^2}$$

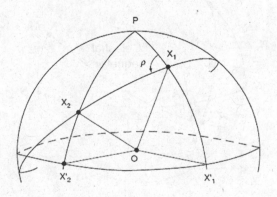

**FIGURE 3.4** Computing the angular distance between two stars $X_1$ and $X_2$.

2. As a second example, consider a telescope mechanically mounted in Alt-Az. Its field of view is in continuous rotation with variable angular speed, because the celestial sphere rotates around a direction that does not coincide with the mechanical axes. For a given star X, call *parallactic angle q* the angle $q = Z\hat{P}X$, which can be expressed as

$$q = Z\hat{P}X = \frac{\sin A \, \cos\phi}{\cos\delta} \qquad (3.18)$$

Its angular velocity (time derivative with $t = ST$, $\dot{A} = 1$) is

$$\dot{q} = \frac{\cos\phi \, \cos A}{\cos\delta \, \cos q} \dot{A} = \frac{\cos\phi \cos A}{\cos h} \qquad (3.19)$$

Field rotation is also encountered in equatorially mounted telescopes if part of the structure is fixed with respect to the ground, e.g., in the so-called Coudè focus, where the light is brought by several mirrors to a large instrument on the floor of the observatory. When the parallactic angle is 90°, namely, when the hour and vertical circles through the star are perpendicular to each other, the star is at its digression, or elongation, as discussed before.

3. Apply the transformation between equatorial and ecliptic coordinates to the Sun (assuming that $\beta_{\odot} = 0°$, an approximation valid for the present purpose). From Equation 3.6, it is immediately seen that

$$\sin\alpha_{\odot} = \tan\delta_{\odot} / \tan\varepsilon \qquad (3.20)$$

so that, with due consideration to the date (namely to the quadrant), the measurement of $\delta_{\odot}$ gives at any time the origin of the right ascension, namely, the position of point $\gamma$. This consideration underlines the fundamental role played by the Sun in the definition of the sidereal time.

4. The *astronomical night* is by definition that period of time when the Sun is 18° or more below the local horizon (see Figure 3.5) of an observer in latitude $\phi$ and longitude $\Lambda$. The period when the Sun has $0° < h_{\odot} < -6°$ is the *civil twilight* and when $-6° < h_{\odot} < -12°$ is the *nautical twilight*. The onset of the civil twilight can be ascertained by the appearance of first magnitude stars and that of the astronomical twilight by the visibility of fourth magnitude stars. The legislation of several countries refers to the start and end of the civil night for obligations such as switching on and off the road lamps, the car lights, etc.

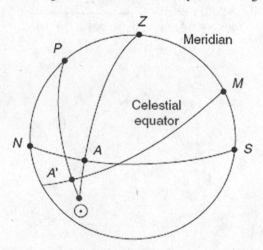

**FIGURE 3.5**   The astronomical night. The arc $A$-Sun ($\odot$) must be $\geq 18°$.

Following the *Astronomical Almanac*, Section A, the terms "sunrise" and "sunset" refer to the instant when the geocentric Zenith distance of the Sun is 90°50′ (16′ solar semi-diameter +34′ of conventional atmospheric refraction), because then the upper limb of the Sun will appear above the horizon of an observer at sea level. Solar parallax is so small that can be ignored. In reality, the precise instant depends on the true atmospheric refraction and height of the observer.

Ignoring the diurnal eastward motion of the Sun, the relation

$$\sin h_\odot = \cos HA_\odot \cos \delta_\odot \cos \phi + \sin \delta_\odot \sin \phi$$

determines the length of the above arcs. By the same formulae, one can calculate the duration of the arc where the Sun is always above the horizon. Those durations obviously depend not only on the latitude of the site but also on the season (namely on $\delta_\odot$). For example, for a site at $\phi \approx +46°$, around the summer and the winter solstices the durations are $15^h44^m$ in summer, $8^h39^m$ in winter, respectively. Notice also that for the Sun, the instant of transit in meridian is not the same as the instant of upper culmination, the difference depending on the date and amounting to few seconds.

Table 3.1 shows examples of the duration of the astronomical night for sites at $|\phi| = 0°$, 23°.5, 45°, 66°.6 for the three cases $\delta_\odot = 0°, \pm\varepsilon$. It is customary to give the durations only to the minute. The total number of astronomical hours is larger for an observatory at lower latitudes. Furthermore, it can be seen that the southern hemisphere has slightly more dark hours than a northern one, because of the different duration of the seasons (see also in Chapters 4 and 10).

The same considerations apply to the Moon. However, the lunar parallax and orbital eccentricity cannot be ignored, so that moonrise and moonset are calculated as the instants when the Zenith distance is $90° + 34′ + s - \pi$, being $s$ the lunar semi-diameter and $\pi$ the lunar parallax at the moment of calculations. As in the case of the Sun, the observed instant may differ according to the real refraction and height of the observer (see also Chapter 8).

The quoted Section A of the *Astronomical Almanac* gives also examples of the calculation of the time when those phenomena take place for different observers, and how to interpolate for other locations. A simpler way to calculate the approximate Universal Time, UT, of rising or setting of a body (be it the Sun, or the Moon or a star) having equatorial coordinates $(\alpha, \delta)$ for a given place of latitude $\phi$ and longitude $\Lambda$ is the following:

$$UT = 0.99727\left[\alpha - \Lambda \pm \cos^{-1}(-\tan\phi\tan\delta) - GMST\,\text{at}\,0^h\,UT\right] \qquad (3.21)$$

where $\alpha$ and $\Lambda$ are in time unit, and GMST is the Greenwich mean sidereal time (see Chapter 4 for the definition of UT and GMST and the reason for the constant 0.99727,

---

**TABLE 3.1**

**Duration of the Astronomical Night**

| $|\phi|$ | $\delta_\odot$ | | | $\delta_\odot$ | | |
|---|---|---|---|---|---|---|
| | $-\varepsilon$ | $0°$ | $\varepsilon$ | $-\varepsilon$ | $0°$ | $\varepsilon$ |
| | $HA_\odot$ (h) | | | Duration Astronomical Night (h) | | |
| 00°.0 | 7.31 | 7.20 | 7.31 | 9.39 | 9.60 | 9.38 |
| 23°.5 | 6.69 | 7.31 | 8.25 | 10.62 | 9.38 | 7.50 |
| 45°.0 | 6.16 | 7.73 | 10.38 | 11.68 | 8.54 | 3.24 |
| 66°.5 | 5.41 | 9.39 | — | 13.18 | 5.24 | — |

which in essence is the ratio between the mean solar time and the *ST*). The positive sign corresponds to setting and the negative to rising. The formula ignores the atmospheric refraction, the semi-diameter, the parallax and variation of coordinates during the day. If $\tan \phi \tan \delta \geq 1$, the phenomenon does not take place. The large diurnal parallax of the Moon requires a correction to Equation 3.21, discussed in Chapter 8.

## EXERCISES

1. *Azimuth counted from the north*: Suppose that the Azimuth *A* is counted from N toward E; find the matrix rotation to derive $(HA, \delta)$ from $(A, h)$; answer: $\mathbf{R} = \mathbf{R}_2(\phi - 90°)\mathbf{R}_3(180°)$.
   The inverse needed to obtain $(A, h)$ from $(HA, \delta)$ is

$$\mathbf{R}^{-1} = \mathbf{R}_3(-180°)\mathbf{R}_2(90° - \phi)$$

   In practice, change sign to $\sin A$ and $\cos A$.
   In the same way, if the Zenith distance *z* is used instead of the altitude *h*, replace $\cos h$ with $\sin z$.

2. *Matrix conversion from* $(\alpha, \delta)$ *to* $(HA, \delta)$: In this case, the full machinery of matrix conversion is surely overabundant. However, when it is set up, it will also work in simple cases. A star has equatorial coordinates $\alpha = 17^h20^m36^s.622$, $\delta = +40°00'00''.00$; find $(HA, \delta)$ when the sidereal time is $ST = 6^h10^m16^s.550$. The rotation angle bringing the hour angle in coincidence with the right ascension has module *ST* around the polar axis:

$$\mathbf{R}_3(ST) = \begin{pmatrix} \cos ST & \sin ST & 0 \\ -\sin ST & \cos ST & 0 \\ 0 & 0 & 1 \end{pmatrix} = \begin{pmatrix} -0.04482 & +0.99899 & 0 \\ -0.99899 & -0.04482 & 0 \\ 0 & 0 & 1 \end{pmatrix}$$

However, $(HA, \delta)$ is a left-handed system, and $(\alpha, \delta)$ is a right-handed one, so the correct transformation by Equation 3.2 is

$$\begin{pmatrix} X \\ Y \\ Z \end{pmatrix}_{HA, \delta} = \begin{pmatrix} 1 & 0 & 0 \\ 0 & -1 & 0 \\ 0 & 0 & 0 \end{pmatrix} \mathbf{R}_3(ST) \begin{pmatrix} x \\ y \\ z \end{pmatrix}_{\alpha, \delta}$$

$$= \begin{pmatrix} 1 & 0 & 0 \\ 0 & -1 & 0 \\ 0 & 0 & 0 \end{pmatrix} \begin{pmatrix} \cos ST & \sin ST & 0 \\ -\sin ST & \cos ST & 0 \\ 0 & 0 & 1 \end{pmatrix} \begin{pmatrix} \cos\alpha \cos\delta \\ \sin\alpha \cos\delta \\ \sin\delta \end{pmatrix}$$

$$= \begin{pmatrix} -0.748103 \\ -0.164705 \\ +0.642817 \end{pmatrix}$$

Finally,

$$HA_1 = \arctan\frac{Y}{X} = \arctan 0.2201637 = 12°24'58''.91,$$

$$\delta = \arctan\frac{Z}{\sqrt{X^2 + Y^2}} = \arcsin Z = \arcsin 0.6428175 = 40°00'08''.05$$

However, $HA_1$ cannot be the correct answer, because $X$ and $Y$ are both negative, so we have to add $180°$ to $HA_1$, finally obtaining $HA - 192°24'58''.92 = 12^h49^m39^s.928$, a result which could have been immediately obtained by $(ST - \alpha)$.

The calculations made for this exercise were purposely carried out with only five significant digits, to show that the final numbers are slightly in error. Repeat with seven digits.

3. *Calculation of* $(HA, \delta)$ *from* $(A, h)$: A given star X has $(A, h) = (317°.6, 32°.4)$ as observed from a site having $\phi = 45°51'$; determine its $(HA, \delta)$. Disregarding the atmospheric refraction, from Equation 3.8 we immediately obtain $\delta(X) = -3°.0$, $HA(X) = 325°.3 = 21^h41^m$ $(= -34°.7 = -2^h19^m)$. Therefore, the star X will reach upper culmination after $2^h19^m$.

4. *Determine* $(A, z)$ *from* $(HA, \delta)$. From a site having $\phi = 28°.755$, observe a star X having $(HA, \delta) = (-9^h30^m, 85°18')$; determine its $(A, z)$.

   Notice that the star is circumpolar. Again, without considering the atmospheric refraction, from

$$\cos z = \sin \delta \sin \phi + \cos \delta \cos \phi \cos HA$$

we get $z = 65°.01$ $(h = 24°.99)$; then, from

$$\cos A = (\sin \delta - \cos z \sin \phi)/\sin z \cos \phi,$$

we get $A = 184°.2$.

   However, the star has a fairly high declination, so it would be advisable to make recourse to the expression of $\tan A$ from Equation 3.9 for better precision.

5. *Errors in transformations*: Assume small errors $(d\alpha, d\delta)$ in your knowledge of the equatorial coordinates of a given star. Determine the corresponding errors on the ecliptic coordinates if the error on $\varepsilon$ can be considered negligible. Solve the inverse problem. Discuss the behavior of the errors in proximity of the ecliptic and celestial poles. Hint: make recourse to the angle $p$ between the declination and the latitude circles.

# 4 First Notions on the Movements of the Earth and the Astronomical Times

This chapter will expound first notions on the diurnal rotation and annual revolution of the Earth, in order to allow an initial understanding of the several definitions of time used in astronomy. Chapter 10 will provide more explanations on the different time scales, including the atomic time of non-astronomical origin. The dynamics of the Earth's rotation and revolution will be discussed in Chapters 6 and 12, respectively. Chapters 14 and 15 will give further notions on the time scales based on the lunar orbital motion.

## 4.1 THE MOVEMENTS OF THE EARTH

The Earth's diurnal rotation takes place around a polar axis whose direction, with respect to the distant stars (namely, in an inertial reference frame), will be considered in this chapter as invariable and with constant angular velocity. In other words, the rotation is expressed by a constant vector $\Omega$. Furthermore, the direction of $\Omega$ is assumed to coincide with the polar axis $c$ of the ellipsoid that mathematically describes the Earth's figure. These assumptions approximate the reality, but they are convenient for the present purposes.

The apparent direct movement of the Sun with respect to the fixed stars, amounting to approximately 1° per day, eastward on the ecliptic, reflects the annual revolution of the Earth around the Sun. According to the first two Kepler's laws, expressed for the geocentric observer:

1. the apparent orbit of the Sun is an ellipse of eccentricity $e$ and semi-major and semi-minor axes $a$ and $b$, respectively, having the Earth in one of the two foci. The equation of the ellipse is

$$\frac{1}{r} = \frac{1}{p}[1 + e\cos v], \quad a = \frac{p}{1-e^2}, \quad b = a\sqrt{1-e^2} \tag{4.1}$$

where $r$ is the distance between Earth and Sun. The argument $v$ is named *true anomaly* of the Sun. By definition, its initial direction $v = 0$ coincides with that of the semi-major axis $a$, at the instant when the Sun passes through the perigee $\Pi$. Given that the civil year begins when the longitude of the Sun is approximately 280°, while the longitude of the perigee at the present time is around $\lambda_\Pi \approx 283°$, this passage occurs few days after the beginning of the year. Approximate values of $\lambda_\Pi$ can be obtained by the formula: $\lambda_\Pi = 282°.940 + 0°.017\Delta t$, with $\Delta t$ in years since J2000; for instance, during year 2019, it is $\lambda_\Pi \approx 283°.264$.

It is worth noticing that the precise *date* of the perigee has appreciable variations from year to year, due to the gravitational pull of the Moon. Indeed, it is the barycenter of the Earth–Moon system that follows the elliptical orbit (ignoring the planetary perturbations), while the perigee is defined as the *minimum distance* between the center of the Earth and the center of the Sun. Moreover, the eccentricity $e$ is a slowly varying quantity under the gravitational influence of the other planets. At the present epoch, its value is approximately 0.0167;

2. the areal velocity (not the angular one!) of the Sun along the ecliptic is constant:

$$\dot{A} = \frac{dA}{dt} = \frac{1}{2}r^2\frac{dv}{dt} = \frac{C}{2} \tag{4.2}$$

where $A$ is the area of the sector between the perigee, and the Sun and C/2 is the so-called *area constant*. Therefore, the angular velocity of the Sun on the ecliptic is greater at the perigee $\Pi$ (at the beginning of January) than at the apogee A (at the beginning of July).

The diurnal rotation of the Earth and the annual revolution of the Sun on the ecliptic provide the basis for two astronomical time scales, the *day* and the *year*.

## 4.2 THE SIDEREAL TIME *ST*

In Chapter 2, the sidereal time *ST* was defined as the hour angle of the vernal equinox $\gamma$: $ST = HA(\gamma)$. At each rotation of the Earth, *HA* increases by one sidereal day of $24^h$. Notice that *HA* is an angle along the celestial equator, representing the position of the equinox with respect to the local meridian. However, the equinox itself is not directly visible as a point, its position being defined by the declination of the Sun $\delta_\odot$ through the relation 3.20:

$$\sin\alpha_\odot = \cot\varepsilon\tan\delta_\odot$$

In other words, *ST* is operationally referred to the Sun, not to the stars, so that the adjective "sidereal" is somewhat misleading. In practice, the operation of referring the equinox directly to the Sun is seldom carried out. In order to determine *ST*, it is much easier to utilize the upper meridian transit of a set of fundamental stars, whose right ascensions also define the origin of the equatorial system. A word of caution is in order here, because each particular set of fundamental stars determines a slightly different equinox. Presently, the best realization of the fundamental catalogue is the already quoted ICRS (adopted by resolution of the IAU starting January 1, 1998), whose right ascension origin can be taken to define the position of $\gamma$. At any rate, irrespective of its origin, the sidereal time is as uniform as the rotation of the Earth, as already pointed out in Chapter 2 and detailed in Chapter 10.

## 4.3 THE SOLAR TIME $T_\odot$ AND THE EQUATION OF TIME *E*

A second rotational time scale can be defined by using the Sun, which for the everyday life is certainly much more important than the equinox. Let us call *solar day* the interval of time between two successive upper transits of the Sun on the meridian of a particular site. Accordingly, the solar time $T_\odot$ is the hour angle of the Sun, augmented by $12^h$, so that the solar day starts at midnight, not at noon. This conventional origin of the solar day was adopted in 1925. However, for the three following years not all Nations or Observatories conformed to the resolution, so that care is needed when using the dates preceding 1928. Therefore,

$$T_\odot = HA_\odot + 12^h$$

This is the time indicated by a sundial (apart from the effects of the atmospheric refraction that can be ignored in this context) in that particular place. Notice that while the sidereal time derives only from the rotation of the Earth, the solar time reflects both the diurnal rotation and the yearly revolution; these two movements do not have any fundamental connection (apart a very slight influence through the constants of precession, that here we can ignore). This independence is also at the root of the difficulties for building yearly calendars based on counting integer numbers of solar days, as discussed later. It is worth noticing that the independence of the rotational state from the revolution is vividly represented by the variety of situations of the planets (see Table 14.2).

Furthermore, we must pay attention that the Sun does not move along the equator but along the ecliptic, according to Kepler's first two laws; those two factors affect both the duration and the uniformity of the solar time. Regarding the duration, the Sun appears to move in direct sense (eastward) on the ecliptic by approximately 1° each day (more precisely, on average by 360°/365 days ≈ $3^m56^s$/day) with respect to the fixed stars and therefore also with respect to the equinox (at least in this approximation). These $3^m56^s$ represent the average extra time the Sun takes to pass the following day in meridian with respect to the equinox. The solar day (indicated with j; more precisely, j is the *mean* solar day) is then, on average, $3^m56^s$ longer than the sidereal day. In the same manner, all units of solar time are correspondingly longer than the units of sidereal time having the same name.

Regarding the uniformity, the solar time: $T_\odot$ is grossly non-uniform, as already known to ancient astronomers. Given that $HA_\odot = ST - \alpha_\odot$, then

$$T_\odot = HA_\odot + 12^h = ST - \alpha_\odot + 12^h \tag{4.3}$$

Equation 4.3 implies that $T_\odot$ and $\alpha_\odot$ have the same degree of non-uniformity. Let $\lambda_\odot$, $\alpha_\odot$ and $\delta_\odot$ be, respectively, the ecliptic longitude, right ascension and declination of the Sun at a given date; from Equation 3.6, the following relations can be derived:

$$\sin\delta_\odot = \sin\lambda_\odot \sin\varepsilon, \quad \tan\alpha_\odot = \tan\lambda_\odot \cos\varepsilon \tag{4.4}$$

Taking the time derivative of the second equation and inserting the first one, we obtain

$$\dot{\alpha}_\odot = \frac{\cos\varepsilon}{1 - \sin^2\lambda_\odot \sin^2\varepsilon}\dot{\lambda}_\odot = \frac{\cos\varepsilon}{\cos^2\delta_\odot}\dot{\lambda}_\odot \tag{4.5}$$

Formula 4.5 comprises both the abovementioned effects, namely, the variable angular speed on the ecliptic and the projection on the equator. To quantify the non-uniformity of $\dot{\alpha}_\odot$, we take into account both factors:

- according to Kepler's II law, the Sun has a daily motion greater at the perigee than at the apogee: $\dot{\lambda}_\odot \approx 61'.1 \approx 4^m4^s\, j^{-1}$ around the beginning of January and $\dot{\lambda}_\odot \approx 57'.2 \approx 3^m49^s\, j^{-1}$ around the beginning of July. This factor alone would produce a solar day $15^s$ longer at the perigee than at the apogee;
- the same motion of $\dot{\lambda}_\odot$ on the ecliptic corresponds to different arcs on the equator according to the declination, from a minimum value of $\dot{\lambda}_\odot \cos\varepsilon \approx 3^m37^s\, j^{-1}$ at the equinoxes, to a maximum value of $\dot{\lambda}_\odot / \cos\varepsilon \approx 4^m16^s\, j^{-1}$ at the solstices. Due to this effect of projection, a constant solar motion along the ecliptic would result in a day approximately $39^s$ longer at the equinoxes than at the solstices.

Therefore, the duration of the true solar day is continuously variable, for two different reasons, which are out of phase and combine with each other with different signs. As a result, the longest solar day occurs around mid-December and lasts approximately $24^h00^m30^s$, about $52^s$ longer than the shortest day, which occurs few days before the autumn equinox. Those seemingly small differences steadily accumulate with the passage of the days, reaching several minutes before changing sign, as we will discuss later on in this paragraph (see Equation 4.12, *Equation of Time E*).

In order to construct a uniform solar time, S. Newcomb (Bibliography section) introduced two hypothetical suns having constant angular velocity, namely, a fictitious sun $F_\odot$ moving on the ecliptic (called by some authors dynamic mean Sun) and a mean sun $M_\odot$ moving on the equator. Both bodies move with the same constant daily motion $n = 3548''.325\, j^{-1}$. This value of $n$ derives from the length of the tropical year, namely, from the time between two consecutive passages of the Sun through the equinox $\gamma$ (a more precise definition will be given later on).

By construction, the fictitious sun $F_\odot$ coincides with the true Sun at perigee $\Pi$ and apogee A. It trails behind the true Sun between $\Pi$ and A; it precedes it between A and $\Pi$. The longitude of $F_\odot$, $\lambda(F_\odot)$, is called the *mean longitude* of the Sun; it must not be confused with the longitude of the mean sun $\lambda(M_\odot)$. From its definition, it follows that the non-uniformity of the right ascension of $F_\odot$, namely, the daily variation of $\dot\alpha(F_\odot)$, is caused only by the projection effect due to its varying declination.

The difference between the ecliptic longitude of the true Sun and that of the fictitious sun, namely, the quantity $EC = \lambda_\odot - \lambda(F_\odot)$, is called *equation of the center*. $EC$ can be calculated from the equation of motion of the Sun in its orbit, whose elements are, as already stated, the eccentricity $e$, the instant $t_0$ of passage of the Sun through the perigee and the true anomaly $v$. It is convenient to introduce an auxiliary quantity named mean anomaly $M = n(t - t_0)$ which increases uniformly with time. Expressing $t$ in mean solar days j and $n$ in time units, the value of the constant is $n = 236^s.555$ j$^{-1}$. Leaving the demonstration to Chapters 12 and 13, the following first-order relation gives the longitude of the true Sun:

$$\lambda_\odot = \lambda_\Pi + v \approx \lambda_\Pi + M + 2e\sin M \tag{4.6}$$

where the last term expresses in an approximate manner the deviation from a circular orbit. By definition, the longitude of the fictitious sun is a linear function of time:

$$\lambda(F_\odot) = \lambda_\Pi + n(t - t_0) = \lambda_\Pi + M \tag{4.7}$$

Finally,

$$EC = \lambda_\odot - \lambda(F_\odot) \approx v - M = 2e\sin M = 0.03345R'\sin M \approx 115'\sin M \tag{4.8}$$

where $R'$ is the constant introduced in Chapter 1 to transform radians to arcminutes. Therefore, at the first order, the *equation of the center EC* is a periodic function of time, with a period of 12 months and amplitude of approximately 115', namely, $7^m40^s$, roughly corresponding to the arc described by the Sun in 2 days. The phenomenon is so evident that Claudius Ptolemaeus could already ascertain it, although with an excessive value. Copernicus is credited with a determination very close to the true one.

Notice the different origins of the angles; the longitudes start from the vernal point $\gamma$ and the anomalies from the perigee $\Pi$. From Equation 4.7, $M = \lambda(F_\odot) - \lambda_\Pi$, so that, using Equation 1.1, $EC$ can also be written as

$$EC = 2e\sin\lambda(F_\odot)\cos\lambda_\Pi - 2e\cos\lambda(F_\odot)\sin\lambda_\Pi + \cdots \tag{4.9}$$

Let us take the time derivative of Equation 4.6, using the mean solar day j as unit of time:

$$\dot\lambda_\odot = n(1 + 2e\cos M + \cdots) = 3548''.325 + 118''.7\cos M + \cdots j^{-1} \tag{4.10}$$

Notice that $\dot\lambda_\odot = n$ on two occasions, namely, when the Sun passes through the semi-minor axes of its orbit. Knowing the date when $M = 0°$ (e.g., Jan. 2.58 in 2002, Jan. 4.62 in 2006, see Table 10.3 for values at other years), we can easily calculate the variation of angular velocity along the ecliptic at each date.

To determine the relationship between $\alpha_\odot$ and $\lambda_\odot$ (a procedure called *reduction to the equator*), let us examine the second of the Equation 4.4, which according to Equation 1.10 can be written as

$$\alpha_\odot = \lambda_\odot + \frac{\cos\varepsilon - 1}{\cos\varepsilon + 1}\sin 2\lambda_\odot + \frac{1}{2}\left(\frac{\cos\varepsilon - 1}{\cos\varepsilon + 1}\right)^2\sin 4\lambda_\odot + \cdots$$

$$= \lambda_\odot - 148'.1\sin 2\lambda_\odot + 3'.2\sin 4\lambda_\odot + \cdots$$

Therefore, the right ascension of the true Sun is connected to its longitude by a series of multiples of $2\lambda_\odot$, with coefficients depending from $c$. Inserting Equations 4.6 and 4.9, we finally obtain

$$\alpha_\odot \approx \lambda_\Pi + M + 2e\sin M - \tan^2\frac{\varepsilon}{2}\sin 2(\lambda_\Pi + M) + \cdots \tag{4.11}$$

a series with periods of 12, 6, 4 months, etc. Taking the time derivative, we can find that $\dot{\alpha}(F_\odot) = n$ four times a year, when $\lambda(F_\odot) \approx 46°14'$, $133°46'$, $226°14'$, $313°46'$. The corresponding dates can be determined by knowing the date of the passage through $\Pi$.

When the fictitious sun $F_\odot$ encounters the equator in $\gamma$ coming from $\Pi$ (later than the true Sun), let the mean sun $M_\odot$ start from $\gamma$ with the same uniform motion $n$. The two hypothetical suns will coincide again at the autumn equinox; in this way, at any instant, $\lambda(F_\odot) = \alpha(M_\odot)$. Finally, calculate the *Equation of Time E*, namely, the difference:

$$E = \alpha_\odot - \alpha(M_\odot) = 460^s.3\sin n(t-t_0) - 592^s.2\sin 2(\lambda_\Pi + M) + \cdots$$

$$= A\sin\lambda(F_\odot) + B\cos\lambda(F_\odot) + C\sin 2\lambda(F_\odot) + D\cos 2\lambda(F_\odot)$$

$$+ E\sin 3\lambda(F_\odot) + F\cos 3\lambda(F_\odot) + G\sin 4\lambda(F_\odot) + \cdots \tag{4.12}$$

The equation of time $E$ is therefore a complicated, but well-known, function of time (see Figure 4.1).

Its value is zero four times a year, namely, in mid-April, mid-June, beginning of September and around December 25; the maximum value of approximately $+16^m$ is reached in early November and the minimum value of approximately $-14^m$ in mid-February. Notice that the exact values at a particular date will vary by few seconds from one year to the next, in a periodic behavior due to the presence of the leap year. For instance, in 2019 the values of the coefficients were $A = -109^s.9$, $B = -427^s.5$, $C = +595^s.9$, $D = -2^s.1$, $E = +4^s.5$, $F = +19^s.2$, $G = -12^s.7$ and $\lambda(F_\odot) = 279°.375 + 0.985647d$, $d$ being the number and fraction of days since January 0, 2019. The precision of this expression is of few seconds.

For any particular site, the difference between the hour angles of the true Sun and mean sun will also equal the difference, changed in sign, between their two right ascensions:

$$HA_\odot - HA(M_\odot) = -\alpha_\odot + \alpha(M_\odot) = E$$

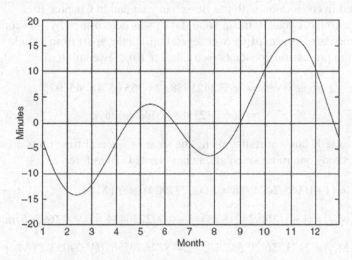

**FIGURE 4.1** The equation of time $E$ for J2000. The convention of sign is that adopted by the *Astronomical Almanac* (apparent minus mean) in Section C2; some texts use the opposite sign.

The hour angle of the mean sun, augmented by $12^h$ in order to have the day starting at midnight, is called the local mean solar time $T(M_\odot)$:

$$T(M_\odot) = HA(M_\odot) + 12^h \qquad (4.13)$$

The interval of time between two passages through the local meridian of the mean sun is called *mean solar day* (the unit already indicated with j) which is divided into $24^h$ of $3600^s$ of mean solar time, whose length is not the same of the sidereal second, as already said.

## 4.4   THE UNIVERSAL TIME UT

In particular, the mean solar time at Greenwich is called universal time UT:

$$UT = HA(M_\odot)(\text{Greenwich}) + 12^h \qquad (4.14)$$

Sometimes, the letter Z is used to indicate Greenwich's mean time; for example, 1433Z means $14^h33^m$ mean solar time at Greenwich.

For another site having longitude $\Lambda$, the local *mean* solar time at $UT = t$ is $T_\Lambda (UT = t) = t \pm \Lambda$, where the sign is $+$ if east of Greenwich and $-$ if west. Therefore, the difference in longitude between two sites can be expressed as a difference in either solar or sidereal time, because $\Lambda$ is an angle.

## 4.5   THE TROPICAL YEAR AND THE RATES OF *ST* AND UT

From their definitions, it follows that the sidereal time *ST* and the mean solar time $T(M_\odot)$, including its particular case UT, have the same degree of uniformity of the Earth's rotation. However, the two times differ in both rate and origin.

The ratio between the two rates can be easily determined. Let us call *tropical year* the interval of time between two consecutive passages of the mean sun through the vernal equinox, in other words the interval of time needed for the longitude of the fictitious sun to increase by 360° with respect to the equinox. At the level of precision of the present discussion, this length of time is the same as an increase of 360° of the right ascension of the mean sun. However, we will see in the following chapters that the equinox is not fixed among the stars, resulting in a slight difference between the two preceding definitions. The correct definition of the tropical year is the period of one complete revolution of the mean longitude of the Sun with respect to the dynamical equinox; this matter will be further discussed in connection with the Besselian year and in Chapter 10.

The value of the tropical year in mean solar days j was determined by Newcomb (*op.* cit.) with utmost precision, thanks to its recording over several millennia; apart from a slight secular variation due to changes in the precession constants (see Chapter 10), Newcomb found

$$1 \text{ tropical year} = 365^j.2421988\ldots = 365^j05^h48^m45^s.975\ldots$$

$$= 366.2421988\ldots \text{sidereal days}$$

The difference of exactly one entire day using the solar or sidereal time units arises because after one tropical year exactly one more sidereal day has elapsed. Therefore,

$$\text{rate} ST = (1 + 1/365.2421988 = 1.002737909)\text{rate} T(M_\odot)$$

$$\text{rate} T(M_\odot) = (1 - 1/366.2421988 = 1 - 0.002730434 = 0.997269566)\text{rate} ST$$

$$24^h T(M_\odot) = 24^h3^m56^s.55537 ST, \quad 24^h ST = 23^h56^m04^s.09053\ T(M_\odot)$$

$$1^s T(M_\odot) = 1^s.0027379 ST, \quad 1^s ST = 0^s.9972696\ T(M_\odot)$$

Regarding the origin, we anticipate an approximate expression, which will be completed in Chapter 10. At any given date $T$, the Greenwich sidereal time at UT = $0^h$ (mean midnight) is

$$ST_{\text{Greenwich}}\left(0^h\,UT\right) = 6^h41^m50^s.815481 + 8640184^s.812866T$$

where $T$ (with its proper sign and all needed decimals) is in Julian centuries before or after the fundamental epoch January 1, 2000 at $12^h$ UT (namely, $T$ is the number of days and fractions from that epoch, divided by 36,525).

For another site having longitude $\Lambda$, the local sidereal time at UT = $t$ is

$$ST_\Lambda\left(UT = t\right) = ST_{\text{Greenwich}}\left(0^h\,UT\right) \pm 1.0027379t$$

where the factor of the last term takes into account the different rates of the two time scales.

## 4.6   THE YEAR AND THE JULIAN CALENDAR

The civil calendars adopted in many countries find their roots on the length of the tropical year, because the seasons follow the course of the Sun along the ecliptic. However, this length cannot be expressed by an integer number of days, not even by a rational fraction. Several remedies were adopted by the different cultures. In Rome, approximately in the year 46 B.C., Julius Caesar agreed to the proposal of the astronomer Sosigenes of adding 1 day each fourth year to the shortest month (February). In this Julian calendar, the fourth year is called *bi-sextus*, or *leap*, year. In the first application of this rule, some 90 days had to be suppressed from the calendar. However, the situation was very confused at least for the following 30 years, until the times of Emperor Augustus. The extension of the Julian calendar into the past (namely, before its adoption) is called the *proleptic* calendar. Because in Chronology year 0 does not exist, passing directly from 1 B.C. to 1 A.D., in performing calculations of intervals of time between two events happened one before and one after Christ, year 1 B.C. is year 0, year 2 B.C. is year −1 and so on. Year 0 is considered a leap year.

With the Julian reform, the duration of the year, averaged over a 4-year period, became exactly 365.25 mean solar days, and the Julian Century 36,525 mean solar days. However, this round number is longer than the duration of the tropical year: after 1000 years, the difference amounts to approximately 8 days. Some adjustments to the calendar were made during the Council of Nicea (325 A.D.). Finally, in 1582, Pope Gregorius XIII decreed to suppress 10 days, jumping from Thursday October 4 directly to Friday October 15. As a further refinement, it was stated that only the secular years divisible by 400 would be leap years. Therefore, following the Gregorian reform, years 1600 and 2000 were leap years, but not 1700, 1800 and 1900; in a cycle of 400 years, there are only 97 such leap years, and the average duration over each cycle is 365.2425 mean solar days. Because in 400 years, there are 146,097 days, which is evenly divisible by 7, the Gregorian civil calendar exactly repeats at each cycle of 400 years. However good, the average value 365.2425 is still an approximation to the true value, so that the Gregorian calendar precedes the Sun by approximately 1 day every 2500 years. As a remedy, year 4000 could be considered a normal year, not a leap one, but no agreement has been reached.

## 4.7   THE BESSELIAN YEAR OR ANNUS FICTUS

Both the tropical and the Julian year are essentially durations, no precise origin being associated with their definitions. Following Bessel, the year starts when the longitude of the fictitious sun, affected by aberration and referred to the mean equinox of date, is exactly $\lambda(F_\odot) = 280°$ and therefore $\alpha(M_\odot) = 18^h40^m$. Such an instant, named *epoch*, is always within 1 day from midnight of December 31. In most applications, for instance, in calculating the amount of precession, this slight difference between the start of the Besselian year and of the civil year is entirely negligible.

The Besselian epoch is indicated by the notation B followed by the year, e.g., B1950.0; any other instant of time during that year is indicated by the fraction of year, e.g., B1950.45678.

The following year will start when again $\lambda(F_\odot) = 280°$, namely, when $\alpha(M_\odot)$ has increased exactly by 360°. As we have indicated before, this interval of time is not quite the same of the tropical year, fixed by the mean longitude of the Sun with respect to the dynamical equinox. Therefore, there is a slight secular acceleration deriving from the different definition of $\lambda(F_\odot)$ and $\alpha(M_\odot)$: the duration of the Besselian year is equal to that of the tropical year minus $0^s.148$ per tropical century since 1900; for instance, B2000.0 started $7^s.4$ before the instant calculated from the duration of the tropical year.

In order to refer the Besselian year to the Julian calendar, it must be recalled that the fundamental epoch B1900.0 corresponds to 1900 January $0^d.813$ = 1899 December 31, $19^h31^m$ (notice the astronomical convention, year first, then month, day, hours and the utilization of the 0 for the last day of the preceding year). The calendar date of another epoch, say B1950.0, is obtained by considering 50 tropical years since then, namely 18,262.110 days, or else 12.110 days more than 50 years of 365 days. Taking into account that 1900 was not a leap year, subtract 12 days and add 0.110 to 0.813 to finally obtain B1950.0 = 1950 January $0^d.923$ = 1949 December 31, $22^h09^m$. This is the civil date of the epoch of many stellar Catalogues of the past, such as the AGK3.

The convention of Bessel remained valid until 1984, when the IAU decreed to move the fundamental epoch to J2000.0 noon (not midnight!) = 2000 January $1^d.5$ UT (actually UT1, see Chapter 10), and to adopt Julian years of $365^j.25$ (or Julian centuries of $36,525^j$). Therefore, the beginning of year 1950.0 corresponds exactly to $18,262^j.5$ days before the fundamental epoch, namely, to 1950 January $1^d.0$, differing by $1^h51^m$ from B1950.0.

## 4.8 THE SEASONS

We have affirmed that the tropical year determines the succession of the seasons. Consider, for instance, the following values valid at epoch 1950.0:

$$\lambda_\odot = 282°04'30'' + M + 115' \sin M + \cdots, \quad M = 3548''.3(t - t_0),$$

$$t_0 = 1950 \text{ January } 3.02 \text{ (date of passage through the perigee)}$$

(4.15)

The seasons started when $\lambda_\odot = 0°$ (spring), $= 90°$ (summer), $= 180°$ (autumn), $= 270°$ (winter). Because no high precision is required, in order to find the corresponding values of $M$ and of $t$, we can ignore the term in $\sin M$, obtaining the values of Table 4.1.

The last two columns were calculated with the appropriate values of Equation 4.15 for the leap years 2000 and 2096 (when the start date will be the earliest of the 21st century). The starting times retard $6^h$ each year, in a cycle of 4 years.

Table 4.1 shows that in the northern hemisphere, the two warm seasons last 7 days longer than in the southern hemisphere; on the other hand, the Earth is closer to the Sun in the southern summer. Averaged over the globe, the sunlight falling on Earth at aphelion is approximately 7% less intense than at perihelion. However, the average temperature of the whole Earth at aphelion is approximately 2.3C

### TABLE 4.1
### Dates of Start of the Seasons

| Season | Start (1950) | Duration (days) | Start (2000) | Start (2096) |
|--------|--------------|-----------------|--------------|--------------|
| Spring | 21.2 March | 92.81 | 20.3 March | 19.5 March |
| Summer | 22.0 June | 93.62 | 21.0 June | 20.1 June |
| Autumn | 23.1 September | 89.82 | 22.7 September | 21.9 September |
| Winter | 22.4 December | 89.00 | 21.5 December | 20.9 December |

higher than at perihelion: this happens because there is more land in the northern hemisphere and more seawater in the southern one. During the month of July, the northern hemisphere is tilted toward the Sun, and the Earth's overall temperature (averaged over both hemispheres) is slightly higher because the Sun shines mostly on continents, which have low heat capacity and warm up more easily. January is the coolest month, because the Earth presents its water-dominated, high heat capacity hemisphere to the Sun. Southern summer in January is therefore cooler than northern summer in July. In order to satisfy the thermal balance of the Earth as a whole, averaged over 1 year, efficient mechanisms of heat transport are required, such as regular winds and sea currents. The observed global warming of about 1C in the last century is not discussed here.

## 4.9   THE JULIAN DATE

In astronomy, it is customary to count the passage of time in mean solar days j starting from an initial arbitrary date. In 1583, based on historical reasoning Joseph Justus Scaliger established that the initial date is midday (not midnight!) of January 1, 4713 B.C. (= −4712 when counting). Such numbering system of days is named Julian Day. Be careful not to confuse the Julian Day with a date in the Julian calendar. According to historians, J. J. Scaliger named Julian Day his system of counting days in honor of his father Julius, and not of Julius Caesar, so that the adjective Julian here has nothing to do with the Julian calendar!

The Julian Date JD at a particular instant of time is obtained by adding to the Julian Day its fractional part. Thus, January 1, 1950, $12^h$ UT, corresponds to JD = 2433283.0, while the old fundamental epoch B1950 corresponds to JD = 2433282.423.

As already stated in paragraph 4.7, the modern fundamental epoch J2000.0 starts at 2000 January $1^d.5$ UT, therefore,

$$J2000.0 = JD2451545.0$$

The value of JD at a given calendar date can be calculated simply by counting the number of intervening days after or before J2000, paying attention to the possible presence of leap years in that interval. For instance, $J2019.5 = 2019$ July $2.875 = JD\ 245\ 8667.375$. To calculate how many days separate two dates, the correct procedure is to calculate the two corresponding JDs, and then make the difference. Practical examples are given in the "Exercises" section at the end of this chapter.

In order to avoid carrying too many decimals, and to start the day at midnight, a Modified Julian Date (MJD) has been introduced, having its zero date on 1858 Nov. 17.0:

$$MJD = JD - 240000.5$$

The JD or MJD scales are a continuous reference of time; however, these scales are as uniform as the mean solar day itself. We will see in the following chapters (in particular in Chapters 6 and 10) that the duration of the day has a secular decrease and erratic small variations, so that JD is not entirely satisfactory for dynamical purposes.

## NOTES

- Gregorius XIII had the advice of several contemporary astronomers, in particular of Luigi Lilius, who also made a reform of the lunar calendar (important for the determination of Easter). The objectives and details of the new calendar were described in 1603 by Christoph Clavius in his letter *Romani Calendarii a Gregorio XIII P.M. restituti explicatio*. The Gregorian reform was not adopted immediately by all countries; for instance, it came to be used in the UK in 1752 (this is the reason for the widespread but wrong statement that the birth of Isaac Newton and death of Galileo Galilei happened in the same year); in Turkey only in 1927. For the historically oriented reader, the following two books can be recommended: Coyne et al. (1983), Heilbron (1999) in the Bibliography section.

- The determination of the date of Easter (the first Sunday after the full Moon happening the 21st of March or the first since that date, as established by the Nicea Council of 325 A.D.), was always a difficult computational problem, because for the Church, the 21st of March is not the spring equinox in the astronomical significance. Therefore, the date of Easter can vary between the 22nd of March and the 25th of April, although the most frequent interval is between March 25th and April 19th. Carl F. Gauss found a tabular expression which is almost always, but not strictly so, correct. See Meeus (1991, in the Bibliography section), for Gauss's and other expressions.

- Another way of representing the equation of time is by means of the so-called *analemma*, an eight-shaped curve often found on celestial globes and sundials. The analemma can be vividly represented by photographing the Sun, ideally each day at mean noon, for an entire year. See an example in http://vrum.chat.ru/Photo/Astro/analema.htm.

## EXERCISES

1. Calculate the mean anomaly of the Sun at the beginning of 2019.

2. An Almanac gives the $ST$(Greenwich) at $0^h$ UT of a given day. Calculate the $ST$ in a particular place of longitude $\Lambda$ at any instant $t$ of $T(M_\odot)$ of the same day. *First solution*: If you don't know UT, calculate the local $ST$ at $0^h$ $T(M_\odot)$, which is equal to $ST$ (Greenwich, $0^h$ UT) + 0.0027379 $\Lambda$ (for instance, for Asiago 0.0027379 $\Lambda = -7^s.58$). This term is often referred to as the local constant. Then, calculate

$$ST\left(t^h T(M_\odot) = ST\left(\Lambda, 0^h T(M_\odot)\right) + 1.0027379t\right.$$

*Second solution*: In your observatory, there is already a clock giving $t$ in UT (which is the common situation today), so that

$$ST\left(t^h UT, \Lambda\right) = ST\left(\text{Greenwich}, 0^h UT\right) + 1.0027379\left(t^h UT\right) + \Lambda$$

(in the IAU convention, subtract west longitude and add east longitude). Inversely, if one has to calculate $T(M_\odot)$ starting from the local $ST$, the following relation will be used:

$$T(M_\odot) = 0.9972696\left(ST - ST\left(0^h T(M_\odot)\right)\right)$$

3. Let a place be at longitude $75° = 5^h$ west, and the local sidereal time be $ST = 23^h30^m$. The date is July 8, 1985. Find the corresponding UT.

   The first operation is to add the west longitude to find the Greenwich $ST$:

$$ST(\text{Greenwich}) = 4^h30^m$$

The Almanac gives, for that date, the value $ST$(Greenwich, $0^h$ UT) = $19^h03^m34^s.376$, which must be subtracted from $4^h40^m$, giving a time difference (in $ST$ units) of $9^h29^m25^s.624$; multiply by 0.9972695663, and finally, obtain UT = $9^h27^m52^s.337$.

   The $ST$ used in Exercises 1, 2 and 3 is the mean sidereal time; to have the apparent one, the small correction named *equation of the Equinox*: $EE = \Delta\psi\cos\varepsilon$ must be added. See also Chapters 5 and 10.

4. Calculate the Julian Day corresponding to a calendar date.

A formula used by many calculators, but valid only in the Gregorian calendar (after October 15, 1582), is the following:

$$JD(Y, M, D, H) = 367Y - INT\left[7 \times \frac{Y + INT((M+9))/12}{4}\right]$$

$$+ INT\left(\frac{275M}{9}\right) + D + 1721013.5 + \frac{H}{24} + 0.5$$

$$\times SIG(190002.5 - 100\ Y - M) + 0.5$$

where INT is the function integer, and SIG is the function *signum*.

After February 28, 1900 and until 2099, the last two terms can be omitted.

For instance, JD(1996,1,1,0) = 2450083.500, JD(2004,1,1,4.75) = 2453009.250.

5. Calculate the civil date corresponding to JD = 2446108.5, knowing that the JD at 0$^h$ UT of January 1, 1900 was JD1900 = 2415020.5. *Solution*: The number of days elapsed since the initial epoch is ND = JD − JD1900 = 31088. The number of years NY elapsed since the initial date is (NY) = INT(JD − JD1900)/365 = 85, INT as function integer is already defined in exercise 4, and the remainder is 0.1726 × 365 = 63 days. However, in 85 years, there are 21 leap days, so that the number of days elapsed since the beginning of the year 1985 is actually 63 − 21 = 42. Subtracting the 31 days of January, we finally obtain the wanted civil date, i.e., February 12, 1985, 0$^h$ UT.

To solve this problem in a general way, one formula is the following

- Let Z = INT(JD + 0.5) and F = (JD + 0.5) − Z
- If Z < 2299161 let A = Z
- If Z ≥ 2299161 determine: A′ = INT(Z − 1867216.25)/36524.25, and let A = Z + 1 + A$^1$ − INT(A′/4)

Then calculate:

$$B = A + 1524, \quad C = INT\left(\frac{B - 122.1}{365.25}\right), \quad D' = INT(365.25C),$$

$$E = INT\left(\frac{B - D'}{30.6001}\right)$$

The day of the month (with decimals) is then D = B − D′ − INT(30.6001E) + F.

The number of the month is M = E − 1 if E < 14, M = E − 13 if E = 14 or 15. The year is Υ = C − 4716 if M > 2, C > −4715 if M = 1 or = 2 (namely, January and February are considered the 13th and 14th months of the preceding year).

For other formulae, and similar problems connected with the date and day of the week, see the book by Meeus (1991) and in several astronomical almanacs.

# 5 The Movements of the Fundamental Planes

The equatorial and ecliptic coordinates are based on the planes of the celestial equator and of the ecliptic, respectively, having a common origin in the vernal point $\gamma$. Every movement of these planes with respect to the fixed stars will result in a variation with time of the equatorial and ecliptic coordinates. The stars will provide an ideally fixed reference frame against which the complex movements of the equator, the ecliptic and their intersections can be determined. For fundamental dynamical reasons, the ecliptic plane is much more stable in the inertial system than the equator, whose movements are larger and known with some residual imprecision even today. Those minute uncertainties, however small, are of great interest for the astronomer, because they somewhat hamper the precise knowledge of the system of motions and of the overall field of forces of the Milky Way. Hence, the efforts made, not only by geophysicists but also by astronomers, to know the movements of the terrestrial observer with ever greater precision.

The discussion of these movements developed in this chapter is based on the traditional approach to this intricate subject. The important revisions implemented by the IAU in 2006 for the definition of precession, nutation and ecliptic (see the "Notes" section), already mentioned in Chapter 2, will be further discussed in Chapters 6 and 10.

## 5.1 FIRST DYNAMICAL CONSIDERATIONS

The equatorial system $(\alpha, \delta)$ is the only one used in high-precision positional catalogues. However, it depends on the orientation in space of the Earth and its rotation (position of the equatorial plane with respect to the fixed stars, meridian, sidereal time), and on the revolution around the Sun (ecliptic, vernal point $\gamma$). The motion of the Earth can be considered, at least in a first approximation, as a combination of two unrelated motions (although the constants of precession weakly tie one to the other, as we will see later), namely, a translation of the center of mass and a rotation of the figure around an axis passing through the barycenter.

In a first, rough approximation, the barycenter of the Earth revolves around the Sun as a point-like particle subject to the gravitational pull of the Sun and of the other planets. More rigorously, it is the Earth–Moon (E–M) barycenter that follows Kepler's laws with respect to the barycenter of the Solar System. Therefore, it is preferable to identify the ecliptic with the orbital plane of the E–M barycenter, and free it from the periodic perturbations mostly due to Venus and Jupiter. The Sun is never more than $2''$ above or below this mean ecliptic, a fact that justifies the simple treatment given in the previous chapters.

Much more complex is the description of the orientation and rotation of the Earth, for several different reasons, among which we quote the following:

- the mass of the whole Earth is $M_\oplus = 5.976 \times 10^{27}$ g; the mass of the oceans is approximately $10^{-4} M_\oplus$ and that of the atmosphere approximately $10^{-7} M_\oplus$. Even considering the solid Earth alone, the distribution of its mass does not possess spherical symmetry (indeed, to the precision of the measurements, not even an azimuthal symmetry);
- the position of its rotation axis is influenced by the presence of the Moon and of the Sun (lunisolar precession and nutation);
- its rotation axis does not coincide with the minor axis of the geometrically best-fitting ellipsoid (Euler nutation).

The variable orientation of the Earth's instantaneous rotation axis with respect to the fixed stars causes a variation of their equatorial coordinates. The variation of its position with respect to the axis of the geometric ellipsoid causes a small variation of the astronomical latitude and longitude of each observatory, and a wandering of the astronomical poles, confined to a circle approximately 18 m in diameter, around the geodetic poles (free or Eulerian nutation, see Chapter 6). Furthermore, the solid Earth is not perfectly rigid, and the distribution of masses can change both at the surface (winds, tides, currents) and in the interior (plate tectonics, earthquakes). Therefore, the rotation of the Earth cannot be uniform for arbitrarily long times; at the present epoch, we witness a secular decrease of its angular velocity, with over-imposed periodic fluctuations and abrupt changes. To be sure, all these complications are small, but well measurable with today's instruments.

The general dynamical problem presents great theoretical complexity. Observationally, there are challenging difficulties to disentangle the minute motions of the observer from the motions of the stars themselves and to keep a uniform time system for centuries or millennia. However, the different effects have different amplitudes and time scales. Therefore, we are justified, at least for the present needs, to follow an almost historical description. At the end, we will sum the different effects one over the other, although the validity of this superposition will break down at very high precisions. To set a limit, a precision of a few tenths of arcsec can be achieved in this way; to reach the thousandth of arcsec, as for Hipparcos or better yet for GAIA, a more rigorous approach is needed.

In the following, we shall call *epoch* the initial time when all constants and their time variations are assumed to be well known, and *date* the generic instant of observation. In several classical texts, the elementary time unit is the year set equal to 1, so that the words "arc" and "motion" mean angular velocities, not movements. Here, for better clarity, we shall distinguish the two meanings, by explicit indication of the time units, derivatives and differentials. When discussing the stellar case, *proper motion* will always maintain its rigorous meaning of angular velocity on the plane tangent to the celestial sphere.

Our treatment can be traced back to the books of Newcomb (1906), Smart (1965) and Danjon (1980) quoted in the Bibliography section. Since 1984, the IAU has enforced a profound revision of several fundamental constants and variables, including time. Therefore, in several places we have provided both pre- and post-1984 definitions and values. Another revision has been applied by IAU starting in 2006, as will be discussed in the "Notes" section to this chapter and in Chapter 6.

## 5.2   THE PRECESSION OF THE EQUINOX

The dominant variation of the stellar coordinates was discovered around 129 B.C. by Hipparchus, comparing his own determination of the ecliptic coordinates of Spica ($\alpha$ Vir) with those derived 144 years earlier by Timocharis: while the ecliptic latitude had remained constant, the longitude had increased by approximately 2° (namely, by 50″.4 per year). The discrepancy of 2° was too large to be attributed to measurement errors. Soon after, the same variation of longitude was found on all stars. To explain it, Hipparchus imagined a rotation in direct (anticlockwise) sense of the whole sphere of the fixed stars around the ecliptic pole. The Sun would therefore encounter the vernal point $\gamma$ each year at a somewhat earlier time. In the words of Hipparchus, the ingress of the Sun in $\gamma$ would precede the previous one by the time taken to describe an arc of 50″.4, hence the expression "precession of the equinox". Over the centuries, the constellation where the Sun is seen to ingress in $\gamma$ changes: it was Aries before the time of Hipparchus, it moved soon after in the Pisces, where it is still today; after another 2000 years, the ingress will be in Aquarius.

Some 1600 years elapsed before Copernicus made the right interpretation of precession: the Earth's rotation axis describes a retrograde cone of semi-aperture $\varepsilon$ around the ecliptic pole E, in a period of approximately 25,800 years (= 360°/50″.4 per year, often called a *platonic year*). Therefore, the celestial pole P is seen at each time on a point of the small circle distant $\varepsilon$ from E, as shown in Figure 5.1. In other words, the parallel of ecliptic latitude $\beta = 90° - \varepsilon$ is the locus described

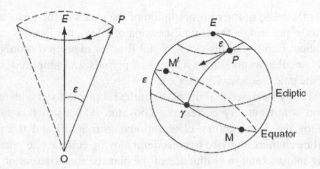

**FIGURE 5.1** Two equivalent representations of the precession of Hipparchus. (a) The vector from the center of the Earth to the celestial north pole P describes a cone of fixed semi-aperture $\varepsilon$ around the ecliptic north pole E, in retrograde direction. (b) The instantaneous motion of the celestial pole is a vector tangent both to the small circle distant $\varepsilon$ from E and to the equinoxial colure. Arc $M\gamma$ = arc $\gamma M'$ = 90°.

by P in 25,800 years. The south celestial pole is seen to move on the corresponding circle having $\beta = -(90° - \varepsilon)$.

During this movement, the celestial poles will be seen in different constellations; today, the north pole approaches the bright star $\alpha$ UMi (Polaris). The present distance of about 45′ will decrease to a minimum value of 27′ in 2102, and then, it will progressively augment. As already said, there is no correspondingly bright star near the present position of the celestial south pole (the one used by the *Astronomical Almanac* as pole star for obtaining southern latitudes is $\sigma$ Octanctis, of visual magnitude 5.5).

The precession, correctly interpreted by Copernicus, was explained on dynamical bases by Newton in his *Principia* (1687): the Earth cannot be a sphere; it must have an elliptical shape, like having a massive annulus around the equator of a spherical Earth. Therefore, the Moon, which is usually not on the equatorial plane, will exert a torque on the Earth's rotation axis. This torque causes a movement of the axis perpendicular to the instantaneous plane passing through the axis itself and the Moon. Quite often, an analogy is made with a spinning top; however, the sense of precession and that of rotation are opposite, the spinning top precesses and rotates in the same sense and the Earth precesses in the opposite sense of the diurnal rotation. The Sun exerts the same effect of the Moon; however, because the amplitude of the torque is proportional to the mass of the responsible body and to the inverse *cube* of its distance from the Earth's barycenter, the magnitude of the lunar effect is more than twice that of the Sun. Therefore, the term "lunisolar" precession was traditionally employed to describe the phenomenon discovered by Hipparchus.

However, Hipparchus's description and Newton's explanation must be only partially correct, for two reasons: first, the Moon's orbit is inclined by 5°9′ to the ecliptic plane; second, the distance of the Moon and to a far lesser extent that of the Sun change during the lunar month and the tropical year, respectively. Consequently, the true movement of the celestial poles cannot rigorously follow the small circles at a constant distance $\varepsilon$ from the ecliptic poles: a series of cyclic terms of different amplitudes and periods, also affecting the instantaneous obliquity, must be present. However, these effects have a periodic behavior with short time scales and small amplitudes; they do not accumulate for centuries as the lunisolar precession does, so they could not be detected until the advent of telescopic observations. The credit for their discovery goes to the English astronomer James Bradley in the 18th century, thanks to a long series of measurements of the declination of the bright star $\gamma$ Dra, which is not too distant from the ecliptic pole E. The declination of the star, once corrected for the lunisolar precession, appeared to increase by 18″ from 1727 to 1736 and to decrease by the same amount from 1736 to 1745, as if the celestial pole has an oscillatory movement. Bradley called *nutation* this oscillation, the same term used for the oscillation of a ship's mast. The observed amount of such nutation was ±9″ with a period of 18.6 years around a mean position. The period is the same of the retrogradation of the nodes of the lunar orbit on the ecliptic. The same effect had to be present on the right ascension of γ Dra, but Bradley could not measure it because of the insufficient stability of his clock. What Bradley measured is only the principal term of the nutation; many smaller terms with a variety of periods and amplitudes are present in the overall oscillation of the pole (see also Chapter 6).

Until now, we have tacitly assumed that the plane of the ecliptic is fixed with respect to the distant stars. This assumption is not entirely correct, as Giovanni B. Cassini had suspected. The level of accuracy reached by the 18th-century observational astronomy and the parallel theoretical progresses of celestial mechanics imposed to consider a moving ecliptic (Leonhard Euler calculated such motion by taking into account the influence of the planets, in particular of Venus and Jupiter). Comparing the values of $\varepsilon$ obtained by Copernicus and then by Tycho Brahe at the end of the 16th century, with the determinations performed during the 19th century by many authors (e.g., Bessel, Struve, Le Verrier, Oppolzer, Newcomb), and finally with the value measured today ($\varepsilon \approx 23°26'$), a secular decrease of approximately $0''.47$ per year is well established. The comparison of the observed values with the planetary perturbation theory first expounded by Euler required a reliable set of planetary masses. Until the middle of the 20th century, this was not the case; the very existence of Pluto was not known before 1938. Today, the masses of the planets have been very well determined, thanks to the accurate measurements of the accelerations of spacecraft that have navigated all planets and many other bodies of the Solar System, including several asteroids (Vesta being the largest) and the dwarf planets Ceres and Pluto (see the "Notes" section). Theoretical developments and computational improvements provide precise orbital elements for millions of years. The decisive influence of the Moon on the stability of $\varepsilon$ has been confirmed (see, e.g., Laskar et al., 1993). The present-day diminution is actually part of a variation with an amplitude of approximately $\pm 1°$ and a period of 44,000 years. Planet Mars has no heavy moon. Therefore, its obliquity can change by much larger values, so its climate can have extreme variations on time scales of the order of 100,000 years (see the papers by Laskar et al., 2004a,b and 2011).

The planets affect the obliquity of the ecliptic with a number of long periods whose overall effect will accumulate for many centuries, just as the lunisolar precession does. Therefore, we are justified to call *planetary precession* the *effect of such perturbations* and add its magnitude to the lunisolar precession. The result of such addition is the *general precession*; because of its negative sign, the planetary precession makes the constant $50''.4$ become slightly smaller. A planetary nutation must also be present; however, the movements of the ecliptic with respect to the equator do not change the stellar declinations, but only the common origin of the right ascensions, so that the planetary nutation goes unnoticed in differential measurements.

## 5.3   THE MOVEMENTS OF THE FUNDAMENTAL PLANES

Figure 5.2 shows the two fundamental planes, celestial equator and ecliptic, at two dates $t_1$ and $t_2$ (with $t_2$ later than $t_1$ in order to fix the sense) intersecting in $\gamma_1$ and $\gamma_2$ with obliquities $\varepsilon_1$ and $\varepsilon_2$, respectively. Each element in this figure, e.g., the angle $j$, can be split in two parts, a secular one $S(t)$ (actually a periodic one but with a very long period) and a short-period one $N(t)$, with the longest period equal to 18.6 years and many shorter ones:

$$j(t) = S(t) + N(t) = at + bt^2 + ct^3 + \cdots + n(t) - n(t_0) \tag{5.1}$$

While the secular terms $S(t)$ (*precession*) are zero at the initial epoch $t_0$, not so are the short-period ones $N(t)$ (*nutation*). The instantaneous elements are called *true* elements, those freed by nutation are called *mean* elements (mean equator, mean equinox, mean obliquity, etc.).

As already pointed out, the definition of the ecliptic requires some caution. In the literature, there is some confusion of terminology, dating back to two different definitions: the one by Newcomb and the one by Le Verrier. The former refers to a rotating observer and the latter to an inertial one.

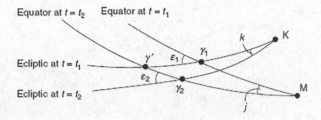

**FIGURE 5.2** The movements of the fundamental planes.

See the discussion by Standish (1981). While the celestial equator is the same for any observer, the ecliptic is only approximately so, because of the finite distance of the Sun. From the knowledge of the shape of the Earth, the apparent ecliptic of a particular observer (the topocentric ecliptic) can be accurately referred to an ideal geocentric observer. However, this translation is not entirely satisfactory: it would be better to call ecliptic the plane passing for the E–M barycenter and freed from all periodic perturbations (due mostly to Venus and Jupiter, which amount to $\pm 0''.4$ and $\pm 0''.2$, respectively). With respect to this ecliptic, the geocentric latitude of the Sun is not strictly zero; it can reach $\pm 0''.6$ according to the instantaneous position of the Moon, which changes with a period of approximately 29 days. Summing up the three different effects, the geocentric latitude of the Sun can amount to $\pm 1''.2$, an angle sufficiently small that we shall ignore it in almost all subsequent considerations. Furthermore, the instantaneous ecliptic is essentially never used; the plane called ecliptic is rather the mean ecliptic in the abovementioned sense of mean planes and mean coordinates. The mean equinox would then be the intersection of the mean equator with the mean ecliptic and the true equinox the intersection of the true equator with the mean ecliptic. Finally, after the introduction of the ICRS, we could call mean equinox the zero point of that catalogue, thus apparently avoiding the problem. However, the ambiguity of the concept "ecliptic" remains, as pointed out, for instance, by Fukushima (2003, see also the "Notes" section). In 2006, the IAU recommended that the ecliptic be precisely defined as the plane perpendicular to the mean orbital angular momentum vector of the Earth–Moon barycenter passing through the Sun.

## 5.4  FIRST-ORDER EFFECTS OF THE PRECESSION ON THE STELLAR COORDINATES

As remarked at the beginning of this chapter, the moderate precision sought for in our treatment permits to consider each effect one after the other and then to sum their amounts; therefore, we shall consider first the lunisolar, then the planetary and then the sum of the two, namely, the *general* precession.

We shall first proceed in the pre-1984 way, when the elementary unit of time was the tropical year and the role played by the vernal point $\gamma$ appears more clearly. The procedure allows the determination of the mean coordinates at any date, starting from the known mean coordinates at a given fundamental epoch. We have called lunisolar precession, without other adjectives, that due to a constant lunisolar torque on a rigid Earth, whose effect would be a strictly periodic rotation of the celestial pole P around the ecliptic one E, with constant velocity $P_0$ and obliquity $\varepsilon$, namely, a progressive, uniform increase of all longitudes expressed by

$$P_0 \cos \varepsilon = \dot{\lambda} = \psi \approx +50''.37, \quad P_0 \approx +54''.91 \ \left( \text{per tropical year} \right) \tag{5.2}$$

where $\psi$ is the notation favored by many authors and used in the following; $\psi$ and $P_0$ are arcs per unit time, in this case the tropical year. The value $P_0 = 54''.91$ derives for 1/3 from the Sun and for 2/3 from the Moon; it is a function of the orbital elements of the two forcing bodies, of the dynamical

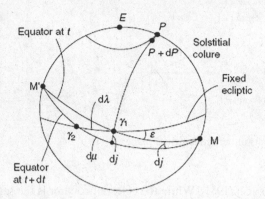

**FIGURE 5.3**  The elementary lunisolar precession. The equator performs an elementary rotation around the line MM′.

figure of the Earth (moments of inertia) and of the obliquity of the ecliptic. In principle, it could be determined by the theory, but not at the level of precision reached by the observations (see also Chapter 6). A variation of any of these elements causes a variation of $P_0$ and $\psi$, but for the moment, this case is not considered. With reference to Figures 5.1 and 5.3, in the assumption of a fixed ecliptic, after an elementary time $dt = 1$ tropical year, the moving equator has performed an elementary rotation $dj$ around the direction MM′. During the same time, the celestial pole has moved from P to P+dP along a great circle which is perpendicular to the solstitial colure and which is also the hour circle of the initial equinox $\gamma_1$, so that $dP = dj$.

The intersection between the ecliptic and the equator moves from $\gamma_1$ to $\gamma_2$, describing the elementary precession in longitude:

$$d\lambda = \psi\,dt = \frac{dj}{\sin\varepsilon} \approx +50''.37$$

The projection of this elementary arc on the moving equator amounts to the elementary lunisolar precession in right ascension:

$$d\mu = (\psi\,dt)\cos\varepsilon = \frac{dj}{\sin\varepsilon}\cos\varepsilon \approx +46''.21$$

The perpendicular component along the hour circle of $\gamma_1$ is the elementary lunisolar precession in declination:

$$dj = (\psi\,dt)\sin\varepsilon \approx +20''.05$$

It is customary to put

$$m = \psi\cos\varepsilon = +46''.21/\text{year} = +3^s.08/\text{year}, \quad n = \psi\sin\varepsilon = +20''.34/\text{year} = +1^s.34/\text{year}$$

where y is the tropical year (do not confuse this constant $n$ with the term in Equation 5.1 nor with the solar annual motion).

Let us take the differentials of the transformation between ecliptic and equatorial coordinates:

$$d\alpha = \frac{\partial\alpha}{\partial\lambda}d\lambda + \frac{\partial\alpha}{\partial\beta}d\beta + \frac{\partial\alpha}{\partial\varepsilon}d\varepsilon, \quad d\delta = \frac{\partial\delta}{\partial\lambda}d\lambda + \frac{\partial\delta}{\partial\beta}d\beta + \frac{\partial\delta}{\partial\varepsilon}d\varepsilon$$

Introduce the initial simplifying assumptions $\dot{\beta} \equiv 0, \quad \dot{\varepsilon} \equiv 0$. From Equation 3.7, we derive

$$\dot{\alpha} = \dot{\lambda}\left(\cos\varepsilon + \sin\varepsilon\sin\alpha\tan\delta\right) = m + n\,\sin\alpha\tan\delta$$

$$\approx \left[3^s.08 + 1^s.34\sin\alpha\tan\delta\right]/\text{year} \tag{5.3}$$

$$\dot{\delta} = \dot{\lambda}\sin\varepsilon\cos\alpha = n\cos\alpha \approx 20''.05\,\cos\alpha/\text{year} \tag{5.4}$$

Equation 5.4 means that the declination increases with time on the whole hemisphere having the vernal point as its pole ($\cos\alpha \geq 0$) and decreases in the other one. However, point $\gamma$ makes a full revolution on the ecliptic in approximately 25,800 years, so that the declination of a star is confined at all times between a maximum and a minimum value. This systematic behavior gives the possibility of determining $n$ with great precision.

More complex is the time derivative of the right ascension expressed by Equation 5.3: due to the greater importance of term $m$, the right ascension will increase on most part of the celestial sphere. However, there is a locus where

$$\dot{\alpha} = m + n\sin\alpha\tan\delta = 0$$

This locus is a spherical triangle having two vertices in the ecliptic and celestial poles, and the third in those stars for which the angle to the two poles is 90°. In each hemisphere, this locus is approximately a small circle of diameter $\varepsilon$ passing through the two poles; inside those two small circles, the right ascensions will decrease with time. In the northern hemisphere, the locus $\dot{\alpha} = 0$ occurs when $12^h \leq \alpha \leq 24^h$ for declinations larger than $90° - \varepsilon$, for instance, at ($16^h$, +70), ($18^h$, $+90-\varepsilon$), ($20^h$, +70) and so on. In the southern one, this locus occurs when $0^h \leq \alpha \leq 12^h$ and the declinations are smaller than $(-90° + \varepsilon)$, for instance, at ($6^h$, $-90+\varepsilon$) and so on. Notice also that $\tan\delta$ becomes very large in the proximity of the celestial poles, so that the calculations there need particular care.

It is instructive to derive the same results from a rotation of the Cartesian equatorial system having the $x$-axis toward $\gamma$ and the $z$-axis toward the celestial north pole at the initial epoch $t_0$. At the time $t_0 + dt$, the position of the system is given by a retrograde rotation $dj$ around $Oy$, followed by a retrograde rotation $d\mu$ around $Oz$. Applying the rotations to the point representing the initial position of the star, namely, to the direction,

$$\begin{cases} a = \cos\alpha\cos\delta \\ b = \sin\alpha\cos\delta \\ c = \sin\delta \end{cases}$$

after simple passages, we obtain the result:

$$\begin{cases} d\alpha = d\mu + \sin\alpha\,\tan\delta\,dj \\ d\delta = \cos\alpha\,dj \end{cases} \tag{5.5}$$

$$\begin{cases} \dfrac{d\alpha}{dt} = \left(\dfrac{d\mu}{dt}\right)_{t_0} + \left(\dfrac{dj}{dt}\right)_{t_0}\sin\alpha\tan\delta\,dj \\[4mm] d\delta = \left(\dfrac{dj}{dt}\right)_{t_0}\cos\alpha \end{cases}$$

These expressions contain the possibility that the "constants" $m$ and $n$ can vary with the epoch $t_0$.

Let us now consider the planetary precession, namely, the effect of the gravitational perturbations of the planets on the ecliptic, and consequently on the precession rate of $\gamma$ along the equator and on the value of the obliquity $\varepsilon$. The previous discussion has given the mathematical possibility to fix the equatorial plane and its poles in the inertial frame. Therefore, we need to consider only the small movement of the ecliptic poles around the so fixed celestial ones. This movement can be represented by an elementary rotation of the ecliptic plane around a given line KK′ (Figure 5.4, for simplicity K′ is not shown), where K and K′ are the poles of the arc EE′. In the notation of the *Astronomical Almanac*, the longitude of K, namely, the arc $\gamma$K, is indicated with $\Pi_A$; at the present epoch, its value is approximately $\Pi_A = 174°.8$. The elementary rotation of the ecliptic moves $\gamma$ in direct sense along the equator, by $g = 0''.13$/year. The projection of this motion on the ecliptic amounts to $g \cdot \cos \varepsilon = 0''.11$/year, which must be subtracted from the value of the lunisolar precession to obtain the value of the general precession $G = \psi - g \cos \varepsilon = +50''.26$/year. Consequently, the value of $m$ (precession in right ascension) slightly decreases to approximately $46''.07$/year $= 3^s.07$/year. The precession in declination, $n$, is not affected.

There is a second consequence of this rotation of the ecliptic around the line KK′, namely, the slight decrease of the obliquity itself. Again with reference to Figure 5.4, consider the elementary spherical triangle $\gamma K\gamma'$, where $\gamma'$ is the position of $\gamma$ on the fixed equator after 1 year, $\Pi_A$ is the longitude of K and $\pi_a$ is the angle in K (namely, the inclination of the mobile ecliptic on the fixed one after 1 tropical year). From Equation 1.6, we obtain

$$\sin \Pi_A \sin \pi_A = \sin g \sin(\varepsilon + \dot{\varepsilon})$$

$$\cos \varepsilon \cos g = \sin g \cot \Pi_A + \sin \varepsilon \cot(\varepsilon + \dot{\varepsilon})$$

where $g, \dot{\varepsilon}, \pi_A$ are three very small angles, so that

$$g \approx \pi_A \sin \Pi_A \operatorname{cosec} \varepsilon, \quad \varepsilon \approx \pi_A \cos \Pi_A \approx -0''.47/\text{year}$$

After $\Delta t$ years, the inclination $k$ of the mobile ecliptic on the fixed one will be

$$k = \pi_A \Delta t \approx \dot{\varepsilon} \Delta t / \cos \Pi_A$$

For instance, at the date J2020.5 the inclination has changed by $-9''.7$ with respect to the mean ecliptic of epoch J2000.0.

Due to the variation of the obliquity of the ecliptic, none of the elements $\psi$, $m$, $n$, $G$, $g$, $\Pi_A$, $\pi_A$, would be constant even if $P_0$ did not vary (as indeed it does).

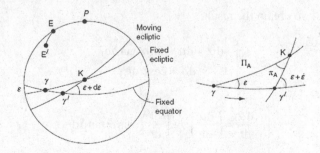

**FIGURE 5.4** The planetary precession. On the right, the situation after 1 year. Notice that arc $\gamma K = \Pi_A$ is approximately $174°.8$, not as drawn.

We come now to the post-1984 set of constants and procedures. Since 1984, the Julian year and the new system of constants adopted by IAU in 1976 (see also Kaplan, 1981) have been enforced in all Almanacs. Here are some of these values:

$$G = 50''.290966 + 0''.02222\ T, \quad \varepsilon = 23°26'21''.448 - 0''.00468150\ T$$

$$m = 46''.124362 + 0''.02793\ T\ , \quad n = 20''.043109 - 0''.008533\ T$$

$$\Pi_A = 174°.8764 + 0°.9137\ T, \quad \dot{\varepsilon} = -0''.46815 - 0.0007\ T,$$

$$g\cos\varepsilon = 0''.1055 - 0''.0189\ T$$

where $T$ is the number of Julian centuries of 36,525 days of $86,400^s$, starting from the fundamental epoch J2000.0. Notice that the ratio between the duration of the Julian year and that of the tropical year, namely, 1.00002136, does not entirely account for the strong revision of the value of $G$. Using centuries, the precessional constants become (see the *Astronomical Almanac* ed. 2019, page B54):

$$M = 1°.28115567\ T + 0°.00038655\ T^2 + 0°.0000101\ T^3 + \cdots$$

$$N = 0°.55671997\ T - 0°.00011930\ T^2 - 0°.00001162\ T^3 + \cdots$$

Let us retain for the moment only the linear term in $t$, and set us the task to derive the equatorial coordinates of a star, given by a catalog at the mean equinox J2000.0, at a date $t = t_0 + \Delta t =$ J2000.0$+ \Delta t$. In a first approximation, the following formulae will suffice:

$$\alpha_t = \alpha_{2000} + \left(m_{2000} + n_{2000} \sin\alpha_{2000} \tan\delta_{2000}\right)\Delta t \tag{5.6}$$

$$\delta_t = \delta_{2000} + n_{2000}\ \sin\alpha_{2000}\ \Delta t \tag{5.7}$$

with $\Delta t$ in Julian years (of course, one has to pay attention to the different units used in right ascension and declination).

A small refinement will improve the precision, considering the time variation of $m$ and $n$:

- calculate the values $m_{1/2}$, $n_{1/2}$ of the "constants" at the epoch $t_{1/2}$ intermediate between J2000.0 and $t$, $t_{1/2} = (t_1 + t_2)/2$;
- calculate the two terms $m' = m_{1/2}\Delta t$, $n' = n_{1/2}\Delta t$;
- using these values of $m'$, $n'$, derive from Equations 5.6 and 5.7 the values of the coordinates $(\alpha_{1/2}, \delta_{1/2})$ at the intermediate epoch:

$$\alpha_{1/2} = \alpha_0 + \frac{1}{2}m' + \frac{1}{2}n'\sin\alpha_0 \tan\delta_0, \quad \delta_{1/2} = \delta_0 + \frac{1}{2}n'\cos\alpha_{1/2}$$

- then, inserting $(\alpha_{1/2}, \delta_{1/2})$ again into Equations 5.6 and 5.7, derive the final ones:

$$\alpha = \alpha_0 + m' + n'\sin\alpha_{1/2}\tan\delta_{1/2} \tag{5.8}$$

$$\delta = \delta_0 + n'\cos\alpha_{1/2} \tag{5.9}$$

These approximate methods, however, fail for stars close to the celestial poles or for long periods. In the first case, we resort to the rigorous formulae expounded in the following (see also the

"Exercises" section). In the second case, Equations 5.6 and 5.7 can be regarded as the first terms of a development in series of $\alpha(t)$, $\delta(t)$. Adding the second derivatives, we obtain

$$\alpha_t = \alpha_{t_0} + \dot{\alpha}_{t_0} \Delta t + \frac{1}{2!} \ddot{\alpha}_{t_0} \Delta t^2 + \cdots,$$

$$\delta_t = \delta_{t_0} + \dot{\delta}_{t_0} \Delta t + \frac{1}{2!} \ddot{\delta}_{t_0} \Delta t^2 + \cdots, \qquad (5.10)$$

For full precision, the derivatives must contain also the variations of the "constants" (see the "Exercises" section). Notice also that it has become customary to calculate the mean coordinates not for the beginning of the year, but for its midpoint.

We shall discuss again the problem of deriving the coordinates at a wanted date after having discussed the nutation.

## 5.5   THE NUTATION

We have called *nutation* the collection of the short-period movements of the equator. The principal part, namely, that discovered by Bradley, is due to the influence of the Moon, whose orbital plane is inclined by $i = 5°9'$ to that of the ecliptic (see Figure 5.5; here, we ignore some smaller movements of the Moon described in Chapters 14 and 15).

Let us call N the ascending node of this orbit on the ecliptic (ascending meaning that node where the lunar latitude passes from negative to positive values) and N′ that on the equator. The two nodes are not fixed in inertial space. Due to the solar perturbation on the lunar orbit, N precesses along the ecliptic, in a retrograde sense, by approximately 191″ each day, or else approximately three lunar diameters westward per lunation, thus making an entire turn in 18.61 years (6798 days). The path of the Moon is indeed a very complex one, a fact that must be taken into account with great accuracy in studies such as the lunar occultations, described in Chapter 15. Correspondingly, in the same time the pole L of the lunar orbit describes a small circle of radius $i$ around the ecliptic pole E. As a consequence, the inclination $\omega$ of the lunar orbit on the celestial equator (and so the declination of the Moon) varies between approximately ±18°.8 and ±28°.8, according to the longitude of the node $\lambda(N)$, whose expression is

$$\lambda(N) = 125°.04452 - 1934°.1363\, T + 0°0000.2971\, T^2 \qquad (5.11)$$

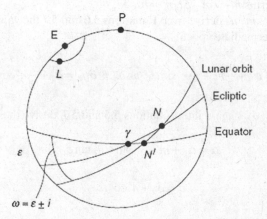

**FIGURE 5.5**   The lunar orbit and Bradley's nutation. $L$ is the pole of the lunar orbit.

with $T$ in Julian centuries from J2000.0. The variation of $\omega$ implies a variation of the perturbing torque exerted by the Moon on the Earth.

To see the effect on the declination of the Moon, consider the spherical triangle $N'\gamma N$; the angle in $\gamma$ is $\varepsilon$; the angle in $N'$ is $180° - \omega$; the arc $\gamma N$ is $\lambda(N)$; and the angle in N is $i$. Therefore,

$$\cos\omega = \cos\varepsilon\cos i - \sin\varepsilon\sin i\cos\lambda(N), \quad \varepsilon - i \leq \omega \leq \varepsilon + i$$

When $\lambda(N) = 0°$, as in 2006.46, the node coincides with $\gamma$, the inclination assumes the maximum possible value ($\omega = +28°36'$) and the declination varies between $\pm 28°36'$ during a lunation. When $\lambda(N) = 180°$, namely, 9.3 years later, the node coincides with the autumn equinox, the inclination on the equator is the minimum possible ($\omega = +18°18'$), and the declination varies between $\pm 18°18'$.

Now, consider the ascending node $N'$ on the equator and its right ascension $\alpha(N')$, namely, the arc $\gamma N'$. From the same spherical triangle $N'\gamma N$ (see again Figure 5.5), one can see that

$$\sin\gamma N' = \sin\alpha(N') = \sin\gamma N\frac{\sin i}{\sin\omega} = \sin\lambda(N)\frac{\sin i}{\sin\omega},$$

$$\left|\alpha(N')\right| \leq \arcsin\left(\frac{\sin i}{\sin\omega}\right) \approx 13°.0$$

Therefore, $N'$ oscillates along the equator around $\gamma$ in 18.6 years, with an amplitude of approximately $\pm 13°$.

Summarizing these considerations, the nutation discovered by Bradley can be described in the following way: the instantaneous movement of the celestial north pole is no longer along the great circle $P\gamma$, but along $PN'$, so that the nutation periodically changes not only the origin of the longitudes but also the obliquity, with amplitudes:

$$\Delta\lambda = -17''.2\sin\lambda(N), \quad \Delta\varepsilon = 9''.2\cos\lambda(N)$$

This movement can be visualized as the instantaneous true pole P describing a retrograde cone around the mean pole $P_m$, which in turn describes a cone of aperture $\varepsilon$ around E. Imagine looking at this movement from the outside of the celestial sphere, as in Figure 5.6.

On the plane tangent to the celestial sphere through $P_m$, the locus occupied by P is an ellipse, of semi-major axis $\Delta\eta = 9''.2$ and semi-minor axis $\Delta\xi = 17''.2\sin\varepsilon = 6''.9$, described with a period

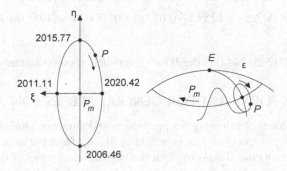

**FIGURE 5.6** Nutation of the instantaneous pole P around the mean pole $P_m$. The $\eta$-axis points toward the ecliptic pole E.

of 18.61 years in the retrograde sense. By dynamical arguments, the French mathematician J. d'Alembert showed that the ratio of the two axes must be equal to ($\cos 2\varepsilon/\cos \varepsilon$).

The complete phenomenon of nutation contains many other terms of smaller but non-negligible amplitudes and different periods; the second most important term in longitude has an amplitude of $1''.32$ and a period of 183 days and that in obliquity has an amplitude $0''.57$ and the same period (these values slightly change with the epoch).

Conventionally, the nutation terms have been divided into two groups, those having long periods (>90 days) and those having short periods (<35 days), which have often been treated separately in the derivation of the true coordinates. This argument is further expanded in Chapter 6. In the following, we shall indicate the complete nutation in longitude with $\Delta\psi$ and that in obliquity with $\Delta\varepsilon$.

The nutation can also be regarded as a retrograde rotation $\mathbf{R_3}$ of the equatorial mean Cartesian frame $O(x_M, y_M, z_M)$ around the $z$-axis by $\Delta\psi \cos \varepsilon$, followed by a direct rotation $\mathbf{R_2}$ around the $y$-axis by $\Delta\psi \sin \varepsilon$. The nutation in obliquity is then a retrograde rotation $\mathbf{R_1}$ around the $x$-axis of amplitude $\Delta\varepsilon$. By using the same procedure as seen in Equations 5.4 and 5.5, one can show that the nutation in longitude has the same structure of the lunisolar precession. Therefore, it will cause a variation of the equatorial coordinates given by

$$\Delta\alpha = \Delta\psi\left(\cos\varepsilon + \sin\varepsilon\sin\alpha\tan\delta\right), \quad \Delta\delta = \Delta\psi\left(\sin\varepsilon\cos\alpha\right) \tag{5.12}$$

The term in obliquity causes a variation of $\varepsilon$, but it does not affect the position of $\gamma$. By taking the derivatives of Equation 3.6, after few simple calculations we obtain

$$\Delta\alpha = -\Delta\varepsilon\cos\alpha\tan\delta, \quad \Delta\delta = \Delta\varepsilon\sin\alpha$$

Therefore, the total effect of the nutation is

$$\begin{cases} \Delta\alpha = \Delta\psi\left(\cos\varepsilon + \sin\varepsilon\sin\alpha\tan\delta\right) - \Delta\varepsilon\cos\alpha\tan\delta \\ \Delta\delta = \Delta\psi\sin\varepsilon\cos\alpha + \Delta\varepsilon\sin\alpha \end{cases} \tag{5.13}$$

whose numerical values depend on the particular date. For instance, for the year 2000 and to a precision of $1''$, the *Astronomical Almanac* gives the following numerical expressions:

$$\begin{cases} \Delta\psi = -0°.0048\,\sin(125°.1 - 0.053d) - 0°.0004\,\sin(198°.0 + 1.971d) \\ \Delta\varepsilon = +0°.0026\cos(125°.1 - 0.053d) + 0°.0002\cos(198°.0 + 1.971d) \end{cases} \tag{5.14}$$

being $d = \text{JD} - 2451543.5$ (2451544.0 is the JD at Greenwich noon on January 0, 2000).

## 5.6  APPROXIMATE FORMULAE FOR GENERAL PRECESSION AND NUTATION

Let us complete the problem of allowing for precession and nutation, again at the first order only. We already know how to precess the mean coordinates ($\alpha_0$, $\delta_0$) from the mean equinox at the epoch $t_0$ of the catalog to the mean coordinates ($\alpha$, $\delta$) at the beginning of year t of the date of observation, using Equations 5.6 and 5.7, or a more refined procedure. Call $\Delta t$ the remaining fraction of the year. Now, we have to add the general precession for this fraction and the nutation terms given in Equation 5.13, obtaining

$$\Delta\alpha = (m + n \, \sin\alpha \tan\delta)\Delta t + \Delta\psi (\cos\varepsilon + \sin\varepsilon \sin\alpha \tan\delta) - \Delta\varepsilon \cos\alpha \tan\delta$$

$$= m\left(\Delta t + \frac{\Delta\psi}{m}\cos\varepsilon\right) + n\left(\Delta t + \frac{\Delta\psi}{n}\sin\varepsilon\right)\sin\alpha \tan\delta - \Delta\varepsilon \cos\alpha \tan\delta$$

which can be written as

$$\Delta\alpha = A'(m + n\sin\alpha \tan\delta) + B'\cos\alpha \tan\delta + E' = A'a' + B'b' + E' \qquad (5.15)$$

$A'$, $B'$, $E'$ are named *Bessel's daily numbers*; their values are rapidly changing functions of the date but not of the star's coordinates. The quantities $a'$ and $b'$, which depend only on the star coordinates, are called Bessel's *star's constants*. In a similar way for the declination,

$$\Delta\delta = A'n\cos\alpha - B'\sin\alpha = A'a'' + B'b'' \qquad (5.16)$$

Actually, the "constants" ($a'$, $b'$, $a''$, $b''$) are slowly varying functions of the epoch, because of the slow variation of $m$ and $n$.

Another way of computing the combined effect of nutation and precession is by means of the so-called *independent day numbers f, g, G*:

$$\alpha = \alpha_0 + f + g\sin(G + \alpha_0)\tan\delta_0 \qquad (5.17)$$

$$\delta = \delta_0 + g(G + \alpha_0)$$

More precise expressions and values of Bessel's daily numbers, star constants and independent day numbers, valid also when the star is close to the celestial pole, can be found in Section B of the *Astronomical Almanac*. As already pointed out, the current practice is to calculate the precession and nutation not for the beginning of the year, but for its midpoint.

At this stage, we have obtained coordinates that are called the "true" coordinates of the star, but they are not yet the observed ones, because other phenomena contribute to the observed values, namely, the aberration, the gravitational deflection of light, the parallax, the proper motions and the atmospheric refraction, which we discuss in later chapters. There is an additional small effect due to general relativity, namely, the *geodesic precession* $P_g$. The inertial reference system in the neighborhood of the geocentric observer has a slight rotation with respect to the heliocentric frame, of the order of $+0''.0192$ per year. Such small correction can be absorbed in $\psi$, so that rigorously: $\psi = P_0 \cos\varepsilon - P_g$. See also Chapter 9.

## 5.7   NEWCOMB'S ROTATION FORMULAE FOR PRECESSION

After having discussed several approximate formulae, we describe the rigorous procedure introduced by Newcomb to derive the mean coordinates at an arbitrary date $t$ from an initial epoch $t_0$. Let us consider the star X (see Figure 5.7), and let $P_0$ and P be two successive positions occupied by the celestial north pole at times $t_0$ and $t$. In the spherical triangle $P_0 PX$, consider the arc $\theta_A$ and the angles $\zeta_{A,} z_A$.

The angle $\zeta_A$ will be very small for small $(t - t_0)$, and so will be the angle $z_A$, because the arc $(P_0 P + P\gamma)$ differs very little from a great circle; furthermore, $\zeta_A \approx z_A$. The arc $\theta_A$ is not exactly the path described by the true pole, which is actually a somewhat irregular curve, but this fact is irrelevant here. At the second order in $T$, the elements $(\zeta_A, z_A, \theta_A)$ are given by

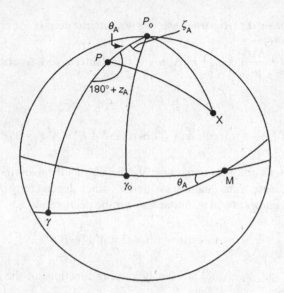

**FIGURE 5.7**  Newcomb's precessional angles. Rigorously, arcs $P_0P$ and $P\gamma$ are not part of the great circle $P_0\gamma$.

$$\left\{ \begin{array}{l} \zeta_A = +0°.0000736 + 0°.6405787T + 0°.0000830T^2 + \cdots \\ z_A = -0°.0000736 + \zeta_A + 0°.000202T^2 + \cdots \\ \quad \theta_A = 0°.5567530T - 0°.0001185T^2 + \cdots \end{array} \right. \qquad (5.18)$$

As usual, $T$ is given in Julian centuries since J2000.0. For instance, the values for reduction from J2000.0 to J2019.5 are

$$\left\{ \begin{array}{l} \zeta_A = 0°.12565278 \\ z_A = 0°.12418889 \\ \theta_A = 0°.10855556 \end{array} \right.$$

From Equations 1.5 and 1.6, we can derive the following rigorous formulae:

$$\left\{ \begin{array}{l} \cos\delta \, \sin(\alpha - z_A) = \cos\delta_0 \sin(\alpha_0 + \zeta_A) \\ \cos\delta \, \cos(\alpha - z_A) = \cos\theta_A \cos\delta_0 \cos(\alpha_0 + \zeta_A) - \sin\theta_A \sin\delta_0 \\ \quad \sin\delta = \sin\theta_A \cos\delta_0 \cos(\alpha_0 + \delta_A) + \cos\theta_A \sin\delta_0 \end{array} \right. \qquad (5.19)$$

and their inverse expressions:

$$\left\{ \begin{array}{l} \sin(\alpha_0 + \zeta_A)\cos\delta_0 = \sin(\alpha - z_A)\cos\delta \\ \cos(\alpha_0 + \zeta_A)\cos\delta_0 = \cos\theta_A \cos\delta \cos(\alpha_0 - z_A) + \sin\theta_A \sin\delta \\ \quad \sin\delta_0 = -\sin\theta_A \cos\delta \cos(\alpha_0 - z_A) + \cos\theta_A \sin\delta \end{array} \right. \qquad (5.20)$$

The transformations 5.19 and 5.20 can be expressed as a rotation matrix $\mathbf{P}$ applied to an initial Cartesian system $O(x_0, y_0, z_0)$ to derive $O(x, y, z)$ and vice versa, namely, $\mathbf{r} = \mathbf{P}\mathbf{r}_0$, or $\mathbf{r}_0 = \mathbf{P}^{-1}\mathbf{r}$. The elements of the matrices can be found by the above equations; for instance,

$$P_{11} = -\sin\zeta_\mathrm{A} \sin z_\mathrm{A} + \cos\zeta_\mathrm{A} \cos z_\mathrm{A} \cos\theta_\mathrm{A}$$

For brevity, we omit the complete expressions, which can be found in the *Astronomical Almanac*. The diagonal elements are very close to one, and all others are very close to zero, so that calculations must be carried out with at least eight significant decimal digits.

To allow for nutation, the rotation $\mathbf{P}$ will be followed by rotation $\mathbf{R}_\mathrm{N}$:

$$\mathbf{R}_\mathrm{N} = \begin{pmatrix} 1 & -\Delta\psi\cos\varepsilon & -\Delta\psi\sin\varepsilon \\ \Delta\psi\cos\varepsilon & 1 & -\Delta\varepsilon \\ \Delta\psi\sin\varepsilon & \Delta\varepsilon & 1 \end{pmatrix} \tag{5.21}$$

whose numerical values are given in Equation 5.14.

Although the method is accurate from the numerical point of view, it cannot be forgotten that the instantaneous values of the constants are not known with infinite precision, and, actually, some correction to their values is always possible with an *a posteriori* analysis of the observations.

## 5.8 PRECESSION AND POSITION ANGLES

Precession and nutation are rigid rotations of the celestial sphere; as such, they do not alter the angular distance between the stars. The apparent shape of a constellation or of a nebula will change in time, because of the proper motions of stars (described in Chapter 8) and/or expansion of filaments, not because of precession and nutation. However, the position angle $p$ between two objects defined in Chapter 3 will change because it is measured from the variable direction of the north celestial pole. After some manipulation, it can be seen that

$$\Delta p / \Delta t = n \sin\alpha \sec\delta = 0°.0056 \sin\alpha \sec\delta \text{ per year} \tag{5.22}$$

Therefore, the position angle must be given in conjunction with an epoch. Its variation is very large in the proximity of the celestial poles. In a similar way, the differential coordinates of the two nearby objects $(\alpha_1 \approx \alpha_2 \approx \alpha, \delta_1 \approx \delta_2 \approx \delta)$ will change by

$$\frac{d\Delta\alpha}{dt} \approx n \sec\delta \left[ \Delta\alpha \cos\alpha \sin d\delta + \Delta\delta \sin\alpha \sec\delta \right], \qquad \frac{d\Delta\delta}{dt} \approx -n \, \Delta\alpha \sin\alpha$$

## 5.9 SOLAR SYSTEM OBJECTS

Allowance for precession and nutation in the case of objects of the Solar System presents several peculiarities. Their Cartesian coordinates, velocities and accelerations are usually known, often with such high precision to require a relativistic treatment. Their positions rapidly change with respect to the background of the "fixed" stars. Topocentric, geocentric, heliocentric and barycentric systems might be required.

Therefore, we must postpone a fuller treatment to later chapters, after having discussed their orbits, the planetary aberration and the different time systems. However, some considerations about the large planets can be made here. From the astrometric point of view, the significant plane is that of the ecliptic, whose stability is much greater than that of the equator. Therefore, in a first approximation, of all elements characterizing the position in space of a planet, only those connected with the position of the equinox on the ecliptic will be affected by precession. We shall have to worry, for instance, about the longitude of the ascending node, very much less about the inclination.

More complicated is the case of the Moon or of periodic comets. See, for instance, the paper by Simon et al. (1994) discussing precession formulae for the Moon and the planets.

## NOTES

- It will be useful to underline that precession and nutation do not alter the geographic coordinates, nor the position of the cardinal points on the horizon, a fact all too often overlooked by science fiction writers.
- The best method Hipparchus and the ancient astronomers had available in order to measure the positions of the stars without precise clocks was to refer their positions to the Sun, using as intermediary a bright planet such as Venus, which can be seen in full daylight. An ingenious variant of the method was to take advantage of a lunar eclipse, when the Sun is at 180° from the Moon, to measure the position of a star near the ecliptic such as Spica, with respect to the center of the Earth's shadow on the Moon.
- As already pointed out, we proceeded in this book by adding one after the other the different effects affecting the true coordinates, so that the order in which the different corrections will be applied is immaterial. This procedure is sufficient for most purposes, but it is not rigorous; see, for instance, the discussion by Soma and Aoki (1990). Standardized rigorous computer programs are available that follow generally agreed procedures, see, for instance, Wallace (1994). See also the site of the HST Science Institute in the Web Sites section.
- Among the many papers useful to follow the evolution of the determination of the precessional constant, see Fricke (1971). The paper by Lieske et al. (1977) contains cross-references to the notations used by several authors. The *1976 IAU Resolution on Astronomical Constants* is given, e.g., by Kaplan (1981). A most important paper on precession formulae is the one by Fukushima (2003), based on the dynamical treatment given by Williams in 1994 (see Chapter 6); in particular, a careful discussion is given of the different meanings of ecliptic. Another discussion of the motions of the ecliptic is given by Harada and Fukushima (2004). This paper also contains an interesting nonlinear method of harmonic analysis.
- The IAU 2006 resolution B1 (Adoption of the P03 Precession Theory and Definition of the Ecliptic). The XXVIth International Astronomical Union General Assembly,

    *Noting*

    (1) the need for a precession theory consistent with dynamical theory, (2) that, while the precession portion of the IAU 2000A precession-nutation model, recommended for use beginning on January 1, 2003 by resolution B1.6 of the XXIVth IAU General Assembly, is based on improved precession rates with respect to the IAU 1976 precession, it is not consistent with dynamical theory, and (3) that resolution B1.6 of the XXIVth General Assembly also encourages the development of new expressions for precession consistent with the IAU 2000A precession-nutation model, and

    *Recognizing*

    (1) that the gravitational attraction of the planets makes a significant contribution to the motion of the Earth's equator, making the terms lunisolar precession and planetary precession misleading, (2) the need for a definition of the ecliptic for both astronomical and civil purposes, and (3) that in the past, the ecliptic has been defined both with respect to an

observer situated in inertial space (inertial definition) and an observer comoving with the ecliptic (rotating definition),

*Accepts*

the conclusions of the IAU Division I Working Group on Precession and the Ecliptic published in Hilton et al. (2006, *Celest. Mech.* 94, 351), and

*Recommends*

(1) that the terms lunisolar precession and planetary precession *be replaced by precession of the equator and precession of the ecliptic, respectively,* (2) that, beginning on January 1, 2009, the precession component of the IAU 2000A precession-nutation model be replaced by the P03 precession theory of Capitaine et al. (2003b) for the precession of the equator (Eqs. 37) and the precession of the ecliptic (Eqs. 38); the same paper provides the polynomial developments for the P03 primary angles and a number of derived quantities for use in both the equinox based and CIO-based paradigms, (3) that the choice of precession parameters be left to the user, and (4) that the ecliptic pole should be explicitly defined by the mean orbital angular momentum vector of the Earth–Moon barycenter in the Barycentric Celestial Reference System (BCRS), and this definition should be explicitly stated to avoid confusion with other, older definitions.

- For examples of numerical methods efficient over long time spans applicable to Earth and Mars, see Laskar et al. (2004a,b, 2011)
- The flyby of Pluto and Charon by the NASA spacecraft New Horizons was carried out in 2015. In January 2019, the spacecraft visited the Kuiper Belt object (486958) Arrokoth (nicknamed Ultima Thule). Vesta and Ceres were visited by the NASA spacecraft DAWN. The third largest asteroid, Lutetia, the small asteroid Steins and comet 67P/C-G were visited by the ESA spacecraft Rosetta from 2008 to 2016. Information about these missions can be found in the Web Sites section.

## EXERCISES

1. Apply Equations 1.5 and 5.7 to a body of the Solar System, in order to transform its Cartesian coordinates $(x, y, z)$ at the epoch $t_0$, to those at another date $t_0 + \Delta t$, ignoring the real movement of the body between the two dates. Provided $\Delta t$ is not too large, so that the precession constants do not vary, we obtain

$$\begin{cases} x(t_0 + \Delta t) = x - (my + nz)\Delta t \\ y(t_0 + \Delta t) = y + mx\Delta t \\ z(t_0 + \Delta t) = z + nx\Delta t \end{cases}$$

where the constants must be expressed in radians per year ($m = 0.0002234$, $n = 0.000097$).

2. Calculate the second derivatives in Equation 5.10. Call $\tau$ the initial epoch, and write

$$\begin{cases} \alpha = \alpha_\tau + a_1 \Delta t + a_2 \Delta t^2 + \cdots \\ \delta = \delta_\tau + b_1 \Delta t + b_2 \Delta t^2 + \cdots \end{cases}$$

After simple passages, we have

$$a_1 = \left( \frac{d\alpha}{dt} \right)_\tau = m_\tau + n_\tau \sin\alpha_\tau \tan\delta_\tau$$

$$a_2 = \frac{1}{2}\left(\frac{d^2\alpha}{dt^2}\right)_\tau = \frac{1}{2}\left[\left(\frac{dm}{dt}\right)_\tau + \left(\frac{dn}{dt}\right)_\tau \sin\alpha_\tau \tan\delta_\tau\right]$$

$$+\frac{1}{4}n_\tau^2\left[1+2\tan^2\delta_\tau\right]\sin 2\alpha_\tau + \frac{1}{2}m_\tau n_\tau \cos\alpha_\tau \tan\delta_\tau$$

$$b_1 = \left(\frac{d\delta}{dt}\right)_\tau = n_\tau \cos\alpha_\tau,$$

$$b_2 = \frac{1}{2}\left(\frac{d^2\delta}{dt^2}\right)_\tau = \frac{1}{2}\left[\left(\frac{dn}{dt}\right)_\tau \cos\alpha_\tau - n_\tau \sin\alpha_\tau\left(m_\tau \sin\alpha_\tau \tan\delta_\tau\right)\right]$$

3. Derive the mean coordinates of Polaris for the year 2100, knowing its J2000.0 position $(2^h31^m46^s.3, +89°15'50''.6)$. Ignore the proper motion of the star.

This exercise has been adapted from Newcomb and Danjon (Bibliography section). The time interval is very long, and the star is very close to the celestial north pole; therefore, we must use the formulae 5.18 and 5.19, with $T = 1$. The result is

$$\text{J2100.0} \left(5^h53^m14^s.9, +89°32'26''.3\right).$$

The result given by Danjon, who, however, started from B1900.0 and with the pre-1994 constants, is B2100.0 $(5^h53^m36^s.7, +89°32'22''.7)$.

# 6 Dynamics of Earth's Rotation

This chapter is dedicated to a very simplified dynamical analysis of the Earth's lunisolar precession, nutation and free (Eulerian) nutation. After Bradley's measurements, many mathematicians of the 18th century, like J. d'Alembert, A.C. Clairaut and L. Euler, accepted the challenge of explaining these observations. Let us start by examining the intuitive explanation of the lunisolar precession put forward by Newton in his *Principia*. We shall give first an approximate description, sufficient to clarify the basis of the overall phenomenon, and then a more quantitative analysis. As in Chapter 5, the traditional approach and denominations of precession, nutation and ecliptic are maintained. In addition to the "Notes" section of the present chapter, see Chapters 2, 5 and 10 for references to recent changes.

## 6.1 NEWTON'S LUNISOLAR PRECESSION

We have already seen that the figure of the Earth is well approximated by an oblate ellipsoid of revolution, with equatorial axis $a_\oplus$ and polar axis $c$; for the moment, the rotational axis is supposed to coincide with $c$. The reason for this shape is the centrifugal force due to the diurnal rotation. This figure can be thought of as an inner sphere (not necessarily homogeneous, but having density with radial symmetry), plus an equatorial bulge. Consider the gravitational attraction exerted by the point-like Sun S on each element of mass of such supposedly *rigid* body (see Figure 6.1); the attraction is proportional to the mass of the Sun and to the inverse square of the distance between the particular element and the Sun. The overall effect on the barycenter of the Earth, responsible for the annual revolution, already discussed in Chapter 4, will be examined in more depth in Chapter 12.

Consider two mass points inside the spherical inner body, such as A and B in Figure 6.1. The force in A will be slightly different from the force in B, because of different distance and direction to S, but this slight imbalance will be perfectly compensated by the two symmetric points, A′, B′. However, this compensation is not possible for points of the equatorial bulge, e.g., points V and W. The net result will be a torque parallel to CS, perpendicular to the plane of the figure and directed inside. As is well known, a gyroscopic axis under the influence of an external torque will tend to orient itself in the plane orthogonal to that containing axis and torque. Therefore, the combination of the angular momentum due to the diurnal rotation and of the solar torque will cause a gyroscopic movement of the Earth axis in the plane containing $c$ and the perpendicular to CS. The solar torque is zero at the equinoxes; it would be zero again if the Sun reached declinations ±90°; it would be maximum if it reached declinations ±45°, but these conditions never happen.

The same reasoning can be applied to the Moon, taking into account the different mass, distance and position with respect to the equator.

Now, we shall show that the gyroscopic motion forced by the external body S has amplitude proportional to the mass of S and to the *inverse cube of its distance* CS, *not to the inverse square*.

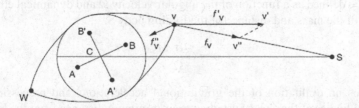

**FIGURE 6.1** Solar precession, winter solstice. The non-spherical shape of the Earth has been greatly exaggerated.

The explanation is intuitively simple, and the imbalance can be considered due to the *gradient* of the gravitational force across the figure of the Earth. For this reason, the effect of the Moon is 2.2 times greater than that of the Sun. To see this conclusion in a quantitative form, refer again to Figure 6.1. Calling $r_S$ the distance between the barycenter of the Earth and of the forcing body S, we get

$$r_V^2 = a_\oplus^2 + r_S^2 - 2a_\oplus r_S \cos\delta_S, \quad r_W^2 = a_\oplus^2 + r_S^2 + 2a_\oplus r_S \cos\delta_S$$

In the limit $a_\oplus/r_S \ll 1$ (certainly true for the Sun; for the Moon: $a_\oplus/r_\mathbb{D} \approx 1/60$),

$$r_V \approx r_S\left(1 - \frac{a_\oplus}{r_S}\cos\delta_S\right), \quad r_W \approx r_S\left(1 + \frac{a_\oplus}{r_S}\cos\delta_S\right)$$

Let $f_V'$ be the component of the gravitational force $f_v$ parallel to CS (and similarly $f_W'$):

$$f_V'/f_V = r_S/r_V, \quad f_W'/f_W = r_S/r_W, \quad \Delta f' = f_V' - f_W' = -K\frac{M_S}{r_S^3}a_\oplus \cos\delta_S$$

Therefore, the key factors for the amplitude of the torque are the mass of S and the inverse cube of its distance, the radius of the Earth and the mass of the equatorial bulge, which enters in the still unspecified constant $K$.

Regarding the mass contained in the equatorial bulge, a rough estimate can be obtained by calculating the volume of the ellipsoid and subtracting that of the inner sphere. Calling $\rho_\oplus$ the mean density, we would have

$$m_{\text{bulge}} \approx \frac{8}{3}\pi\rho_\oplus c^3 \frac{a_\oplus - c}{c} \approx 2fM_\oplus \approx \frac{2}{300}M_\oplus \approx f_{\text{bulge}}M_\oplus$$

where $f$ is the flattening factor defined in Chapter 2. However, the value $f_{\text{bulge}} \approx 2/300$ is certainly overestimated, because the density of the crust of the Earth is smaller than the average (3.3 instead of 5.5), and because at any instant the effective mass is less than that of the whole bulge.

Regarding the order of magnitude of the precessional velocity, from the basic equations of precessional motions we recall that the torque is proportional to the product of the moment of inertia $C$ of the forced body (Earth) around its polar axis, times its rotational (diurnal) velocity $\Omega$ and times the precessional velocity $\omega_{\text{prec}}$, so that

$$\omega_{\text{prec}} \approx \frac{\Delta F \cdot a_\oplus}{C \cdot \Omega} \approx G\frac{m_{\text{bulge}} \cdot a_\oplus \cdot a_\oplus}{M_\oplus a_\oplus^2 \cdot \Omega}\frac{M_S}{r_S^3} \approx G\frac{f_{\text{bulge}}}{\Omega}\frac{M_S}{r_S^3}$$

(notice the cancellation of $\delta_S$, due to the presence of $\sin\delta_S$ on both terms). In general, a precessional parameter $\alpha$ can be defined as a function of the angular velocity $\Omega$ and dynamical ellipticity $\sigma$ of the forced body, and of the mass and distance of the forcing body S:

$$\alpha = G\frac{\sigma}{\Omega}\frac{M_S}{r_S^3}$$

Table 6.1 provides an evaluation of the gravitational accelerations and precessional effects $F_P$ exerted by several bodies of the Solar System at their minimum distance from the Earth.

Examine first the fourth and fifth columns, where the distance enters with the second power. The gravitational acceleration exerted by the Sun on the Earth is $6.1 \times 10^{-4}$ the gravity at the surface

**TABLE 6.1**

**Gravitational and Precessional Effects of Bodies of the Solar System on the Earth**

| Body | Mass (in Earth Masses $M_\oplus$) | Minimum Distance (in Earth Radii $a_\oplus$) | $g/g_\oplus$ | $(g/g_\oplus)_\odot$ | $F_{pr}/F_\odot$ |
|---|---|---|---|---|---|
| Sun | $3.33 \times 10^5$ | $2.354 \times 10^4$ | $6.1 \times 10^{-4}$ | 1 | 1 |
| Moon | $1.23 \times 10^{-2}$ | 60.2 | $3.4 \times 10^{-6}$ | $5.6 \times 10^{-3}$ | 2.19 |
| Venus | $8.15 \times 10^{-1}$ | $6.49 \times 10^3$ | $1.9 \times 10^{-8}$ | $3.2 \times 10^{-5}$ | $1.2 \times 10^{-4}$ |
| Mars | $1.08 \times 10^{-1}$ | $1.23 \times 10^4$ | $7.2 \times 10^{-10}$ | $1.2 \times 10^{-6}$ | $2.3 \times 10^{-6}$ |
| Jupiter | $3.18 \times 10^2$ | $9.85 \times 10^4$ | $3.3 \times 10^{-8}$ | $5.4 \times 10^{-5}$ | $1.3 \times 10^{-5}$ |
| Saturn | $9.52 \times 10^1$ | $2.00 \times 10^5$ | $2.4 \times 10^{-9}$ | $3.9 \times 10^{-6}$ | $4.6 \times 10^{-7}$ |

(indicated with $g_\oplus$). The gravitational attraction by the Moon is 200 times smaller than that of the Sun. The attractions exerted by Venus and Jupiter are two orders of magnitude below that of the Moon. The fifth column normalizes the gravitational attractions to that if the Sun. The last column, where the distance enters with its cube, shows that the precessional effect of the Moon is 2.19 times higher than that of the Sun, whereas those of Venus and Jupiter are, respectively, four and five orders of magnitude below the solar one. The values in the last column enter also into the discussion of the sea and solid Earth tides.

## 6.2 THE LUNISOLAR TORQUE

We examine in this section the influence of the astronomical coordinates of the Sun and of the Moon on precession and nutation. Let us consider first the Sun (see Figure 6.2).

At the summer solstice, the moment of the torque is oriented toward the vernal point $\gamma$, and therefore, the Earth's axis will tend to fall toward $\gamma$; the same will happen 6 months later, at the winter solstice. On the other hand, at the equinoxes the moment will be zero.

Thus, we can consider the solar moment $\mathbf{K}_\odot$ as composed of two vectors of constant and equal modules, one always directed toward $\gamma$ and one rotating in the equatorial plane with a period of 6 (*not* 12!) months. The first component forces the north pole P to constantly fall toward $\gamma$. Therefore, the instantaneous velocity of P will be a uniform retrograde vector tangent both to the hour circle of $\gamma$ and to the small circle of radius $\varepsilon$ centered on the ecliptic pole E. The modulus of this vector of solar origin is approximately 3.19 times smaller than the lunisolar precession, namely, $15''.8$ sin $\varepsilon \approx 6''.3$ per year, which would also be the elementary rotation of the equator around MM' in Figure 5.1. The vernal point itself is forced to a retrograde motion on the ecliptic of $15''.8$ per year. This would be the very slow *mean solar* precession.

The second component, rotating on the equatorial plane with a period of 6 months, is responsible for a movement of the north pole P along a circle around the *mean* pole, of very small amplitude. This solar nutation produces (1) a periodic oscillation of $\gamma$ about the mean position with a semi-amplitude of $1''.6$ and (2) a periodic variation of the obliquity with total amplitude of $0''.55$ (the obliquity being maximal at the equinoxes and minimal at the solstices). The *true* celestial poles would therefore appear to describe a cycloid around the mean poles, with arcs $3''.2$ long and $1''.1$ high, every 6 months.

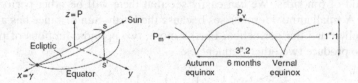

**FIGURE 6.2** The solar precession.

In analytical terms, the previous discussion showed that the solar torque has amplitude $K_\odot$ given by

$$K_\odot = \Delta f' a_\oplus \sin \delta_\odot = k' \sin \delta_\odot \cos \delta_\odot = k \sin 2\delta_\odot$$

At every date, the vector $\mathbf{K}_\odot$ is orthogonal both to the polar axis and to the line joining the Earth and the Sun, therefore pointing to direction $(\alpha_\odot - 90°)$ in the equatorial plane; in the geocentric equatorial Cartesian reference system $(x, y, z)$, its components will be

$$\begin{cases} K_x = k \sin 2\delta_\odot \cos(\alpha_\odot - 90°) = 2k \sin \delta_\odot \cos \delta_\odot \sin \alpha_\odot \\ K_y = k \sin 2\delta_\odot \sin(\alpha_\odot - 90°) = -2k \sin \delta_\odot \cos \delta_\odot \cos \alpha_\odot \\ K_z = 0 \end{cases} \quad (6.1)$$

or else indicating as usual with $\lambda_\odot$ the longitude of the Sun:

$$\begin{cases} K_x = k \sin \varepsilon \cos \varepsilon (1 - \cos 2\lambda_\odot) \\ K_y = k \sin \varepsilon (-\sin 2\lambda_\odot) = k \cos \varepsilon \tan \varepsilon (-\sin 2\lambda_\odot) \\ K_z = 0 \end{cases} \quad (6.2)$$

Given that $\mathbf{K}_\odot$ cannot alter the module of the diurnal rotation velocity vector $\mathbf{\Omega}$, but only its direction, its components will be directly proportional to the Cartesian components of the velocity of the pole, namely, $K_x \propto \dot{x}$ and similar for the other components. Now, assume that $\dot\lambda_\odot$ is a uniform function of time (this assumption is not strictly correct, as discussed in Chapter 4), namely, that $\dot\lambda_\odot = 2\pi$ if the time unit is the tropical year. Then, by integrating Equation 6.2 from the initial point $\lambda_\odot = 0$, we get

$$\begin{cases} x \mathrm{cosec}\, \varepsilon = P_\odot \cos \varepsilon \left( t - \frac{1}{2\dot\lambda_\odot} \sin 2\lambda_\odot \right) = p_\odot \left( t - \frac{1}{2\dot\lambda_\odot} \sin 2\lambda_\odot \right) \\ y = \frac{1}{2\dot\lambda_\odot} P_\odot \cos \varepsilon \tan \varepsilon \cos 2\lambda_\odot = \frac{p_\odot}{2\dot\lambda_\odot} \tan \varepsilon \cos 2\lambda_\odot \\ z = 1 \end{cases} \quad (6.3)$$

where $P_\odot$ is a "constant" whose value (17″.2 per tropical year) depends on the angular momentum of the Earth and on the average solar torque. The projection of $P_\odot$ on the ecliptic, $p_\odot = P_\odot \cos \varepsilon$, is the solar contribution to the total precession (its value is still less constant than that of $P_\odot$ because of the variable obliquity). Disregarding these minute variations, the term $x \, \mathrm{cosec}\, \varepsilon$ represents the motion in longitude $\Delta \lambda$ of the north pole; the term $y$ gives the variation of the obliquity $\Delta \varepsilon$; and the $z$-component stays constant. We have confirmed, by this simplified analytical treatment, the existence of a progressive term in $\Delta \lambda$ increasing as $p_\odot t$ and of two periodic terms in $\Delta \lambda$ and $\Delta \varepsilon$, both with a period of 6 months. We can easily see that there will be other periodic terms in the solar precession. A small annual term arises because the Earth–Sun distance has a slight variation between the aphelion and the perihelion. Furthermore, two periodic motions of period $T_1$ and $T_2$ couple together to produce two other frequencies:

$$\frac{1}{T_+} = \frac{1}{T_1} + \frac{1}{T_1}, \quad \frac{1}{T_-} = \frac{1}{T_1} - \frac{1}{T_2} \quad (6.4)$$

so that two other periods of 4 and 12 months (and others of smaller amplitude) will be present in the complete solar nutation expression.

The above treatment cannot be extended so easily to the Moon, because of the much greater intricacies of the lunar orbit, the larger precessional torque and the larger variation in declination caused by the inclination of 5°9′ of the orbital plane on the ecliptic. Nevertheless, a first conclusion can be reached almost immediately: owing to the inverse proportionality of the amplitude to $d\lambda/dt$, which varies 13 times more rapidly than that of the Sun (13 lunar months in 1 year), the period of the lunar semi-month (13.7 days) has amplitude smaller than that of the 6-month period in the solar one. Instead, the very long time (18.61 years, 6789 days) associated with the retrogradation of the nodes will be extremely important, say approximately $18.6/2.19 \approx 8.5$ times larger than the semi-annual solar term. Any variation of the other orbital elements will give rise to corresponding terms in the lunar nutation, and to their coupling frequencies, so that the total number of measurable terms rapidly grows.

The 1980 IAU Theory of Nutation was computed by determining the nutation in longitude and obliquity of a rigid Earth (Kinoshita, 1977) and introducing some modifications to allow for the non-rigidity. Table 6.2 reports the nine terms with amplitude larger than 0″.03 (the values are at J2000.0, all have minute variations).

Although the 1980 theory contained 106 terms in longitude and obliquity, it still failed to represent the observations. See the "Notes" section to Chapter 5 for the IAU resolution of 2006 with the new model of precession and nutation.

## 6.3 THE PRECESSIONAL POTENTIAL

Let us consider a *rigid* Earth as an ellipsoidal body of center C. Let $C(X, Y, Z)$ be the geocentric equatorial Cartesian system connected to the body and therefore rotating in the inertial frame. Let $a, b, c$ be the three principal semi-axes, and $A, B, C$ the three corresponding moments of inertia:

$$A = \int \left(Y^2 + Z^2\right) dM, \quad B = \int \left(X^2 + Z^2\right) dM, \quad C = \int \left(X^2 + Y^2\right) dM \qquad (6.5)$$

where $dM$ is the mass element; the integration must be performed over the volume of the body. For symmetry reasons, it is

$$\int dM = M_\oplus, \quad \int X \, dM = \int Y \, dM = \int Z \, dM = \int XY \, dM = \cdots = \int X^2 Y \, dM = \cdots = \int X^3 \, dM = \cdots = 0$$

---

**TABLE 6.2**
**High-Amplitude Terms in the 1980 IAU Theory of Nutation**

| Term Number | Period (days) | Longitude (″) | Obliquity (″) |
|---|---|---|---|
| 1 | 6798.4 | −17.1996 | 9.2025 |
| 2 | 3399.2 | 0.2062 | −0.0895 |
| 9 | 182.6 | −1.3187 | 0.5736 |
| 10 | 365.3 | 0.1426 | 0.0054 |
| 11 | 121.7 | −0.0517 | 0.0224 |
| 31 | 13.7 | −0.2274 | 0.0977 |
| 32 | 27.6 | 0.0712 | −0.0007 |
| 33 | 13.6 | −0.0386 | 0.0200 |
| 34 | 9.1 | −0.0301 | 0.0129 |

---

For the simplified model of homogeneous and rotationally symmetric ellipsoid with $c$ slightly smaller than $a$ and $b$, we would have

$$A = \frac{1}{5}M_\oplus\left(b^2 + c^2\right), \quad B = \frac{1}{5}M_\oplus\left(a^2 + c^2\right), \quad C = \frac{1}{5}M_\oplus\left(a^2 + b^2\right)$$

$$a = b = a_\oplus, \quad A = B, \quad \frac{C}{A} = \frac{2a_\oplus^2}{a_\oplus^2 + c^2} \approx 1.003, \quad \frac{C-A}{A} \approx \frac{C-A}{C} \approx 0.003$$

With reference to Figure 6.3, let Q be a massive point external to the Earth, at a distance $r$ from C much greater than $a_\odot$. Let $v$ be the angle PCQ.

For a generic point P inside the Earth, of elementary mass d$M$, distant $R$ from C and $r_{PQ}$ from Q, we have

$$r_{PQ}^2 = R^2 + r^2 - 2rR\cos v = r^2\left(1 + \frac{R^2 - 2rR\cos v}{r^2}\right)$$

where $0 \le R \le a_\oplus$.

With the simplification $a_\oplus \ll r_{PQ}$ and $r$, we apply the series expansion seen in Equation 1.12:

$$\frac{1}{r_{PQ}} = \frac{1}{r}\left[1 + \frac{R}{r}\cos v + \frac{R^2}{2r^2}\left(3\cos^2 v - 1\right) + \frac{R^3}{2r^3}\left(5\cos^3 v - 3\cos v\right) + \cdots\right]$$

$$= \frac{1}{r}\left[1 + \frac{R}{r}P_1(\cos v) + \left(\frac{R}{r}\right)^2 P_2(\cos v) + \cdots\right] \tag{6.6}$$

where $P_n$ are Legendre's polynomials of order $n$, defined by the general expression:

$$P_n(x) = \frac{1}{2^n n!}\frac{d^n}{dx^n}\left[\left(x^2 - 1\right)^n\right] \tag{6.7}$$

$$P_0(x) = 1, \quad P_1(x) = x, \quad P_2(x) = \frac{1}{2}\left(3x^2 - 1\right)$$

$$P_3(x) = \frac{1}{2}\left(5x^3 - 3x\right), \quad P_4(x) = \frac{1}{8}\left(35x^4 - 30x^2 + 3\right), \cdots$$

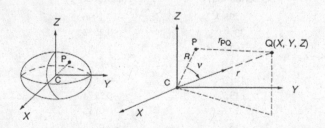

**FIGURE 6.3**  Q is a generic body (in the present case, the Sun or the Moon) outside the Earth. P is a point-like mass inside it.

The elementary potential energy $dU$ in P due to Q is

$$dU = GM_Q \frac{dM}{r_{PQ}}$$

which can be integrated over the whole body of the Earth to give

$$U = GM_Q \int \frac{dM}{r_{PQ}} = G \frac{M_Q}{r} \left( U_0 + \frac{U_1}{r} + \frac{U_2}{2r^2} + \cdots \right) \tag{6.8}$$

In the same manner, the potential of the Earth on a distant point Q can be expressed in terms of *spherical harmonics*, first derived by Laplace, as

$$U = \frac{GM_\oplus}{r} \left[ 1 - \sum_{n=2}^{\infty} \left( \frac{a_\oplus}{r} \right)^n J_n P_n (\sin \phi') \right] \tag{6.9}$$

where $\phi'$ is the geocentric latitude of Q. The constants $J_n$ can be derived with high precision from the observations of the orbits of artificial satellites. This expression will be used in Chapter 14 to examine how the non-spherical Earth influences the motion of a nearby artificial satellite. $J_1 = 0$ when the center of mass is at the origin; the value of the dominant term $J_2 = (C - A)/M_\oplus a_\oplus$ is $1.0826 \times 10^{-3}$. Its time rate of change in the IAU 2006 model is $dJ_2/dt = -0.3001 \times 10^{-9}$/century.

The first term $U_0$ in Equation 6.8 is the mass of the Earth; the second term $U_1$ vanishes because of the definition of the barycenter; all the following uneven terms $U_3$, $U_5$, etc. must be zero for symmetry reasons; therefore, in the present approximations we need only to consider $U_2$:

$$U_2 = \int \left( 3\cos^2 v - 1 \right) R^2 \, dM = A + B + C - 3I \approx 2A + C - 3I \tag{6.10}$$

where $I$ is the moment of inertia of the Earth around the direction to Q. Therefore, to the order $(1/r)^3$

$$U = GM_Q \left( \frac{M_\oplus}{r} + \frac{2A + C - 3I}{2r^3} \right)$$

Let $(X_Q, Y_Q, Z_Q)$ be the coordinates of Q in the Earth-fixed reference system and $(l, m, n)$ its direction cosines:

$$l = \frac{X_Q}{r}, \qquad m = \frac{Y_Q}{r}, \qquad n = \frac{Z_Q}{r}, \qquad l^2 + m^2 + n^2 = 1$$

Then,

$$I = Al^2 + Bm^2 + Cn^2 = A + (B - A)m^2 + (C - A)n^2 \approx A + (C - A)n^2$$

$$U \approx G \frac{M_Q}{r} \left( M_\oplus + \frac{(C - A)}{2r^2} - 3 \frac{(C - A)}{2r^2} \frac{Z_Q^2}{r^2} \right)$$

Only the third term will be effective in precessional and nutational phenomena, being the only one that explicitly depends on the position of the forcing body Q with respect to the equator. Finally, Q being very distant from C, with a slight imprecision we can also put $Z_Q/r = \sin\delta_Q$, so that

$$U_{prec} \approx -3GM_Q \frac{(C-A)}{2r^3} \sin^2\delta_Q$$

Adding together Sun and Moon,

$$U_{prec} \approx -3G \frac{(C-A)}{2} \left[ \frac{M_{\leftmoon}}{r_{\leftmoon}^3} \sin^2\delta_{\leftmoon} + \frac{M_{\odot}}{r_{\odot}^3} \sin^2\delta_{\odot} \right] \tag{6.11}$$

Let us take into account Kepler's third law, namely, that the square of the period of revolution is proportional the cube of the semi-major axis (for a demonstration, see Chapter 12):

$$P_Q^2 \left( M_Q + M_{\oplus} \right) = \frac{4\pi^2}{n_Q^2} \left( M_Q + M_{\oplus} \right) = \frac{4\pi^2}{G} a_Q^3$$

where $P_Q$ is the period of revolution of Q (either Sun or Moon) around the Earth, $n_Q$ its mean motion and $a_Q$ the semi-major axis of its orbit. Approximating the real orbits of the Sun and the Moon with circular ones, the cube of the instantaneous distances in Equation 6.11 can be substituted by the square of the mean motions, so that

$$U_{prec} \approx -3 \frac{(C-A)}{2} \left[ n_{\leftmoon}^2 \frac{M_{\leftmoon}}{(M_{\leftmoon}+M_{\oplus})} \sin^2\delta_{\leftmoon} + n_{\odot}^2 \frac{M_{\odot}}{(M_{\odot}+M_{\oplus})} \sin^2\delta_{\odot} \right]$$

$$\approx -3 \frac{(C-A)}{2} \left[ n_{\leftmoon}^2 \frac{M_{\leftmoon}}{(M_{\leftmoon}+M_{\oplus})} \sin^2\delta_{\leftmoon} + n_{\odot}^2 \sin^2\delta_{\odot} \right] \tag{6.12}$$

where the mass of the Sun has disappeared because it greatly exceeds that of the Earth.

Although many approximations were made in deriving Equation 6.12, some fundamental points have nevertheless been brought to light. Namely, $U_{prec}$ depends on

- the declination and mean motion of the forcing bodies;
- the moments of inertia of the Earth, whose mechanical ellipticity can be derived without any model of the interior;
- the mass of the Moon but not of the Sun.

For a fuller discussion of Equation 6.12, see the "Notes" section.

Notice that before the advent of the Space age (more precisely, before the spacecraft Ranger 5 had flown very near the Moon, in October 1962), the best value of the lunar mass ($\approx M_{\oplus}/80$) was estimated from the constants of precession and nutation.

## 6.4  THE EARTH'S FREE ROTATION

Let us compare the precessional energy given by Equation 6.12 with that of the diurnal rotation $T$:

$$T = \frac{1}{2} C\Omega^2, \quad \Omega = 2\pi / (\text{sidereal day}) \approx 7.292 \times 10^{-5} /\text{s}$$

$U_{prec}$ can be maximized by taking the highest value of the declination of Sun and Moon, and hence,

$$\left|\frac{U_{prec}}{T}\right| \leq 3\frac{(C-A)}{C}\left(\frac{n_\odot}{\Omega}\right)^2\left[1+0.25\frac{M_\circlwithdot}{M_\oplus}\left(\frac{n_\circlwithdot}{n_\odot}\right)^2\right] \tag{6.13}$$

where $n_\circlwithdot/n_\odot \approx 1/13$, $n_\odot/\Omega \approx 1/366.25$, so that $U_{prec}/T \leq 1 \times 10^{-7}$, a very modest fraction indeed. Therefore, in this paragraph we will examine the rotation of a free Earth, as if the Moon and the Sun were not forcing the precession and nutation.

Let us perform again the dynamical analysis of the forced lunisolar precession, with a more formal procedure. Consider two geocentric reference systems, the inertial ecliptic one $(X_0, Y_0, Z_0)$, and the equatorial one $(X, Y, Z)$ fixed in the body and thus rotating with respect to the inertial system. For simplicity, we orient the axes $(X, Y, Z)$ toward the principal axes of inertia. Let $\Omega$ $(\Omega_1, \Omega_2, \Omega_3)$ be the angular velocity, resolved along the three orthogonal components and $I$ the inertia tensor, reduced to the diagonal form $(A, B, C)$ by the choice of axes directions. The rotational energy $T$ of the body and its total angular momentum $\mathbf{M}$ are

$$T = \frac{1}{2}\left(A\Omega_1^2 + B\Omega_2^2 + C\Omega_3^2\right),$$

$$\mathbf{M}(M_1, M_2, M_3) = \mathbf{M}(A\Omega_1, B\Omega_2, C\Omega_3) = \int \mathbf{R} \times (\Omega \times \mathbf{R})\mathrm{d}m$$

where the integral is extended over the mass of the body. Notice that the directions of $\mathbf{M}$ and $\Omega$ do not necessarily coincide unless the body has spherical symmetry, because the direction of $\Omega$ need not coincide with that of the principal axis of inertia.

However, by assuming the spheroidal shape, with $A = B < C$, it can be shown that at all times the directions of $\mathbf{M}$, $\Omega$, and axis $Z$ will be in the same plane. In the absence of external forces, the direction of $\mathbf{M}$ is fixed in the inertial space, so that the inertial observer sees a precession of the rotation axis around $\mathbf{M}$, with angular velocity $\omega_{prec} = M/A$.

In the real case of external forces due to Moon and Sun, the vector equations of motions in the inertial system centered in the barycenter will be

$$\frac{\mathrm{d}\mathbf{P}}{\mathrm{d}t} = \mathbf{F}, \quad \frac{\mathrm{d}\mathbf{M}}{\mathrm{d}t} = \mathbf{K}$$

where $\mathbf{P}$ is the total linear momentum, $\mathbf{F}$ are the total external forces (namely, those of lunar and solar origin, ignoring the planetary perturbations) and $\mathbf{K}$ is their total moment. In the rotating body-fixed system $(X, Y, Z)$, we can connect the three components of $\Omega$ to the three components of $\mathbf{K}$ through the following Euler's equations:

$$\begin{cases} A\dfrac{\mathrm{d}\Omega_1}{\mathrm{d}t} + (C-B)\Omega_2\Omega_3 = K_1 \\[2mm] B\dfrac{\mathrm{d}\Omega_2}{\mathrm{d}t} + (A-C)\Omega_1\Omega_3 = K_2 \\[2mm] C\dfrac{\mathrm{d}\Omega_3}{\mathrm{d}t} + (B-A)\Omega_1\Omega_2 = K_3 \end{cases} \tag{6.14}$$

In the absence of external forces, $K_{1,2,3} = 0$ and with azimuthal symmetry $A = B$, the system reduces to

$$
\begin{cases}
A\dfrac{d\Omega_1}{dt} + (C - A)\Omega_2\Omega_3 = 0 \\[2mm]
A\dfrac{d\Omega_2}{dt} - (C - A)\Omega_1\Omega_3 = 0 \\[2mm]
C\dfrac{d\Omega_3}{dt} = 0
\end{cases}
$$

After an elementary time $dt$, the rotating system $(X, Y, Z)$ has a different orientation with respect to the inertial system $(X_0, Y_0, Z_0)$ (see Figure 6.4), which can be resolved in an elementary rotation with angular velocities $(\Omega_1, \Omega_2, \Omega_3)$.

We have defined the plane $XY$ as the geocentric equator and the plane $X_0Y_0$ as the fixed ecliptic. The two planes intersect along the line of nodes, the ascending one being indicated with N. In the present context, N is the initial vernal equinox $\gamma$. Thus, the instantaneous position of the rotating frame in the inertial system will be specified by three Eulerian angles $(\theta, \psi, \varphi)$; the angle $\varphi$ expresses the diurnal rotation of the Earth. The time derivative of $\psi$ is connected to the precession in longitude; $\theta$ and its time derivative express the obliquity of the ecliptic and its variation, and therefore the nutation. According to the previous discussion, the derivatives are composed of secular and periodic terms, whose coefficients depend on the mechanical ellipticity of the Earth, the orbital elements and the masses of the Sun and of the Moon.

From Figure 6.4, we see that the time derivatives of the Eulerian angles can be written in terms of the angular velocities $(\Omega_1, \Omega_2, \Omega_3)$. We omit the demonstration of these relationships, because of the many elaborated passages required to obtain them, and only recall the result:

$$
\begin{cases}
\dot{\theta} = \Omega_2 \sin\varphi - \Omega_1 \cos\varphi \\[2mm]
\dot{\psi} \sin\theta = \Omega_2 \cos\varphi + \Omega_1 \sin\varphi \\[2mm]
\dot{\varphi} = \Omega_3 + \cot\theta\left(\Omega_1 \sin\varphi + \Omega_2 \cos\varphi\right)
\end{cases}
\tag{6.15}
$$

$$
\begin{cases}
\Omega_1 = \dot{\psi} \sin\theta \sin\varphi - \dot{\theta} \cos\varphi \\[2mm]
\Omega_2 = \dot{\psi} \sin\theta \cos\varphi + \dot{\theta} \sin\varphi \\[2mm]
\Omega_3 = \dot{\varphi} - \dot{\psi} \cos\theta
\end{cases}
$$

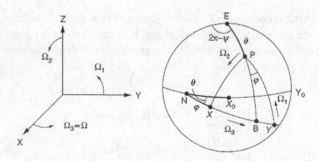

**FIGURE 6.4** The Eulerian rotation angles. E, ecliptic pole; P, celestial north pole; N, ascending node of the ecliptic on the equator; $\theta$ = angle EP = angle $X_0NX$ = nutation angle; $2\pi - \psi$ = angle $NX_0$ = angle $Y_0EP$ = precession angle; $\varphi$ angle $NX$ = angle $YPB$ = angle of diurnal rotation.

For the Earth, both $\Omega_1$ and $\Omega_2$ are much smaller than $\Omega_3$, while $\Omega_3 \approx \Omega$ (conditions not necessarily true in the general case). Thus, in the inertial system, the rotational energy $T$ is given by

$$2T = A\left(\dot{\theta}^2 + \dot{\psi}^2 \sin^2\theta\right) + C\left(\dot{\phi} - \dot{\psi}\cos\theta\right)^2 \approx C\Omega^2$$

Using the Lagrangian formalism, the following relations can be obtained:

$$
\left\{
\begin{array}{l}
A\dot{\Omega}_1 + (C-A)\Omega_2\Omega_3 = \dfrac{\sin\varphi}{\sin\theta}\left(\cos\theta\dfrac{\partial U}{\partial \varphi} + \dfrac{\partial U}{\partial \psi}\right) - \cos\varphi\dfrac{\partial U}{\partial \theta} \\[4mm]
A\dot{\Omega}_2 + (A-C)\Omega_1\Omega_3 = \dfrac{\cos\varphi}{\sin\theta}\left(\cos\theta\dfrac{\partial U}{\partial \varphi} + \dfrac{\partial U}{\partial \psi}\right) + \sin\varphi\dfrac{\partial U}{\partial \theta} \\[4mm]
\qquad\qquad\qquad C\dot{\Omega}_3 = \dfrac{\partial U}{\partial \varphi}
\end{array}
\right.
$$

where $U$ is the total potential due to the external bodies. Notice that the assumption of symmetry permits $\partial U/\partial\varphi = 0$ in the last equation; $\Omega_3$ is constant even in the case of forced precession and nutation (to be sure, if all simplifying assumptions are fulfilled).

   In the hypothetical case of the free Earth, $U$ and its derivatives can be ignored, so that

$$
\left\{
\begin{array}{l}
A\dot{\Omega}_1 + (C-A)\Omega_2\Omega_3 = 0 \\[2mm]
B\dot{\Omega}_2 + (A-C)\Omega_1\Omega_3 = 0 \\[2mm]
\qquad\quad C\dot{\Omega}_3 = 0
\end{array}
\right.
\qquad\qquad (6.16)
$$

The component $\Omega_3$ of the angular velocity remains obviously constant as before, while (remember, $B = A$):

$$\frac{d\Omega_1}{dt} = -\Omega_3\frac{C-A}{A}\Omega_2 = -\omega\Omega_2, \qquad \frac{d\Omega_2}{dt} = \Omega_3\frac{C-A}{A}\Omega_2 = \omega\Omega_1$$

Expressing $\Omega_3 \approx \Omega$ in days$^{-1}$, the constant

$$\omega = \Omega_3\frac{C-A}{A} \approx 0.003$$

corresponds to a period of about 303 sidereal days. Finally, with an appropriate choice of the initial time, one can arrive to the result:

$$\Omega_1 = \chi\Omega_3\cos\frac{C-A}{A}\Omega_3 t, \qquad \Omega_2 = \chi\Omega_3\sin\frac{C-A}{A}\Omega_3 t \qquad\qquad (6.17)$$

where $\chi$ is a constant which has to be determined by the observations.

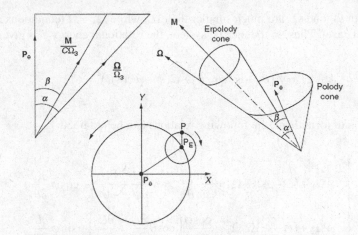

**FIGURE 6.5** The rigid body rotation in the Poinsot representation. $P_e$, unit vector to the pole of the geoid (axis Z); $P_E$, Eulerian position of the rotation pole. The observations prove that angle $P_eP_E \approx 0''.3$, so that $\alpha - \beta \approx 0''.001$ (**M**, **Ω** and $P_E$ essentially coincide). The direction $P_eX$ defines the terrestrial meridian of Greenwich. The true instantaneous rotation pole is in P: the difference $P_EP$ is due to the forced lunisolar precession, varying approximately between 0 and $0''.02$ (Oppolzer's terms). The astronomical meridian of Greenwich passes through P.

To understand the previous results, consider in Figure 6.5 the three vectors **Ω**, **M**, $P_e$, the latter being the unit vector in the direction of the pole of the spheroid, namely, along the $Z$-axis (this representation often goes under the name of the French mathematician L. Poinsot). Let us normalize **Ω** by $\Omega_3$ and **M** by $C\Omega_3$:

$$\Omega/\Omega_3 = \left( \frac{\Omega_1}{\Omega_3}, \frac{\Omega_2}{\Omega_3}, 1 \right),$$

$$\mathbf{M}/C\Omega_3 = \left( \frac{A}{C}\frac{\Omega_1}{\Omega_3}, \frac{A}{C}\frac{\Omega_2}{\Omega_3}, 1 \right) \approx \left( 0.997\frac{\Omega_1}{\Omega_3}, 0.997\frac{\Omega_1}{\Omega_3}, 1 \right)$$

The two vectors $\Omega/\Omega_3$ and $\mathbf{M}/C\Omega_3$ are very similar in direction and amplitude. Furthermore, they must also be almost equal to $P_e$, so that only refined measurements will be able to distinguish between them. The theory cannot fix the value of the angle $\beta$ in Figure 6.5, and it must be determined by the observations, but satisfying the condition: $[(\alpha - \beta)/\beta] = [(C - A)/A] \approx 1/303$.

Recalling that the three vectors must be at all times coplanar, we conclude that the inertial observer notices an almost diurnal rotation of the plane containing **Ω** and $P_e$ around **M**, with expected frequency:

$$\omega_{\text{fp}} = \frac{M}{I_1} = \left( 1 + \frac{C - A}{A} \right)\Omega_3 \approx 1.003\Omega_3$$

The body-fixed observer should notice a rotation of **Ω** around $P_e$ (cone of polody) and around **M** (cone of erpolody). The observations show (see next paragraph) that the angle between **Ω** and $P_e$ is approximately $0''.3$, so that the angle between **Ω** and **M** must be approximately 303 times smaller, namely, of the order of $0''.001$. Projected on the terrestrial surface at the geographic poles, these angles correspond to 9 m and 3 cm, respectively. Therefore, for the terrestrial observer the dominant motion is the polody, which, according to the above results, should have a period of approximately 303 sidereal days (10 months).

This free-body nutation is superimposed to the forced lunisolar precession and nutation, whose presence forces the motion of **M** with respect to the fixed stars and slightly alters the previous considerations by a minute variation of the apertures of the cones (the difference $\mathbf{P_e P}$ in Figure 6.5, Oppolzer's terms), at the level of few hundredths of arc seconds.

In conclusion, if what is exposed is correct, from the Earth's surface we should observe the diurnal rotation axis to perform, every 10 months, a cone around the geodetic north pole, with an amplitude $\sigma$ that one cannot determine *a priori* (being a function of unknown initial conditions), but that hopefully is measurable. Figure 6.6 shows, in a graphical form, the difference between forced and free precession, also called *Chandler's wobble*.

The immediate observational consequence of this wobble of the rotation axis would be a periodic alteration of the astronomical latitude of each observatory, so that the phenomenon is also called *variation of the astronomical latitudes*. Obviously, the longitudes would be affected in the same way, but with the intrinsic difficulties associated with all determinations of the zero point that we have discussed in several places.

Therefore, the astronomical latitudes $\phi$ should vary according to

$$\phi = \phi_0 + \phi_1 \cos\left(\omega_{pol} t + q\right)$$

Many astronomers attempted to detect the phenomenon, until finally, in 1888, Küstner measured variations at the level of $0''.3$. Soon after, Chandler was able to demonstrate the existence of two different periods, one of approximately 366 days (one sidereal year) and one of approximately 435 days. The instantaneous pole would be seen by an observer looking at it from above to perform two counterclockwise revolutions around the principal moment of inertia, with periods of 1 and 1.2 years, respectively. The first term is clearly of meteorological origin, the seasonal variation of the weight of the snow in the polar caps, the sea currents, the atmospheric loading of the regular winds, etc., but the duration of the second term came as a real surprise, being much longer than the expected free Eulerian period. Therefore, one major assumption had to be in error in the previous reasoning. Newcomb identified the responsible factor in the non-rigidity of the Earth. The Earth behaves more like an elastic than a rigid body; indeed, its elasticity is surprisingly good, being comparable to that of a steel sphere. Precise computations to quantify the effects of elasticity are very difficult to make. It is however understandable that if the spheroid is elastic, only a fraction of its mass will be effective at any given time, lowering C and the frequency $\omega$. Another way of looking at this phenomenon is to think that part of the equatorial elastic bulge will be readjusted by the centrifugal force around the rotation axis. In the same manner, the constants of the forced precession and nutation will be affected by the elasticity.

Following the discoveries by Küstner and Chandler, an International Service for the Latitudes (ILS) was instituted at the close of the 19th century. Several stations, all having latitude $+39°08'$, were chosen in Italy, Japan, USA and Russia, and equipped with instruments such as meridian

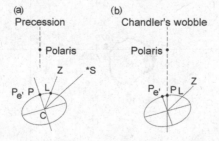

**FIGURE 6.6**   The wobble of the rotation axis $P$ around the geodetic pole $P_e$. L is a site with zenith Z; (a) the forced lunisolar precession in the approximation of the rotational pole coinciding with the pole of the figure produces a slow variation of the equatorial coordinates of the star S; (b) the slight difference between the two poles produces small variations of the latitude of L.

circles, Zenith tubes and Danjon's astrolabes. The ILS brought to light a complex wandering of the poles, which is usually represented in a Cartesian system centered in a mean pole and having the *X*-axis pointing toward Greenwich and the *Y*-axis toward west. The superposition of the two revolutions approximately averages to zero every 6 years; for this reason, averages were made over that interval (initially from 1900 to 1905). See Dick et al. (Bibliography section) for an account of the first years of the ILS.

The results of the ILS were in a sense disappointing. First, the instrumental and human resources were never adequate to perform the difficult task. Furthermore, a great problem was the very definition of the reference and time systems: until clocks much more precise than astronomical phenomena became available, it was difficult to disentangle the minute perturbations affecting the meridian and the vertical of a particular site. An entirely different approach was necessary, which became available in the second part of the 20th century, thanks to atomic clocks, Very Long Baseline Interferometers (VLBI) at radio frequencies and laser ranging of dedicated satellites (e.g., the LAGEOS) and of the Moon, as mentioned in Chapters 2 and 5 (see also the Web Sites section).

Figure 6.7 shows results obtained from 1996 to 2000 with the VLBI, together with a reanalysis of the historical drift of the mean pole. The much higher precision of measurements was immediately demonstrated by the smoothness of the trajectory of the pole.

In addition to the small and periodic variation of latitudes and longitudes, such modern techniques have confirmed and improved the determination of several phenomena proper to the Earth's body. The influence of tides on the rotation of the Earth is described in many books and articles, for instance, by Brosche et al. (1989) and by Ray et al. (1994). Even the oblateness of the Earth can have measurable changes. See, for instance, Dickey et al. (2002).

VLBI data have confirmed, beyond doubt, the existence and importance of the relative movements of the continents. The first ideas were put forward by the German geographer A. L. Wegener in 1911 (see Wegener (1929) in the Bibliography section) and enriched around 1960 by the plate tectonics theory. The distance between two observatories can change at the level of a few centimeters a year, a movement reflected in the relative displacements of the celestial sources amounting to few milliarcseconds per year (see Figure 6.8).

At the present level of precision to measure angular displacements, approaching the microarcsec, the concept of rotation of the Earth, as measured from stations on moving plaques, becomes extremely intriguing, and so do astronomical theories of precession and nutation. An overview of the complexity of the problem can be appreciated by the 2019 Report on the outcomes of the activities of the IAU/IAG Joint Working Group on Theory of Earth Rotation and Validation by Ferrándiz and

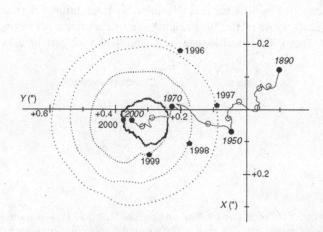

**FIGURE 6.7**   VLBI determination of the wandering of the poles from 1996 to 2000, together with a reanalysis of the historical data on the mean pole movement from 1890 to 2000 (Adapted from IERS data.)

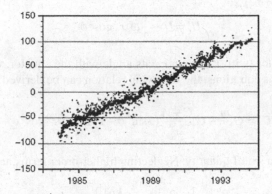

**FIGURE 6.8** The increase of the relative distance between two VLBI stations (Westford in USA and Wetzell in Germany) 1983 to 1995; the ordinate scale is in mm.

collaborators and references therein (see the Web Sites section). The report points out, for instance, the need to update the IAU 2000 theory of nutation; moreover, the IAU2006 precession model may be not entirely correct nor consistent with the IAU2000 nutation theory. The polar motion and time scales (in particular UT1) are affected too by improved geophysical models. Further work is clearly necessary.

Chapter 10 will give a further discussion of Earth's variable rotation on short and secular time scales, affecting, for instance, the length of the day and the number of days in the year in past eras.

## NOTES

- A detailed discussion of Euler's Equation 6.12, covering the entire problems of general precession and nutation, is given in many textbooks. For instance, M. Smart, (*Celestial Mechanics*, Bibliography section) showed that

$$-U_{\text{prec}} = \frac{3}{2}(C-A)n_\odot^2 \left\{ \begin{array}{l} f\left(\dfrac{M_\circleddash}{M_\odot}, r_\circleddash, i, e, e_\circleddash\right)\sin^2\theta \\ +\left[(g_1\cos\psi - g_2\sin\psi)\sin\theta\cos\theta + H\sin^2\theta\right]t + V \end{array} \right\}$$

where $t$ is the time since the initial epoch, and $f$, $g_1$, $g_2$, $H$ and $V$ are functions of the orbital elements of the Sun and of the Moon. From this, the expressions of $\psi(t)$ and $\theta(t)$ become

$$\psi(t) = at + bt^2 + \psi_{\text{per}} = 50''37 - 0''.0001t^2 - 17''.23\sin\Lambda_\circleddash + \cdots$$

$$\theta(t) = \theta_0 + ct^2 + \theta_{\text{per}} = 23°.26 + \frac{1}{2}5''.6\times10^{-6}t^2 + 9''.21\cos\Lambda_\circleddash + \cdots$$

where $\Lambda_\circleddash$ is the longitude of the ascending node of the lunar orbits. These results justify the approximate discussion and validate the results obtained with Equation 6.12.

## EXERCISE

Derive the gravity at the surface of a rotating ellipsoidal body of mass $m$ and radius $R$ (Clairaut's formula). Let $U$ be the potential of the body gravitational field and $\boldsymbol{\omega}$ its angular velocity around the polar axis. On any equipotential equilibrium surface, the quantity $U'$,

$$U' = U + \frac{1}{2}\omega^2 r^2 \cos^2\phi$$

where **r** is the radius vector (with $r \approx R$) and $\phi$ its angle with the equator, will be constant on that surface. From the previous equations, the following relation can be derived:

$$U' = G\frac{m}{r} - G\frac{C-A}{2r^3}\left(3\sin^2\phi - 1\right) + \frac{1}{2}\omega^2 r^2 \cos^2\phi$$

Let $r = R(1 - \eta)$, $\eta$ being a small quantity. Neglecting higher-order terms, it can be proven that

$$\eta = \left[3\frac{C-A}{2mR^2} + \frac{\omega^2 R^3}{2GM}\right]\sin^2\phi$$

Let $\sigma$ indicate the ratio between centrifugal force and gravitational force at the equator:

$$\sigma = \omega^2 R^3 / Gm$$

Recall the equations for the oblate spheroid and of its flattening $f$ given in Chapter 2. Expanding $r^2$ by the binomial theorem and retaining only the terms in $f$, we finally get for the equilibrium surface:

$$r = R\left[1 - \left(3\frac{C-A}{2mR^2} + \frac{1}{2}\sigma\right)\sin^2\phi\right]$$

By measuring $f$ and $\sigma$, we can thus measure $(C - A)$ and derive $U'$ to get the surface gravity $g$:

$$g = -\frac{\partial U'}{\partial r} = G\frac{m}{r^2} - 3G\frac{C-A}{2r^4}\left(3\sin^2\phi - 1\right) - \omega^2\left(1 - \sin^2\phi\right)$$

Finally,

$$g = g_0\left[1 + \left(\frac{5}{2}\sigma - f\right)\sin^2\phi\right] \tag{6.18}$$

where $g_0$ is the gravity at the equator:

$$g_0 = G\frac{m}{R}\left[1 + f - \frac{3}{2}\sigma\right]$$

Equation 6.18 is Clairaut's formula. At the equator of the Earth, $g_0 = 978.049$ cm/s$^2$ (measured), while it would be 981.43 cm/s$^2$ without rotation. At the second order, the more precise formula (Airy, Callandreau) can be applied:

$$g = 978.049\left(1 - 0.005288\sin^2\phi - 0.000006\sin^2 2\phi\right)$$

# 7 Aberration of Light

As seen in the previous chapters, precession and nutation are phenomena due to the variable orientation of the observer's system with respect to the fixed stars. The aberration, instead, is an effect due to the finite velocity of light and to the motions of the observer with respect to the celestial sources. As a consequence, the apparent direction of a celestial source, be it a distant star or a body of the Solar System, is not the same as the geometric direction at the instant of the observation. Following the traditional treatment, we shall call *stellar aberration* that part of the displacement that depends on the motion of the observer around the Sun, namely, the annual revolution and the diurnal rotation. *Planetary aberration* is the sum of the stellar aberration and the part of displacement due to the light time from the planet, comet or asteroid to the observer. The part of displacement due to the motion of the Solar System in the inertial space, a sort of *secular* aberration, is absorbed in the *proper motions* of the stars, to be discussed in Chapter 9.

Many philosophers and physicists of the past suspected the finiteness of the velocity of light, indicated as usual by $c$, long ago. Galileo Galilei suggested a method to measure it by means of the ephemerides of the Medicean moons of Jupiter he had discovered in Padova in January 1610. The credit for obtaining the first reliable value of $c$ by purely astronomical means goes to Öleg Römer, at the end of the 17th century. In Paris, Römer had taken up the task of comparing the observed times of the eclipses of the Medicean moons of Jupiter with the tables compiled by Cassini. Those times could have great practical importance for the determination of the longitude of ships, as suggested by Galileo. The observations proved that Cassini's ephemerides for the innermost moon, Io, were quite accurate at the quadratures of Jupiter with the Sun. However, the observed positions retarded or anticipated by approximately 11 (subsequently corrected in 8) min near opposition or conjunction (see Figure 7.1). Thus, the required correction had the form:

$$\Delta t = \pm \tau_a \cos(\lambda_\odot - \lambda_J)$$

where $\lambda_\odot$, $\lambda_J$ are, respectively, the ecliptic longitudes of the Sun and of Jupiter and $\tau_a \approx 8^m$.

Römer interpreted the time interval $\tau_a$ as the time needed by the light to cross the Earth–Sun distance, namely, the astronomical unit AU, for the epoch a bold idea indeed. Cassini had determined in 1672 the fairly precise value of $9''.5$ for the solar parallax. Values for the velocity of light

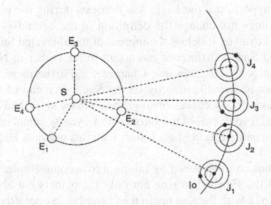

**FIGURE 7.1** Successive positions of Jupiter and Io as seen from the Earth. Position 4 (conjunction) is not directly observable.

ranging from 230,000 to 360,000 km/s could be derived, according to the assumed value of the light time and solar distance.

In 1727, James Bradley discovered that the declination of $\gamma$ Dra (a star not too distant from the ecliptic pole) has a periodic variation during the yearly revolution of the terrestrial observer due to the same cause, namely, the finite velocity of light. More than a century later, around 1850, H. Fizeau and L. Foucault, with their well-known terrestrial experiments employing toothed wheels and rotating mirrors, refined the value of $c$ derived from the astronomical observations.

This chapter discusses the aberration of light, mostly limited to first order in the quantity $V/c$, in the frame of the Newtonian dynamics. Although the derived formulae will not be precise to better than $0''.01$, the advantage of this approximate procedure is to separate all the various factors entering in the aberration, allowing the possibility of summing their effects one over the other. Should greater accuracy be needed, e.g., in reducing the data of the astrometric satellites Hipparcos and GAIA, more precise formulae based on special and general relativity must be employed.

## 7.1 THE SOLAR ABERRATION

For the moment, the Earth's orbit around the Sun is taken as being circular with radius $a_\odot = 1$ AU and uniform velocity vector $\mathbf{V}$. The direction of $\mathbf{V}$ is perpendicular to the radius, and its modulus is given by

$$V = 2\pi \frac{a_\odot}{P} = na_\odot$$

where $P$ is the sidereal year and $n$ ($\approx 3548''.2$/day) the solar mean motion introduced in Chapter 4.

The light crosses the AU in time $\tau_a$ (a quantity referred to as aberration time, or equation of light, or light time for the unit distance). For the geocentric observer, in those $8^m$ the Sun will have moved from its apparent position (when the light left it) to the geometrically correct but *unobservable* position corresponding to the arrival of the light. The angular distance between the two positions is

$$K = 2\pi \frac{a_\odot}{Pc} = n\tau_a = \frac{V}{c} \quad \text{(radians)}$$

The quantity $K$ is named *constant of solar aberration*. In round numbers,

$$V \approx 30 \text{ km/s}, \quad K = V/c \approx 0.0001 \text{ radians} \approx 20''.6$$

Under those simplifying hypotheses, the angle $K$ is constant during the year, even in sign. Notice that the solar aberration does not change the definition of the equinoxes as intersections of the equator with the ecliptic; however, it delays the ingress of the aberrated Sun in the equinoxes with respect to the geometrical Sun, a difference important to remember in Newcomb's definition of the longitude of the mean *geometric* Sun (see Chapter 10). Furthermore, the solar aberration as previously described is not observable directly on the Sun, but it can be noticed from a number of external effects, such as velocity curves of single and binary stars, and light curves of variable stars. For instance, the radial velocities of the stars exhibit a yearly component due to the projection of the Earth velocity $\mathbf{V}$ in the direction of each star; with this method, H. Spencer Jones obtained in 1928 the value $K = 20''.475$.

The above considerations can be refined by taking into account Kepler's first and second laws. The *unperturbed* orbit of the Earth (ignoring not only the planets but also the Moon) would be slightly elliptical ($e \approx 0.0167$), with the Sun not in the center but in one of the two foci. The modulus $V$ of the velocity will therefore change with the date, being maximum at perihelion and minimum at aphelion. It is useful to consider the vector $\mathbf{V}$ as the sum of two components: $\mathbf{V}_t$ perpendicular to the major axis of the ellipse (a direction known as line of apses) and $\mathbf{V}_r$ perpendicular to the radius vector. With simple geometric considerations, it is seen that the moduli of both components are constant and that the *direction* of $\mathbf{V}_t$ is also constant:

$$\mathbf{V}_r = \frac{na_\odot}{\sqrt{1-e^2}}, \quad \mathbf{V}_t = \frac{na_\odot e}{\sqrt{1-e^2}} \approx \frac{1}{60} \mathbf{V}_r$$

where $a_\odot$ is now the semi-major axis of the Earth's orbit, as shown in Figure 7.2.

Therefore, the velocity that properly enters in the definition of the solar aberration is $\mathbf{V}_r$, so that the constant of aberration is

$$K = \frac{\mathbf{V}_r}{c} = \frac{na_\odot}{c\sqrt{1-e^2}} \approx 20''.496$$

Notice that the quantity $K$ is not really constant over long time intervals, because of the secular change of $e$. The component $\mathbf{V}_t$ is responsible for the so-called elliptical *aberration* $K_e \approx 0''.343 \approx 0^s.023$, which changes from day to day and in principle is observable through the *equation of time* (see Chapter 10).

In total, the difference in ecliptic longitude between the aberrated (observable) and the geometric Sun is

$$\lambda_\odot - \lambda = -K\left[1 - e\cos(\lambda_\odot - \lambda_\Pi)\right]$$

where $\lambda_\Pi$ is the longitude of the perigee (at 180° from the longitude of the perihelion; at the present epoch $\lambda_\Pi \approx 18^h 50^m$, see also Chapter 4). A corresponding equation must be applied to the difference in right ascensions. The *Astronomical Almanac* publishes in Section C the *geometric* ecliptic coordinates referred to as the mean equinox of date; the apparent longitude is then given by the formula:

$$\lambda_\odot = \lambda + \Delta\psi - 20''.496/R$$

where $\Delta\psi$ is the nutation term for the date and $R$ the distance in AU. To this level of precision, the latitude is not affected.

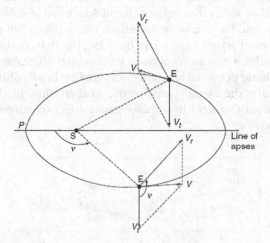

**FIGURE 7.2** A representation of the elliptical motion of the Earth. P is the perihelion (at 180° from the perigee Π).

In these considerations, we have not yet discussed the very small effects due to the latitude of the Sun with respect to the dynamical ecliptic; to the velocity of the Earth with respect to the barycenter of the Earth–Moon system (approximately 12.5 m/s), nor to the planetary perturbations. These components will be discussed later. In the usual treatment, the gravitational deflection of light is taken into account separately.

The value of $K$ has been revised from time to time, according to the different determinations of $a_\odot$, $e$ and $c$. For instance, in the first part of the 20th century the following values were adopted:

$$a_\odot = 149{,}504{,}000 \, \text{km}, \quad c = 299{,}774 \, \text{km/s}, \quad e = 0.0167301 \, (1950),$$

$$K = 20''.487 = 1^s.336, \quad K_e = 0''.343 = 0^s.023$$

The new set of constants adopted by the IAU in 2009/2012 are:

$$1\text{AU} = 149{,}597{,}870.700 \, \text{km}, \quad c = 299{,}792.458 \, \text{km/s},$$

$$e = 0.01670790 (\text{J2000}), \quad K = 20''.49552 = 1^s.366, \quad K_e = 0''.343 = 0^s.023$$

Notice that today the velocity of light is considered the *natural defining constant*; the other constants are called *auxiliary defining constants*. Using these values, the light time for unit distance becomes $\tau_a = 499^s.00478384 = 8^m19^s.00478384$, $1/\tau_a = 173.144632674$ AU/day, corresponding to a solar parallax $\pi_\odot = 8''.794143$. See also Chapter 8.

## 7.2   THE ANNUAL ABERRATION

As previously highlighted, the annual aberration affecting the stars was discovered by Bradley by observing with his meridian circle the second magnitude star $\gamma$ Dra, in an attempt to measure its parallax. During the year, the declination of the star was seen to oscillate by $\approx \pm 20''.5$ around a mean position, reaching the maximum deviation at the solar opposition or conjunction. The motion was too large to be attributable to a distance effect; moreover, the dates were 3 months out of phase with those expected from the annual parallax. Furthermore, Bradley noticed the close numerical coincidence with the solar aberration constant. Thus, he suspected that the cause was the same, namely, the finite velocity of light. Obviously, the apparent Right Ascension of the star had to be affected in the same way, but Bradley could not measure that effect, due to the lack of precision of his clocks.

Figure 7.3 shows a series of positions of $\gamma$ Dra from 1920 to 1941, during a complete revolution of the nodes of the lunar orbit. One can see a small processional effect (small because of the proximity of the star to the ecliptic pole), the nutation discovered by Bradley himself (the sinusoid with period 18.6 years) and finally the annual variation due to aberration. Bradley's discovery conclusively proved the correctness of Römer's hypothesis, gave a direct way to determine $K$ and provided a much more precise value of $c$.

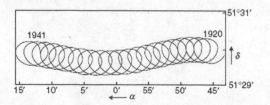

**FIGURE 7.3**   A series of positions of $\gamma$ Dra from 1920 to 1941. (Adapted from Danjon, Bibliography section.)

An intuitive way of understanding the yearly aberration, based on the Galilean transformation of velocities and the simplified hypothesis of circular orbit, is the following (Figure 7.4).

Let C be the center of the objective of the telescope and E the intersection of the optical axis with the focal plane, so that the line EC is the direction of sight. The Earth's velocity vector **V** points toward an instantaneous direction named *apex of motion*, which is on the ecliptic at 90° from the Sun. Let $\theta$ be the angle between the line of sight and the apex, in the plane defined by the two directions. During the time $\Delta t$ employed by the light to travel the distance EC, the Earth moves by $V \Delta t = (V/c)\, EV$.

Therefore, the telescope must be pointed in direction $\theta'$, not $\theta$, inclining it toward the direction of the apex. From the figure, it is easily seen that

$$\sin(\theta - \theta') = \sin \Delta\theta = \frac{V}{c}\sin\theta' \tag{7.1}$$

where $(V/c)\sin\theta'$ is the component of the Earth velocity perpendicular to the apparent direction $\theta'$ of the star.

Since $V \ll c$, the sine of the difference approximately equals the difference:

$$\sin\Delta\theta \approx \Delta\theta = \theta - \theta' \approx \frac{V}{c}\sin\theta', \qquad \theta = \theta' + \frac{V}{c}\sin\theta' \tag{7.2}$$

Using this approximate formula, the geometrical direction $\theta$ of the star, at the time when the light reached the observer (not at the time it left the source), can be derived from the apparent direction $\theta'$. The difference (geometric minus apparent) is zero in the direction of the apex and of the antapex (at 180° from the apex).

With vector notation, let **n** be the unit vector in direction $\theta$ and **n′** the unit vector in direction $\theta'$; the Galilean addition of velocities gives

$$\mathbf{n}' = \frac{\mathbf{n} + \dfrac{\mathbf{V}}{c}}{\left|\mathbf{n} + \dfrac{\mathbf{V}}{c}\right|} \tag{7.3}$$

Taking the scalar part of $n^n'$,

$$\sin\Delta\theta = \frac{\dfrac{V}{c}\sin\theta}{\sqrt{1 + 2\dfrac{V}{c} + \left(\dfrac{V}{c}\right)^2}} = \frac{V}{c}\sin\theta - \frac{1}{2}\left(\frac{V}{c}\right)^2 \sin 2\theta + \cdots \tag{7.4}$$

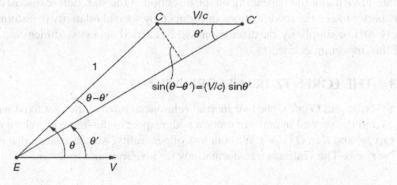

**FIGURE 7.4**  The intuitive explanation of aberration, in the pre-relativistic approximation where $EC = 1$, $EC' > 1$. The Earth's velocity vector **V** points to the apex in the plane of the ecliptic, 90° from the Sun.

which is the pre-relativistic second-order approximation of the aberrational amount. Notice the dependence of the second term from sin $2\theta$ and that the apparent direction is expressed as a function of the geometric one. The maximum value of the second-order term is $0''.001$.

The aberration effect we have described does not dependent on wavelength, focal length of the telescope, distance and velocity of the star with respect to the terrestrial observer. The aberration will be almost identical for both components of a strict binary system of stars or galaxies (apart from that due to the slight difference in relative positions). One might also wonder what the correct value of $c$ to be used in observations with ground telescopes is, either that in air or that in vacuum (the two velocities differ in the visible range by some 67 km/s in normal conditions of temperature and pressure, a difference well measurable). The correct answer is the velocity in vacuum, because the atmosphere partakes of the same translational motion of the Earth barycenter, and no further aberration is introduced by its presence (the atmospheric refraction is one of the main factors limiting the precision of positional measurements, including the determination of the aberration, but this is an entirely different effect). The Astronomer Royal George B. Airy in 1872 obtained the observational proof by filling his telescope with water: the aberration amount did not change.

In the simplifying hypothesis of the terrestrial circular orbit, the velocity vector **V** rotates during the year by 360° in the plane of the ecliptic with constant modulus, always pointing to 90° from the Sun. Consequently, the yearly aberration of a star having ecliptic latitude $\beta$ appears as an elliptical motion with semi-major axis parallel to the ecliptic and equal to $K$. The semi-minor axis, perpendicular to the ecliptic, is equal to $K \sin \beta$. This ellipse degenerates in a circle at the ecliptic poles, in a segment on the ecliptic itself. Notice that the star will never be seen in its geometrical position, except for an ecliptical star twice a year.

The dimensions of this ellipse are the same for all celestial bodies having the same ecliptic latitude (whether they are planets, stars, galaxies, quasars, etc.) and do not reflect the ellipticity of the Earth's orbit, although finally the cause is the yearly motion. In other words, the aberration cannot be noticed by differential measurements of stars in a small area of sky. Over large angles, however, it will cause a (small) distortion of the celestial sphere, distinctively different from precession and nutation, which are rigid rotations of the sphere. Any other periodic motion of the terrestrial observer, for instance, the diurnal rotation or the motion around the Earth–Moon barycenter, will cause a corresponding periodic phenomenon of aberration, suitably scaled for its velocity direction and modulus. Allowing for all these aberrational effects will result in referencing the position of each star to an ideal observer fixed in the barycenter of the Solar System. However, even the latter will be in secular motion with respect to the fixed star system, so that the true coordinates and the true aspect of the celestial sphere are to a certain extent unknown.

The effect of the annual aberration on the equatorial coordinates can easily be visualized: the ellipse in ecliptic coordinates will maintain the same dimension in the equatorial ones, but it will be rotated by an amount depending on the direction of the star. Before discussing the precise formulae, it is better to see the modifications introduced by special relativity (Einstein, 1905, the Bibliography section). For simplicity, the discussion will be carried out in two dimensions, sufficient to emphasize all the important elements.

## 7.3 THE LORENTZ TRANSFORMATIONS

Let O($x$, $y$) and O'($x'$, $y'$) be two inertial reference systems: the first fixed with respect to the distant stars and the second in uniform motion with respect to the first with velocity **V** along $x$. The time is $t$ in O($x$, $y$) and $t'$ in O'($x'$, $y'$). Without loss of generality, we can assume that the two systems coincide at $t = t' = 0$. The Galilean transformations for positions and times are

$$x' = x - Vt, \ y' = y, \ t' = t$$

Special relativity tells us that the correct transformations (named after Lorentz) are instead

$$x' = \frac{x - Vt}{\sqrt{1 - \left(\dfrac{V}{c}\right)^2}}, \quad y' = y, \quad t' = \frac{t - \dfrac{Vx}{c^2}}{\sqrt{1 - \left(\dfrac{V}{c}\right)^2}}$$

Resorting again to the vector notation, the Lorentz transformations give, for the unit vectors **n** in direction $\theta$ and **n'** in direction $\theta'$, the expression:

$$\mathbf{n'} = \frac{\sqrt{1 - \left(\dfrac{V}{c}\right)^2}\,\mathbf{n} + \dfrac{\mathbf{V}}{c} + \left(\mathbf{n}\dfrac{\mathbf{V}}{c}\right)\left(\dfrac{\mathbf{V}}{c}\right) \Big/ \left(1 + \sqrt{1 - \left(\dfrac{V}{c}\right)^2}\right)}{1 + \mathbf{n}\dfrac{V}{c}} \tag{7.5}$$

Taking the modulus of the vector product $\mathbf{n} \char94 \mathbf{n'}$ and expanding in powers of $(V/c)$, we obtain

$$\sin \Delta\theta_{\text{Rel}} = \frac{\left(\dfrac{V}{c}\right)\sin\theta + \dfrac{1}{2\sqrt{1 - \left(\dfrac{V}{c}\right)^2}}\left(\dfrac{V}{c}\right)^2 \sin 2\theta}{1 + \dfrac{V}{c}\cos\theta} \tag{7.6}$$

$$= \frac{V}{c}\sin\theta - \frac{1}{4}\left(\frac{V}{c}\right)^2 \sin 2\theta + \cdots$$

At this level of approximation, the relativistic aberration differs from the Galilean Equation (7.4) by $1/4(V/c)^2 \sin 2\theta$. Therefore, the elementary Equation (7.2) is approximated to terms of the order of $(V/c)^2$ for two distinct reasons: first, the neglect of higher order terms in the trigonometric expansions; second, incorrect transformation rules.

Numerically, the Galilean and relativistic expressions give the same results for the annual aberration to within $0''.0005$. However, there is a profound conceptual difference: even in the vacuum above the terrestrial atmosphere, in the Galilean relativity aberration only in the direction of the group velocity appears, not in that of the phase velocity. According to special relativity, aberration is present in both (see the books by Møller and Böhm quoted in the Bibliography section).

## 7.4   EFFECTS OF ANNUAL ABERRATION ON THE STELLAR COORDINATES

In order to determine the main components of the annual variation of the apparent ecliptic coordinates due to aberration, let us proceed for a moment with the circular approximation of Earth's orbit. On the celestial sphere, the plane defined by the observer, the terrestrial apex T' and the geometric position X of the star of ecliptic coordinates $(\lambda, \beta)$ translates in a great circle passing through X and T' (see Figure 7.5).

The apparent position of the star on this great circle is X', displaced from X toward T' by the amount $XX' = K \sin\theta$. T' is on the ecliptic at 90° behind the Sun, whose true (geometric) longitude at the date is $\lambda_\odot$. Therefore, the displacement is always toward the Sun. The smallness of $K$ allows the utilization of plane trigonometry in the small triangle XX'U; after simple passages, we get

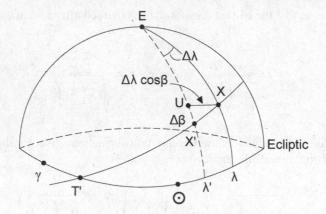

**FIGURE 7.5** The displacement of the apparent position due to annual aberration in ecliptic coordinates, in the approximation of the Earth's circular orbit. E is the pole of the ecliptic. T' is the apex of the Earth's motion, 90° from the Sun.

$$\Delta x = (\lambda' - \lambda)\cos\beta = \Delta\lambda \cos\beta = -K\cos(\lambda_\odot - \lambda)$$

$$\Delta y = (\beta' - \beta) = \Delta\beta = -K\sin(\lambda_\odot - \lambda)\sin\beta \tag{7.7}$$

which are the parametric equations of an ellipse with semi-major axis $K$ and semi-minor axis $K \sin\beta$.

During the course of the year, the star traces an elliptical locus having the true star in its center. This is the annual aberration ellipse. The star describes the ellipse in retrograde sense if $\beta > 0$ and in direct sense if $\beta < 0$, passing through the end points of the semi-major axis when in conjunction with the Sun (westerly end point) or opposition to it (easterly end point). The star passes through the semi-minor axis when at the quadratures (at the westerly quadrature, the star is nearer to the ecliptic). The ellipse degenerates in a circle for a star at the ecliptic pole, in a segment for an ecliptic one, the only case when we can see the star in its true position, twice a year.

Let us add the effect of the slight ellipticity $e$ of the terrestrial orbit, namely, the small and constant (ignoring the secular variation of $e$) velocity component of amplitude $K_e$ perpendicular to the semi-major axis. First, as already discussed for the solar aberration, the value of the constant $K$ in the annual aberration must be understood as

$$K = \frac{V_r}{c} = \frac{na}{c\sqrt{1 - e^2}} \approx 20''.496.$$

Second, the effect of the perpendicular component is as follows: the geometric position of the star is not exactly in the center of the ellipse of aberration, but displaced from it by $0''.343$, in a direction whose longitude is $\lambda = \lambda_\Pi - 90°$, namely, at 90° from the geocentric longitude of the perigee of the Sun. This displacement, called *elliptic aberration*, when projected in longitude and latitude amounts, respectively, to

$$\Delta\lambda_e = -eK\cos\beta\cos(\lambda_\Pi - \lambda), \quad \Delta\beta_e = -eK\sin(\lambda_\Pi - \lambda)\sin\beta$$

These two so-called *E-terms*, which explicitly depend on $e$, must be added with their signs to the already determined $\Delta\lambda$, $\Delta\beta$. Some authors prefer using the longitude of the perihelion ($\lambda_\oplus = 180° - \lambda_\Pi$), thus reversing the sign of the second term.

The value of the elliptic "constant" $e\,K$ has a slight secular variation, for two different reasons: first, the eccentricity decreases with the centuries, and second, the Earth's orbit slowly processes in its plane (for this reason $\lambda_\Pi$ increases by 11".63/year, performing a full sidereal revolution in approximately 110,000 years). The changes resulting from these variations are very small, of the order of 0".02 in 1000 years. In the past, the terms of the elliptic aberration were absorbed or even omitted in the published mean positions. This habit was discontinued by the FK5 and subsequent catalogs, in order not to introduce any systematic error, however small, in the mean coordinates. Therefore, since FK5 the mean places published in the Almanacs do not contain the elliptic terms of aberration. According to the 1976 IAU resolutions, stellar aberration must be computed from the total velocity of the Earth referred to the barycenter of the Solar System. The Earth indeed moves around the barycenter of the Earth–Moon system with a velocity $\approx 4 \times 10^{-4}$ smaller than that of revolution, giving rise to monthly term of amplitude $\approx 0".009$, which in the following we neglect.

We now calculate the corresponding variations in equatorial coordinates. The aberration ellipse will maintain its shape, but its orientation will change in the new reference system. Ignoring for the moment the small elliptic terms, from the general transformation rules given in Chapter 3 we derive the apparent geocentric coordinates $\alpha'$, $\delta'$ from the geometric geocentric coordinates $\alpha$, $\delta$ through the expressions (see Figure 7.6):

$$
\left\{
\begin{aligned}
\Delta\alpha &= \alpha' - \alpha = -K\,\frac{\sin\alpha\sin\lambda_\odot + \cos\varepsilon\cos\alpha\cos\lambda_\odot}{\cos\delta} = \frac{1}{c}\,\frac{-\dot{X}\sin\alpha + \dot{Y}\cos\alpha}{\cos\delta} \\
\Delta\delta &= \delta' - \delta = -K\left(\sin\varepsilon\cos\delta\cos\lambda_\odot + \cos\alpha\sin\delta\sin\lambda_\odot - \cos\varepsilon\sin\alpha\sin\delta\cos\lambda_\odot\right) \qquad (7.8) \\
&= \frac{1}{c}\left(-\dot{X}\cos\alpha\sin\delta - \dot{Y}\sin\alpha\sin\delta + \dot{Z}\cos\delta\right)
\end{aligned}
\right.
$$

where $(\dot{X}, \dot{Y}, \dot{Z})$ are the components of the Earth's velocity vector, whose approximate values in AU/day are

$$
\dot{X} = +0.0172\sin\lambda_\odot, \quad \dot{Y} = -0.0158\cos\lambda_\odot, \quad \dot{Z} = -0.0068\cos\lambda_\odot
$$

Precise values are found in the *Astronomical Almanac*.

Therefore, the corrections take the form:

$$
\alpha' - \alpha = Cc + Dd, \qquad \delta' - \delta = Cc' + Dd'
$$

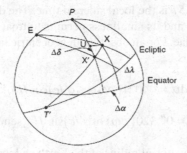

**FIGURE 7.6**  Aberration in the transformation from ecliptic to equatorial coordinates.

in which $C$, $D$ depend on the Sun's longitude and therefore on the date, while $c$, $c'$, $d$, $d'$ depend on the coordinates of the star and on the obliquity of the ecliptic $\varepsilon$. Or else

$$\Delta\alpha = \frac{1}{15\cos\delta}(C\cos\alpha + D\sin\alpha) = \frac{1}{15\cos\delta}h\sin(H+\alpha)$$

$$\Delta\delta = C(\tan\varepsilon\cos\delta - \text{sen}\,\alpha\,\text{sen}\,\delta) + D\cos\alpha\,\text{sen}\,\delta = h\cos(H+\alpha) + i\cos\delta$$

where

$$h\cos H = -K\,\text{sen}\,\lambda_\odot, \quad h\,\text{sen}\,H = -K\cos\varepsilon\cos\lambda_\odot, \quad i = -K\,\text{sen}\,\varepsilon\cos\lambda_\odot$$

$C$ and $D$ are *Bessel's day numbers*, and $H$, $h$, $i$ are the *independent day numbers*. Their values are tabulated for each day in the *Astronomical Almanac*.

To take into account the elliptic aberration, we must add the (almost) constant terms:

$$\Delta\alpha = -Ke[\sin\lambda_\Pi\sin\alpha + \cos\lambda_\Pi\cos\varepsilon\cos\alpha]/\cos\delta,$$

$$\Delta\delta = -Ke[\sin\lambda_\Pi\cos\alpha + \cos\lambda_\Pi\cos\varepsilon\sin\alpha]\sin\delta + \cos\lambda_\Pi\sin\varepsilon\cos\delta$$

where the "constant" actually depends on $e$.

Notice the formal similarity of the expression of the aberration with the expression of the nutation, although the physical bases are so different.

We have already highlighted that the aberration introduces a slight distortion of the celestial sphere; the distance $s$ between two stars and their position angle $p$ will be altered by an amount that can be calculated in advance, and removed before the astrometric reduction of a plate is carried out. To give an order of magnitude, over an arc of $1°$ the maximum effect of the annual aberration is $0^s.02/\cos\delta$ in $\alpha$, $0''.3$ in $\delta$. Bessel's formulae provide a convenient way to carry out this differential calculation:

$$\Delta s = s[h\cos(H+\alpha)\cos\delta - i\sin\delta], \quad \Delta p = h\sin(H+\alpha)\tan\delta$$

## 7.5   THE DIURNAL ABERRATION

The diurnal rotation velocity is responsible for a similar effect, of much smaller amplitude and dependent on the geocentric latitude $\phi'$ of the observer. Indeed, the diurnal velocity is approximately 0.46 km/s at the equator (the angular velocity is $\omega \approx 7.292 \times 10^{-5}$ /s); its apex is on the equatorial plane, at $90°$ from the meridian and toward east, therefore with equatorial coordinates $(\alpha_\oplus = LST + 6^h, \delta_\oplus = 0°)$, where $LST$ is the local sidereal time. The diurnal aberration ellipse is thus parallel to the equatorial system, and its smallness permits to treat the difference (apparent minus geometric) with first-order formulae. For an observer in geocentric latitude $\phi'$, at distance $\rho$ km from the Earth's center, the difference is

$$dHA = -d\alpha = 0^s.021(\rho\omega\cos\phi'/c)\cos HA/\cos\delta,$$

$$d\delta = 0''.320(\rho\omega\cos\phi'/c)\sin HA\,\text{sen}\,\delta$$

No sensible error is made if the equatorial radius of the Earth and the astronomical latitude are used in these formulae. Notice that when the star is on the meridian, the effect on declination is zero, while the eastward displacement causes a later transit. The delay, operationally indistinguishable from a collimation error of the telescope, is equal to

$$\Delta t = 0^s.021 \cos\phi / \cos\delta$$

For a satellite such as the HST, orbiting at approximately 600 km in a plane inclined by 27° to the equator and with a period of 90$^m$, the diurnal velocity has the higher value of about 7 km/s. Therefore, the effect of the orbital aberration must be computed with care. The HST Science Institute discusses the theme "velocity aberration" in several places, for instance, https://hst-docs.stsci.edu/display/FGSDHB/5.2+Observation-Level+Position+Mode+Errors#id-5.2Observation-LevelPositionModeErrors-5.2.7DifferentialVelocity Aberration.

## 7.6 PLANETARY ABERRATION AND PLANETARY PERTURBATIONS

The correction for stellar aberration provides the geometric direction to the star at the time when the light reaches the observer; no allowance is made for the motion of the star in the long and generally unknown time interval between emission and reception. In the case of the bodies of the Solar System, whose orbital elements (and therefore, also the distance $\rho$ from the Earth at the time $t$ of observation) are known with high accuracy, we can explicitly take into account the finite time $\tau$ of propagation of the light from the body to the observer. Notice that the observations provide the topocentric positions. By knowing the geographic location of the observer, the topocentric positions can be reduced to geocentric and finally barycentric positions.

The term *planetary aberration* rigorously means the sum of the finite light time and the stellar aberration, which is independent of the distance and motion of the source and thus affecting in the same manner the apparent positions of the stars surrounding the body. Notice that some authors though call planetary aberration only the first term. Therefore, the measured equatorial coordinates $(\alpha, \delta)$ of a Solar System body contain the same aberrational terms $\Delta\alpha$, $\Delta\delta$ affecting the surrounding stars, so that in differential methods for determining its coordinates, the stellar aberration terms are automatically eliminated.

To determine the planetary aberration depending on the light time, let us assume that we know the barycentric ephemeris of the body in its orbit (coordinates, velocity). Therefore, at the instant $t$ of the observation, we know the geometric barycentric position of the body and that of the Earth; the light time $\tau_t = \rho_t/c$ is also known. However, the observed coordinates $(\alpha_t, \delta_t)$ pertain to the instant the light left the object, namely, the instant $t - \tau$ when the positions of both object and Earth were slightly different. In a first step, we make the simplifying assumption that the light time is short enough that during it, the body is moving in uniform motion, so that a first determination of the apparent ephemeris $\left(\alpha_t^1, \delta_t^1\right)$ can be derived from the geometric position $(\alpha_{\text{geom}}, \delta_{\text{geom}})$ at the same instant by

$$\alpha_t^1 = \alpha_{\text{geom}} - \tau_t \frac{d\alpha}{dt}, \quad \delta_t^1 = \delta_{\text{geom}} - \tau_t \frac{d\delta}{dt}$$

Any difference with the observed values indicates how good the approximation is. For instance, we could presume an error in $\tau_t$, obtaining a second revised value of this parameter and iterate the procedure. Notice that in general, relativistic terms have negligible amplitude.

More intricate would be the reverse problem, of deriving the geometric coordinates from the observed ones, because *a priori* the light time is not known, as it happens, for instance, for a new comet. In this case, a series of observations (at least three) can be used to derive a first tentative orbit and then proceed by successive approximations. See also Chapter 13.

The approximate procedures to remove the several effects of the aberration of light described in the previous paragraphs are quite adequate in most applications. Should greater precision be needed, additional terms must be considered. The ephemerides of the Solar System bodies are usually given in the barycentric reference system. As discussed in more depth in Chapter 8, due to planetary perturbations the Sun moves with respect to the barycenter in a complex path, with a velocity of about 14 m/s (corresponding to an angle $\approx 0''.010$) and an approximate period of 20 years. The total effect is mostly due to Jupiter; Saturn, Venus and the other gaseous planes are from five to ten times less effective than Jupiter. Therefore, a correction to the aberrated coordinates should be added with an amplitude depending from the ecliptic longitude of the plane (the latitude can be neglected).

As already stated, when the positional precision needs to be better than a few hundredths of an arcsec, e.g., in the calculation of the occultation of a star by a planet, the problem becomes quite intricate, and resorting to precise formulae is mandatory.

We will treat the complex case of the occultation of a star by the Moon in Chapter 15. Notice that for the Moon, the *Astronomical Almanac* provides the apparent geocentric ecliptic coordinate referred to the true ecliptic and equinox of date; the apparent right ascension and declination are referred to the true equator and equinox of date. Reduction formulae from topocentric to geocentric equatorial rectangular coordinates are also provided.

Another example of difficult cases is the determination of the instants of the transit of Mercury across the solar disk, where the effect of general relativity on light propagation is important (see the next paragraph). Another interesting case is the correct determination of the periods of eclipsing binaries. See, for instance, Irwin (1959, References section).

## 7.7   THE GRAVITATIONAL DEFLECTION OF LIGHT

The gravitational deflection of light caused by the Sun was not included in the pre-1984 formulae giving the apparent positions of celestial bodies. Such deflection was already foreseen by the Newtonian theory, but with a value twice as small as that based on general relativity (Einstein, 1915, Bibliography section). Dyson et al. (1920), taking advantage of the solar eclipse of May 1919, confirmed (although certainly not in a conclusive way) the correctness of Einstein's prediction. Since then, the gravitational deflection of light, together with the anomalous precession of the perihelion of Mercury and the gravitational redshift of the spectral lines, has been one of the fundamental astronomical proofs of that theory. For further discussion of general relativity effects, see also Chapter 14.

Karl Schwarzschild (1916, References section) introduced a typical radius associated with a spherical mass $M$, the so-called Schwarzschild's radius $r_S$, given by

$$r_S = 2\frac{GM}{c^2}$$

whose value is about 3.0 km for the Sun and 0.9 cm for the Earth. The influence of the mass of the Sun on a grazing light ray will make its path slightly concave toward the Sun; this effect is the manifestation of the curvature of space-time due to mass. Therefore, in first approximation, a star near the limb of the Sun will be seen by the terrestrial observer in a direction slightly displaced, radially outward, by the quantity:

$$\Delta\theta = (1+\Gamma)\frac{r_S}{R_\odot} = (1+\Gamma)\frac{r_S}{a_\odot\alpha_\odot}$$

where $a_\odot$ is the AU, and $R_\odot$ and $\alpha_\odot$ are the linear ($\approx 750,000$ km) and angular ($\approx 16'$) radii of the Sun, respectively. The constant $\Gamma$ is equal to 0 in the Newtonian theory and to 1 in general relativity; therefore, $\Delta\theta \approx 0''.9$ in the first case and $\approx 1''.8$ in the second. Notice that the deflection is independent of the wavelength; in particular, it is the same in the optical and radio domain. Indeed, the radio measurements are much more precise than the optical ones (see, e.g., Fomalont and Shramek, 1975; Jones, 1976).

To understand why the ratio between the Newtonian and the general relativity values is precisely 2, we resort to a reasoning due to Eddington (1920, Bibliography section). Gravitation acts as a refracting medium, having index of refraction $n \approx 1 + 2m/r$, where $m$ and $r$ are a-dimensional mass and radius of the deflecting body. By solving the light path in this medium, it can be shown that the asymptotic total deflection of the ray passing at distance $r$ is $4m/r$ (radians), while in the Newtonian theory, it is only $2m/r$. The specification of asymptotic deflection, in this context, means that both observer and source are considered at infinite distance from the Sun. The discussion is much more complicated if the source of light is not at infinity but inside the Solar System, for instance, in the case of Mercury passing behind the Sun (the gravitational deflection being smaller than the asymptotic value).

In conclusion, the outward radial displacement $\Delta E$ of the apparent direction of the star decreases linearly with the angular distance from the center of the Sun, so that, after some manipulation, the general formula can be easily derived:

$$\Delta E = \frac{r_{\!s}}{a_{\odot}}\frac{1+\cos E}{\sin E} = 0''.00407\,\frac{1}{\tan\dfrac{E}{2}} \tag{7.9}$$

See Figure 7.7.

At grazing incidence, the value of the displacement is $1''.866$. At $45°$ from the center of the Sun, the displacement is still at the level of $0''.01$ and of $0''.004$ at $90°$. An appropriate projection of this angle on the equatorial system will permit the determination of the corrections to be applied to the apparent coordinates:

$$\cos E = \sin\delta\sin\delta_{\odot} + \cos\delta\cos\delta_{\odot}\cos(\alpha - \alpha_{\odot})$$

$$\Delta\alpha = 0^{s}.000271\cos\delta_{\odot}\sin(\alpha - \alpha_{\odot})/(1-\cos E)\cos\delta$$

$$\Delta\delta = 0''.00407\big[\sin\delta\sin\delta_{\odot}\cos(\alpha - \alpha_{\odot}) - \cos\delta\sin\delta_{\odot}\big]/(1-\cos E)$$

Since the displacement depends on the distance from the center of the Sun, the gravitational deflection amounts to a slight distortion of the celestial sphere. Therefore, it appears also in differential measurements of high precision over large arcs.

Notice that at a distance of 1 AU, the velocity $V$ of a body in circular orbit around the mass $M$ is

$$V^2 = 2\frac{GM}{a_{\odot}}$$

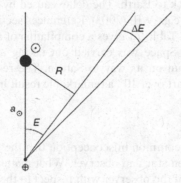

**FIGURE 7.7**   The gravitational deflection of light for a very distant source.

**TABLE 7.1**

**Values of Γ**

| Method | Γ |
| --- | --- |
| Optical solar eclipses | $0.90 \pm 0.22$ |
| Radar echoes from Venus and Mercury | $1.03 \pm 0.06$ |
| Radar echoes from Voyager 2 | $1.00 \pm 0.06$ |
| Signal from radio pulsar $1937 + 21$ | $1.03 \pm 0.05$ |
| Solar occultations of radio sources by VLBI | $0.9996 \pm 0.0017$ |

so that, for a generic distance $R < a_\odot$,

$$\frac{r_S}{R} \geq \frac{V^2}{c^2}$$

In other words, the gravitational deflection is as large, or larger, than the second-order terms of the annual aberration.

What has been said for the Sun could be repeated for Jupiter, Saturn or any other planet. In projects aiming to achieve a positional precision much better than $0''.001$, the gravitational field of all planets of the Solar System must be computed (see the "Notes" section). To achieve this level of precision, allowance must be made for the finite distance of the observer from the deflecting mass and also for its position on the surface of the Earth (where the gravitational deflection can never exceed $0''.0003$).

A relativistic treatment can be found, for instance, in Green (1985, Bibliography section). Beyond the complexity of the formulae, it is shown there that the effect of aberration, light deflection and parallax can still be considered separately, even if *a priori* space and time cannot be separated in general relativity. A more recent treatment, especially connected with the VLBI data, is found in Walter and Sovers (Bibliography section).

At the level of precision achieved by the VLBI and GAIA, the passage of the decades may make the galactic rotation measurable, as *secular* or *galactic* aberration of very distant extragalactic sources. The distance between the Sun and the galactic center is around 8 kpc, while the rotational velocity is approximately 230 km/s. Therefore, the change in direction of this velocity produces a maximum aberrational displacement of few microarsec per year, which accumulates in time and will eventually become detectable (see, e.g., MacMillan et al., 2019).

For completeness, it must be pointed out that the space-time curvature due to gravity also affects the times of a signal received by two antennae of the VLBI network, or sent from Earth to a planet or to a spacecraft and relayed back to Earth. The delay caused by the gravitational field is usually referred to as *Shapiro delay* (see, e.g., Will (2003), References section). Therefore, there are several ways to measure the parameter Γ. Table 7.1 gives a compilation of a few results.

One such measurement from Space was carried out taking advantage of the passage of the Cassini spacecraft beyond the Sun on its way to Saturn. The results have validated the general relativity to an accuracy of one part over $10^5$, a remarkable result indeed (see Bertotti et al., 2003).

## NOTES

- We have tried to avoid the common misconception that the stellar aberration depends on the relative velocity between star and observer. What matters is the change in the direction of the velocity vector of the observer with respect to the star. An account of the traps incurred by many textbooks can be found, for instance, in Liebscher and Brosche (1998).

- Regarding general relativity and celestial mechanics, see, for instance, Kovalevsky and Brumberg eds., (1986) in the Bibliography section. The References section reports two articles published by P. Strumpff from 1979 to 1985 and a paper by Hellings (1986).
- A pictorial representation of the gravitational bending of light by the planets is given, in the frame of the GAIA Project, in the ESA's Report to the 34th COSPAR Meeting, SP-1259, p. 96, 2002.
- Aberration is also present in laser ranging to satellites equipped with retro-reflectors. Therefore, the angle of the retro-reflectors must be slightly changed with respect to a stationary corner cube to achieve identical directions between incoming and outgoing laser rays (see, for instance, Minato et al., 1992). Two more papers can be quoted in this respect: Ragazzoni (1997) and Wing (2003) in References section.

# 8 The Parallax

The phenomenon of the parallax is caused by the finite distance of a planet or star from the observer. Two observers located in two different positions, or the same observer moving from one place to another due to diurnal rotation or annual revolution of the Earth, will see the object in two different positions on the celestial sphere. Of all the phenomena described so far altering the apparent direction of the source, the parallax is the first to give direct information on the distance and nature of the heavenly body, not only on the motions of the observer nor on fundamental properties of light and of space-time. Furthermore, the knowledge of the distance between the observers provides a link between the terrestrial and the cosmic distance scales. Therefore, the determination of the parallaxes is a fundamental astronomical measurement.

Although the present treatment is carried out for a terrestrial observer, the basic concepts could be adapted to an observer on another planet or aboard a spacecraft navigating the Solar System.

## 8.1 THE TRIGONOMETRIC PARALLAX

Given two observers, O and C at a distance $R$ from each other and an object X at distance $d'$ from O and $d$ from C, O will see the object in direction $z'$ and C in direction $z = z' - p$ with respect to the baseline OC (see Figure 8.1). The angle $p$ is the parallax of X with respect to the baseline OC.

The following exact relations hold

$$CX' = d\cos z = R + d'\cos z', \quad XX' = d\sin z = d'\sin z',$$

$$CX = d = d'\cos p + R\cos z, \quad CC' = d\sin p = R\sin z',$$

$$C'X = d\cos p = d' + R\cos z', \quad OO' = d'\sin p = R\sin z,$$

$$\tan p = \frac{R}{d'}\frac{\sin z'}{1 + \frac{R}{d'}\cos z'} = \frac{d}{d'}\frac{R}{d}\frac{\sin z}{1 + \frac{R}{d'}\cos z'}$$

We adapt now these generic relationships, useful in terrestrial triangulations or when observing an artificial satellite in low orbit, to the astronomical case. In the following sections, we shall distinguish between objects in the Solar System, where the Earth's radius is sufficiently large to provide a useful baseline (*diurnal parallax*) and stars, for which this radius is exceedingly small and the baseline is provided by the radius of the orbit of the annual revolution around the Sun (*annual parallax*).

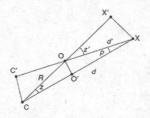

**FIGURE 8.1** The geometric parallax.

## 8.2   THE DIURNAL PARALLAX

When determining the celestial coordinates of an object of the Solar System (for instance, a planet, but the same holds for a comet or an asteroid), it will be necessary to take into account the location of the observer on the terrestrial surface, namely, his topography, in order to translate these coordinates to an ideal geocentric observer. This process will therefore transform *topocentric* into *geocentric* coordinates. Let C be the center of the Earth, O the generic observer on the surface, at a distance $R = \rho a_{\oplus}$ from C ($\rho \approx 1$). Let X be the planet distant $d$ from C and $d'$ from O (see Figure 8.2), and the angle $p = X\hat{C}O$ is the instantaneous parallax of X. Notice that in a generic observation, the plane through X, O and C will not coincide with the meridian of O.

In Figure 8.2, the line OZ is the astronomical vertical and the line CZ′ the geocentric vertical; the observations provide the astronomical zenith distance $z_0$ that can be transformed into the distance $z'$ from the geocentric vertical by means of the deviation of the vertical $v = \phi' - \phi$ given in Equation 2.4:

$$z' = z_0 - (\phi' - \phi) = z_0 - v$$

The angle $p$ is then given by

$$\sin p = \rho \frac{a_{\oplus}}{d} \sin z' \tag{8.1}$$

Finally, if the parallax $p$ can be determined by the observations (see later), or if the distance $d$ is known, the geocentric zenith distance $z$ can be determined by

$$z = z' - p \tag{8.2}$$

Notice that the topocentric zenith distance is larger than the geocentric one and that the change of the line of sight from the topocentric to the geocentric observer takes place in the plane OCX. This plane essentially coincides with the vertical plane of X as seen by O (disregarding the small difference between astronomical and geodetic vertical). Therefore, the influence of the diurnal parallax is practically to increase the zenith distance (or to decrease the elevation above the horizon) measured by the topocentric observer with respect to the geocentric one (hence the other designation, *parallax of elevation*, although rigorously the circle through X′Z′ is not exactly the vertical one). In particular, the observations in meridian give directly the geocentric right ascension. The diurnal parallax is then simply the variation of declination. As already stated, if the observations provide $p$, we can

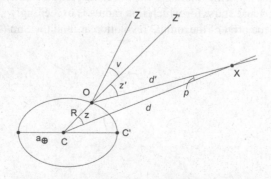

**FIGURE 8.2**   The instantaneous diurnal parallax. The ellipticity of the Earth is greatly exaggerated. The plane COX does not necessarily coincide with the meridian of O.

derive $d$ and $z$ from Equations 8.1 and 8.2. If, instead, we know the geocentric zenith distance $z$ and the distance $d$, we can derive the topocentric parallax $p$ by developing $\sin z' = \sin(z + p)$:

$$\tan p = \rho \frac{a_\oplus}{d} \frac{\sin z}{\left(1 - \rho \dfrac{a_\oplus}{d} \cos z\right)} \tag{8.3}$$

The angle $p$ is variable with the rotation of the Earth, even disregarding the motion of the planet. Therefore, it has been agreed to name horizontal equatorial parallax $\pi$ the angle under which X sees perpendicularly the Earth radius $a_\oplus$. In other words, $\pi$ is the maximum value of $p$ at any given distance of X from C, but $\pi$ itself will vary according to the relative positions of X and C in their heliocentric orbits.

The horizontal parallax of the Moon, $\pi_\mathrm{\mathrm{\mathrm{}}}$, varies between $54'$ and $61'$ due to the strong eccentricity of its geocentric orbit. These values are about twice the apparent diameter of the lunar disk; therefore, the measured lunar parallax has a strong dependence on the position of the observer and of the Moon in its orbit. Consequently, the occurrence of topocentric phenomena such as a lunar occultation (see Chapter 15) depends very critically on the date.

All other celestial bodies are much farther than the Moon (apart from occasional objects passing inside its orbit), so that their diurnal parallaxes are small but vary by great amounts according to their orbital position. For instance, $\pi(\text{Venus})$ varies between $5''$ and $34''$. Therefore, essentially, in all cases (except artificial satellites or near Earth objects) we are justified to assume $d \gg a_\oplus$ and $\pi$ very small, so that

$$d = \frac{a_\oplus}{\sin \pi} \approx \frac{a_\oplus}{\pi}, \quad \sin \pi \approx \pi = \frac{a_\oplus}{d}$$

For the Sun, averaging over the slight ellipticity of the orbit, the horizontal parallax is essentially constant:

$$d_\odot = a_\odot \approx \frac{a_\oplus}{\pi_\odot}, \quad \pi_\odot'' \approx 8''.8$$

The average distance to the Sun $a_\odot$ is the astronomical unit AU, $1\ \text{AU} \approx 1.49 \times 10^8 \text{km}$ (see the precise values in Chapter 7). Expressing the distance $d$ to a generic body in AU and the parallax in seconds of arc,

$$\pi'' = \left(\frac{1}{d(\text{AU})}\right)\pi_\odot'' \tag{8.4}$$

When the body is on the meridian,

$$\sin p = \rho \sin \pi \sin z' \approx \rho \pi \sin z' \approx p,$$

$$\tan p = \rho \sin \pi \frac{\sin z}{1 - \rho \sin \pi \cos z} \approx \rho \pi \frac{\sin z}{1 - \rho \pi \cos z} \approx p$$

where the approximations are justified because the angles $p$, $\pi$ are always so small that sin and tan essentially coincide with their arcs. Using arcsec and AUs as units, we have

$$p'' = \rho \frac{\pi_\odot}{d} \sin z' \approx 8''.8 \frac{\rho}{d(\text{AU})} \sin z' \tag{8.5}$$

Suppose now to observe the same planet from two stations having the same longitude, at the same instant and in meridian (this is an idealized case, but of great historical significance). It is then easy to find the horizontal equatorial parallax of the planet. From $p = \rho\pi \sin z' = z' - z$ written for the two observers 1 and 2, we derive

$$z_1' = z_{01} - v_1, \quad z_2' = z_{02} - v_2, \quad \pi = \frac{(z_1' + z_2') - (z_1 + z_2)}{\rho_1 \sin z_1' + \rho_2 \sin z_2'}$$

In these relations, the topocentric zenith distances are given by the observations, while the sum of the geocentric zenith distances is equal to the difference of the two astronomical latitudes: $z_1 + z_2 = \phi_1 - \phi_2$. To maximize this factor, the two observers should preferably be located on opposite sides of the equator. In realistic cases, appropriate corrections must be applied for the difference in longitudes and the orbital motion of the planet between the two observations. Moreover, to minimize the effects of refraction and other systematic errors, the measurements should be taken with respect to a set of nearby fundamental stars. Let us consider now the generic case of a body observed outside the meridian, with the simplifying assumption of astronomical vertical coinciding with the geodetic one. The geocentric zenith $Z'$ is on the meridian at the small distance $v$ from the astronomical one $Z$, and closer to the horizon (see Figure 8.3).

$X'$ will be the observed topocentric position of the body and $X$ the geocentric one on the great circle through $XX''$ and $Z'$. The angle $Z'\hat{P}X' = Z\hat{P}X'$ is the topocentric hour angle $HA' = HA(X')$, while the arc $X_1'X'$ is the topocentric declination $\delta' = \delta(X')$. The corresponding geocentric quantities are $Z'\hat{P}X = Z\hat{P}X = HA(X)$, arc $X_1X = \delta(X)$. As shown in Chapter 2, it is

$$PZ' = 90° - \phi', \quad Z'X' = z', \quad Z'X = z, \quad XX' = p, \quad z' = z + p$$

Therefore, the diurnal parallax augments the zenith distance along the vertical circle and leaves the azimuth essentially constant (hence the other designation parallax of altitude, although the circle through $X'Z'$ is not rigorously the vertical one).

Now, call $A$ the angle in $Z'$, which is common to the spherical triangles $PZ'X'$ and $PZ'X$. Letting $\Delta HA = HA' - HA$, after some manipulation we find

$$\frac{\cos\phi' \sin p}{\sin z \sin(z + p)} = \frac{\sin A \sin \Delta HA}{\sin HA \sin(HA + \Delta HA)} \tag{8.6}$$

Taking into account that

$$\sin p = \rho \frac{a_\oplus}{d} \sin(z + p), \quad \sin A \sin z = \cos\delta \sin HA$$

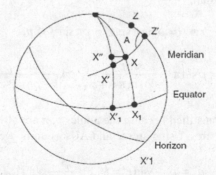

FIGURE 8.3    Effects of the diurnal parallax on the celestial coordinates.

we finally get the rigorous formulae:

$$\begin{cases} \tan\Delta HA = -\tan\Delta\alpha = -\rho\dfrac{a_\oplus}{d}\cos\phi'\ \dfrac{\sin HA}{\rho\dfrac{a_\oplus}{d}\cos\phi'\cos HA - \cos\delta} \\[4mm] \tan\delta' = \dfrac{\tan\delta - \rho\dfrac{a_\oplus}{d}\sin\phi'\sec\delta\cos\Delta HA}{1 - \rho\dfrac{a_\oplus}{d}\cos\phi'\sec\delta\cos HA} \\[4mm] \sin\ p = \sin\pi\ \dfrac{\sin HA'\cos\delta'}{\sin HA\cos\delta} \end{cases} \tag{8.7}$$

Let us introduce the auxiliary quantity $q$:

$$q = \cot\phi'\ \frac{\cos\big[(HA + HA')/2\big]}{\cos\big[(HA - HA')/2\big]}$$

by which

$$\tan\Delta\delta = \frac{\rho\dfrac{a_\oplus}{d}\sin\phi'\big(q\sin\delta - \cos\delta\big)}{1 - \rho\dfrac{a_\oplus}{d}\sin\phi'\big(q\cos\delta + \sin\delta\big)}$$

where $\Delta\delta = \delta' - \delta$.

The ratio between topocentric and geocentric distances is expressed as

$$\frac{d'}{d} = \cos(\delta' - \delta) - \rho\frac{a_\oplus}{a}\sin\phi'\big(q\cos\delta' + \sin\delta'\big)$$

These rigorous formulae seldom need to be applied. For the occasional asteroids coming inside the orbit of the Moon, or for meteors in the upper atmosphere, radar echoes can often be obtained that provide high-precision distances and velocities. However, in these cases it is more efficient to use Cartesian coordinates and apply the refined methods used for tracking satellites.

For the usually much more distant celestial objects, the quantity $\rho a_\oplus/d = \rho\pi$ is so small that Equations 8.7 become

$$\begin{cases} -\Delta HA = \Delta\alpha = -\rho\dfrac{a_\oplus}{a}\cos\phi'\ \dfrac{\sin HA}{\cos\delta} = -\rho\pi\cos\phi'\operatorname{cosec} z\sec\delta \\[3mm] \Delta\ \delta = -\rho\dfrac{a_\oplus}{a}\big(\sin\phi'\cos\delta - \cos\phi'\sin\delta\cos HA\big) \\[3mm] \qquad = -\rho\pi\big(\sin\phi'\operatorname{cosec} z\sec\delta - \tan\delta\cot z\big) \end{cases} \tag{8.8}$$

Recalling Equation 8.5, another way of expressing the same quantities is

$$\begin{cases} -\Delta HA'' = \Delta\alpha'' = -\rho\dfrac{\pi_{\odot}''}{d(\mathrm{AU})}\cos\phi'\,\dfrac{\sin HA}{\cos\delta} \\[4mm] \Delta\delta'' = -\rho\dfrac{\pi_{\odot}''}{d(\mathrm{AU})}\left(\sin\phi'\cos\delta - \cos\phi'\sin\delta\cos HA\right) \end{cases} \tag{8.9}$$

No sensible error is made if the observed quantities $HA'$, $\delta'$ are substituted to the geocentric values in the right side of these equations. Notice, however, that the symmetry between geocentric and topocentric coordinates in Equation 8.9 is illusory: indeed, the knowledge of the geocentric coordinates depends on the previous knowledge of the geocentric distance. For a new comet or asteroid, $d$ is unknown, so that only the trigonometric expressions (also named *parallactic factors*) can be computed. These equations provide a second method for obtaining distances (*diurnal method*), this time using observations taken at the same observatory but at two different hour angles (ideally, $12^{\mathrm{h}}$ apart), provided the parallax of the Sun is known.

If we could follow the apparent path of the comet or asteroid with respect to the distant stars for a full Earth's rotation, we would observe a circle (projected as an ellipse) trailed by the revolutions around the Sun of the Earth and of the body itself. This kind of observation is indeed possible, e.g. from the HST, whose orbital semi-period is approximately $45^{\mathrm{m}}$; an asteroid will therefore leave, on a sufficiently long exposure, an appreciably curved trail (see the "Notes" section).

The diurnal parallax has an interesting consequence on the apparent diameters. Let $s$ be the radius in km of a planet (treated here for simplicity as a disk) and $S$ its apparent geocentric angular radius when at geocentric distance $d$; $S'$, $d'$ are the corresponding topocentric values. The following relations will hold

$$\sin S = \frac{s}{d}, \quad \sin S' = \frac{s}{d'} = \frac{s}{d}\frac{d}{d'} = \sin S\,\frac{\sin z'}{\sin z}$$

The angles are usually small quantities, so that

$$S' = S\,\frac{\sin z'}{\sin z} \tag{8.10}$$

From the Earth's surface, the effect for the Sun never exceeds $0.1''$, but it is noticeable for the Moon, whose geocentric apparent radius is given by $S_{\circleddash} = 0.27245\pi_{\circleddash}$, with a maximum variation of approximately $40''$. Therefore, this change in lunar diameter with the location of the observer must be taken into account when calculating precise circumstances of eclipses and occultations. The diurnal parallax affects the Moon in another way, namely, causing slight variations of the rising and setting instants. The rise retards and the setting anticipates, contrary to the effect of the atmospheric refraction which amounts at the horizon to approximately $34'$ (see Chapters 2 and 11). As already mentioned in Chapter 3, when calculating the times of rise or set for an observer at sea level, the altitude $h$ of the *center* of the disk can be derived from

$$h = -34' - \text{semi-diameter} + \text{horizontal parallax}$$

At these times, the upper limb of the Moon is on the horizon, while $h$ varies from 5 to 11 arcminutes according to the distance of the Moon. Recalling Equations 3.10 and 3.21, the hour angle HA of rising and setting of the upper limb of the Moon can be derived from

$$\cos HA = -\tan\varphi\tan\delta + \sin(\pi - 34' - 16')\sec\varphi\sec\delta' \tag{8.11}$$

Thus, the correction to the apparent hour angle due the lunar diurnal parallax can be larger than that due to the refraction at the horizon.

## 8.3 SOLAR AND LUNAR PARALLAXES

The direct determination of the solar parallax $\pi_\odot$ poses great difficulties, due to its small amount and the lack of precise reference points on the surface. Copernicus and Tycho Brahe, for instance, adopted a value of 3 arcminutes! Soon, it was found advantageous to make recourse to indirect methods, of which we provide two examples:

1. Suppose an accurate table of geocentric positions (namely, the ephemerides) of a given planet is available, and compute

$$\alpha'(t_2) - \alpha'(t_1) = \alpha(t_2) - \alpha(t_1) + \pi_\odot \left[ f_\alpha(t_2) - f_\alpha(t_1) \right]$$

$$\delta\alpha'(t_2) - \delta'(t_1) = \delta(t_2) - \delta(t_1) + \pi_\odot \left[ f_\delta(t_2) - f_\delta(t_1) \right]$$

where the $f$'s are the parallactic factors in Equation 8.9. From a set of well-measured topocentric positions, the unknown $\pi_\odot$ can be derived. Indeed, it is not even necessary to know the geocentric positions if advantage is taken of the moments when the planet appears stationary on the celestial sphere. Furthermore, if the measurements are made at the same time from two different locations, the unknown parallax of the planet is also found. With this method, applied in 1672 by Picard in Paris ($\phi = +49°50'$) and by Lacaille in Cayenne ($\phi = +4°57'$) to planet Mars, a difference of $15''$ in the two topocentric declinations was found, corresponding to an equatorial parallax of $25''.5$. From this value, Cassini was able to derive the first reliable value of $\pi_\odot$, namely $9''.5$, a value he could essentially confirm measuring from Paris the right ascensions of the planet. It is worth recalling that during the same expedition, Lacaille found that his pendulum was slower than in Paris, losing some $2^m$ in $24^h$, the first experimental proof of the non-sphericity of the Earth. A variant of this method was used by Encke (1822, 1824), see the References section, by a new reduction of the data taken on the rare occasion of the transit of Venus on the solar disk in 1761 and 1769; his value of $\pi_\odot$ was, however, decidedly lower than the true one.

2. Consider Kepler's third law (see Chapter 12) for a given planet, in its slightly approximate expression $P^2 = (4\pi/GM_\odot)a^3$, P being the sidereal period, a the semi-major axis and $M_\odot$ the mass of the Sun. Comparing this expression with the corresponding one for the Earth,

$$P^2/P_\oplus^2 = a^3/a_\oplus^3$$

one can derive only the relative dimensions of the orbits. At least one absolute determination in kilometers is needed to fix the scale of the Solar System. To reach this goal, it is advantageous to observe from several locations an asteroid (having a star-like image) whose orbit takes it as close as possible to the Earth. Several asteroids were tried, the one giving the best results being 433 Eros, whose perihelion is at 1.13 AU. Eros came at the opposition from 1900 to 1901 and again from 1930 to 1931. The first event, during which the minimum distance was 0.32 AU, was observed visually and gave the value $\pi_\odot = (8''.806 \pm 0''.004)$. In the second opposition, the small body came much closer, 0.17 AU; moreover, the observations took advantage of photography. The derived value was $\pi_\odot = (8''.790 \pm 0''.001)$. Eros was observed again in 1975, but this time with radar echoes, a technique that had been successfully applied to the Moon, to Venus, to Mercury. The radar data gave not only the distance to Eros, but also the first determination of its dimensions, approximately $16 \times 35 \, km^2$. Finally, Eros was reached in 2001 by the spacecraft NEAR, who crashed on its surface at the end of a very successful mission (see the Web Sites section).

Table 8.1 provides a compilation of historical values of $\pi_\odot$; notice that the variations from time to time usually greatly exceeded the quoted probable errors, an indication of the presence of systematic errors in several methods.

**TABLE 8.1**

**Values of the Solar Parallax**

| Period | $\pi_\odot$ | Method |
|---|---|---|
| 1801–1833 | 9.0″ | Parallax of Mars, transits of Venus in 1761–1769 |
| 1834–1869 | 8.5776″ | New discussion by Encke of previous Venus transits data |
| 1870–1881 | 8.95″ | Le Verrier, from the lunar orbit (solar parallactic inequality) |
| 1882–1900 | 8.848″ | Best value according to Newcomb |
| 1901 | 8.80″ | Official value agreed in 1896 |
| 1930–1931 | 8.790″ | Spencer–Jones, from Eros opposition |
| 1964 (IAU) | 8.79405″ | Radar, corresponding to $149.60003 \times 1011$ m |
| 1976 (IAU) | 8.794148″ | Radar, corresponding to $149.59787 \times 1011$ m |

It is worth recalling that other indirect methods of finding the solar parallax were the radial velocities of ecliptic stars and the aberration of the Medicean moons of Jupiter (see also Chapters 7 and 9).

On closer examination, the methods based on Kepler's third law (e.g., the historical Eros observations) provide the mass of the Moon in addition to the solar parallax, because it is the Earth–Moon barycenter that follows an (almost) Keplerian orbit around the Sun. As also discussed in Chapter 6, the geocentric observer has a monthly inequality with respect to the dynamical ecliptic that can reach 6″.44 when the Moon is in quadrature. This method of obtaining the mass of the Moon was used in addition to those expounded in Chapter 6.

Regarding the lunar parallax, of great historical significance is the method followed by Hipparchus. During a lunar eclipse, be S, E and L the centers of Sun, Earth and Moon, respectively. The shadow of the Earth is a cone of vertex $E_1$ and height HK when intercepted by the Moon in L; as seen from E, the angular semi-aperture of the shadow LH is $\psi/2$, while the angular solar radius is $\alpha_\odot/2$. From simple geometry, we derive

$$\psi/2 = \pi_\odot - \alpha_\odot/2 + \pi_{\mathbb{D}} \approx -\alpha_\odot/2 + \pi_{\mathbb{D}}$$

The angle $\psi/2$ ($\approx 2.5\alpha_\odot/2$) can be determined by the instants of immersion and emersion of the Moon in the shadow. Although the method cannot be very precise, Hipparchus arrived to the value 3489″, fairly close to the true value of 3422″.

Finally, a direct and precise measurement was obtained by Lalande in 1752 from Berlin in Germany and Lacaille from the Cape in South Africa, taking advantage of the close coincidence in longitude and large difference in latitude.

In today's practice, the instantaneous distance and the rotation of the figure can be obtained with excellent precision (around 1 cm) by laser echoes from the retro-reflectors that were placed on its surface by American Apollo and Soviet Luna missions from 1969 to 1976 (see, for instance, the Project Apollo, Web Sites section).

## 8.4   THE ANNUAL PARALLAX

Annual parallaxes are the fundamental method for the direct determination of the stellar distances.

Figure 8.4 shows the Earth in two diametrically opposed positions along its orbit, for the moment taken as circular with radius $a_\odot = 1$ AU. $E_1$ and $E_2$ are two positions of the geocentric observer 6 months apart, S is the ideal heliocentric observer, X is a nearby star, which will be seen from the Earth as projected in $X_1$ and $X_2$ on the celestial sphere (in other words, with respect to the background of the distant stars).

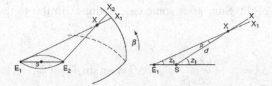

**FIGURE 8.4**  The annual parallax.

Calling ES the direction Earth–Sun, $z$ the angle of the line of sight with that direction from S and $z'$ from E, and $d$ the heliocentric distance of X, we have

$$a_\odot \sin z' = d \sin(z - z') = d \sin p, \quad \sin \pi = \frac{a_\odot}{d}, \quad \sin z' = \frac{\sin(z' - z)}{\sin \pi}$$

The angle $\pi$, under which the AU is seen perpendicularly from the star, namely, the maximum value of $p$, is called the *annual parallax* of X. Until now, the star Proxima Centauri (in the triple system of $\alpha$ Cen) has the largest observed value, $\pi = 0''.76$; therefore, no sensible error is made by using arcs instead of their sine or tangent, and therefore,

$$z' = z - \pi \sin z' \approx z - \pi \sin z, \quad \pi \approx \frac{a_\odot}{d} \tag{8.12}$$

The finite distance has the effect that the geocentric observer sees the star closer to the Sun by the slight amount $(z' - z)$; on the celestial sphere, this apparent movement occurs along the great circle passing through X and S. During the year, S moves along the ecliptic; therefore, the locus X' of apparent geocentric positions will be an ellipse centered on the heliocentric position X. Notice that this ellipse has nothing to do with the ellipticity of the Earth's orbit (that we have assumed circular). It is simply the projection effect depending on the ecliptic latitude $\beta$ of X; the slight ellipticity of the orbit ($e = 0.0167$) will introduce small modifications, which presently we will neglect. Notice also that the annual parallax could be determined, in principle, by measuring the variation of the angular distance of the nearby star from the Sun during the year; in practice, it is much easier to determine its relative positions in respect to the background of the other stars.

Let us put these qualitative conclusions in a more quantitative way, by reasoning in ecliptic coordinates (see Figure 8.5).

If $(\lambda, \beta)$ are the heliocentric coordinates of X, the geocentric coordinates will differ by very small amounts that we can treat as differentials, $(\lambda + \Delta\lambda, \beta + \Delta\beta)$. In other words, the small triangle XX''X' can be considered as a plane triangle with infinitesimal sides:

$$X''X = \Delta\lambda \cos\beta, \quad X''X' = \Delta\beta, \quad XX' = z - z' = \pi \sin z$$

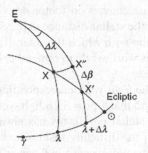

**FIGURE 8.5**  The annual parallax in ecliptic coordinates.

Calling $\lambda_\odot$ the longitude of the Sun, after some calculations (similar to those made for the annual aberration), we get

$$\Delta\lambda = \pi\frac{\sin(\lambda_\odot - \lambda)}{\cos\beta}, \quad \Delta\beta = -\pi\sin\beta\cos(\lambda_\odot - \lambda)$$

or else

$$\cos^2\beta\left(\frac{\Delta\lambda}{\pi}\right)^2 + \left(\frac{\Delta\beta}{\pi\sin\beta}\right)^2 = 1 \qquad (8.13)$$

where $\Delta\lambda\cos\beta$ is a movement parallel to the ecliptic, and $\Delta\beta$ is perpendicular to it; therefore, the annual locus of the star is an ellipse, with semi-major axis $\pi$ and semi-minor axis $\pi\sin\beta$. For an ecliptic star, this locus degenerates in a straight segment, and for a star seen toward the ecliptic pole, the locus is a circle. The star passes through the semi-major axis when its longitude is $\pm90°$ from that of the Sun, for which reason there are two preferred dates in the year for any particular region of sky. If, in addition, observations in meridian are preferred, parallax work is better done at dusk and dawn. Notice also that the ellipse of parallax anticipates by 3 months that of the annual aberration; of course, its dimensions are proportional to the distance and not fixed as those of aberration.

In equatorial coordinates, this ellipse will be rotated by a certain angle. It is not difficult to prove the following relations:

$$\begin{cases} \Delta\alpha\cos\delta = \pi(\cos\varepsilon\cos\alpha\sin\lambda_\odot - \sin\alpha\cos\lambda_\odot) \\ = \pi(Y_\odot\cos\alpha - X_\odot\sin\alpha) \\ \Delta\delta = \pi(\sin\varepsilon\cos\delta\sin\lambda_\odot - \cos\alpha\sin\delta\ \lambda_\odot - \cos\varepsilon\sin\delta\sin\alpha\sin\lambda_\odot) \\ = \pi(Z_\odot\cos\delta - X_\odot\cos\alpha\sin\delta - Y_\odot\sin\alpha\sin\delta) \end{cases} \qquad (8.14)$$

where $(X_\odot, Y_\odot, Z_\odot)$ are the geocentric equatorial coordinates of the Sun. These corrections $\Delta\alpha$, $\Delta\delta$ must be added with their signs to the heliocentric coordinates, subtracted from the geocentric ones.

Using the AU as baseline, it is possible to institute the fundamental unit of astronomical stellar distances, namely, the parsec: 1 parsec (pc) is the distance from where the AU subtends perpendicularly an angle of $1''$. Therefore,

$$1\mathrm{pc} \approx 206264.8\,\mathrm{AU} \approx 3.09\times10^{13}\,\mathrm{km}$$

It is worth remembering that the last conversion factor derives from the solar parallax, but any revision of its value will not change the stellar distance expressed in parsecs. This remark is not so important today, because the precision with which the solar distance is known is much higher than that of stellar distances; however, it is worthwhile to remember how the cosmic distances ladder has been connected to the terrestrial units.

A secondary unit of distance is the *light-year*, corresponding to the distance traveled in vacuum by the light in 1 year; it is easily found that 1 pc $\approx$ 3.26 light-years.

The measurement of the minute stellar parallaxes has always been a difficult task, quite often plagued by systematic errors. Part of the difficulty resides in the proper motion of the nearby stars, which trails the parallax ellipse, as we will discuss in Chapter 9. Until recently, the precision could rarely be claimed to be better than $0''.01$. If we restrict our considerations to stars having a relative error

$\sigma(\pi)/\pi < 1/3$, we then find an astrometric horizon around 30 pc. The astrometric satellite Hipparcos enlarged this horizon by ten times, but still the volume of the Milky Way directly accessible to trigonometric distance determination was a very slight fraction of the total. GAIA, with several hundred times better precision, enlarges the astrometric horizon to reach the nearest galaxies.

The above treatment of stellar parallaxes is approximate for several reasons:

- the orbit of the Earth is slightly elliptical;
- the geocentric observer should be replaced by the Earth–Moon barycentric one; the effect is smaller than the diurnal aberration on that particular star, because the barycenter is inside the body of the Earth;
- the heliocentric observer should be substituted by the one in the barycenter of the Solar System. The distance between the two never exceeds 0.01 AU (two solar radii), with a complex behavior in time according to the longitude of Jupiter, Saturn and the other planets, as shown in Figure 8.6;
- the relativistic deflection of light affects the apparent directions according to the angular distance from the Sun and large planets.

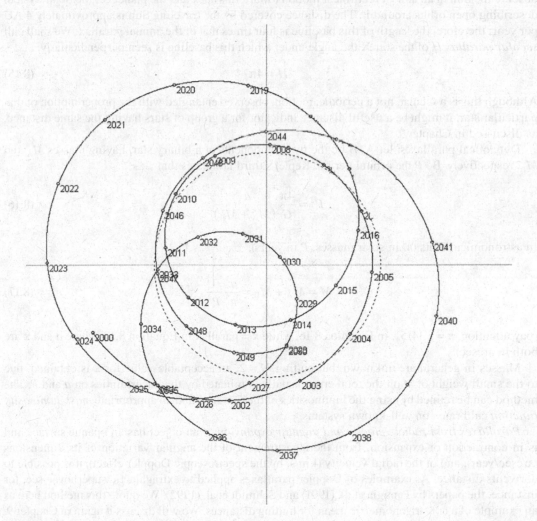

**FIGURE 8.6** Solar System barycenter motion relative to the Sun's disk (the inner light circle), for the years 2000–2051. The vernal equinox is to the right. (Courtesy Larry McNish (RASC)).

The differences between our approximate treatment and a more rigorous one are of the order of $(a_\odot/d)^2$ from the trigonometric point of view alone. Furthermore, a correct treatment in the frame of general relativity would introduce terms depending on the radial velocity $V_r$ and the proper motion $\mu$ of the star, because parallaxes cannot be totally separated from aberration and velocities.

## 8.5  SECULAR AND DYNAMICAL PARALLAXES

The previous discussion has shown the great difficulties and limitations of measuring distances in the Universe by direct trigonometric means. Indirect criteria have been devised, quite often having only a statistical validity, not providing the value for a particular star but only for a group or an association. In Chapters 16 and 17, we will examine some photometric and spectroscopic distance indicators; here, we extend the concept of trigonometric parallaxes to encompass the secular movement of the Solar System and the consequences of Kepler's third law to binary stars.

*Secular parallaxes*: the Solar System moves with respect to the ensemble of the nearby stars with a velocity of approximately 20 km/s toward a point named *apex* of the solar motion, situated in the constellation Lyra. A hypothetic observer, at rest in the frame of this group of nearby stars, would observe the Sun in an almost rectilinear motion toward this apex and the planets of the Solar System describing open orbits around it. The distance covered by the traveling Sun is approximately 4 AU per year; therefore, the length of this baseline is four times that of the annual parallax. We shall call *secular parallax H* of the star X the angle under which this baseline is seen perpendicularly:

$$H \approx 4\pi \tag{8.15}$$

Although this is a secular, not a periodic, motion, observed entangled with the proper motion of that particular star, it might be a useful distance indicator for a group of stars having the same distance, as discussed in Chapter 9.

Dynamical parallaxes: let A, B be the two components of a binary star, having masses $M_A$ and $M_B$, respectively. Be $P$ the orbital period. Kepler's third law states that

$$P^2 = \frac{4\pi}{G} \frac{a^3}{(M_A + M_B)} \tag{8.16}$$

In astronomical units ($M$ in solar masses, $P$ in years),

$$M = M_A + M_B = \frac{a^3}{\pi^3} \frac{1}{P^2} \tag{8.17}$$

(pay attention, $\pi = 3.1415...$ in Equation 8.16, while $\pi$ = parallax in Equation 8.17, where $a$ and $\pi$ are both in arcsec).

Masses in general are unknown, but putting $M = 2$, an acceptable value for $\pi$ is obtained, due to the small weight of $M$ on the total error, usually dominated by the uncertainties on $a$ and $P$. The method can be refined by using the luminosities of the two stars and an appropriate *mass-luminosity function* calibrated on well-known systems.

*Parallaxes from radial velocity and angular expansion*: if an object has an opaque surface and is in contraction or expansion, from the measurement of the angular variation of its dimensions (arcsec/year), and of the radial velocity (km/s, by the spectroscopic Doppler effect), it is possible to derive its distance. As examples of Doppler parallaxes applied to extragalactic supernovae see, for instance, the papers by Panagia et al. (1991) and Schmidt et al. (1992). We quote this method here as an example of a non-trigonometric mean for finding distances. We will discuss it again in Chapter 9.

## NOTES

- Aristarchus of Samos (310–230 B.C.) attempted the determination of the ratio between the lunar and solar distances by measuring the angles Moon, Sun and Earth at the first or last quarter. At quadratures, the angle at the Moon must be 90°, the angle at the Earth can be measured and hence the third angle can be calculated; from the lunar parallax, the solar parallax could then be derived. The measurements of Aristarchus were fairly crude; he produced an angle of 87° instead of 89°51′; nevertheless, for the first time it could be concluded that the Sun must be much more distant (and consequently much larger) than the Moon.

- The first reliable annual parallax was determined by Bessel (1838) for the star 61 Cyg, suggested by Piazzi as likely to be near to the Sun because of its very large proper motion of 5″/year. Bessel at Könisgrad, using a so-called *heliograph*, namely, a telescope with a split objective built by Fraunhofer, obtained the value $\pi = 0''.314 \pm 0''.020$, in very good agreement with the modern value of $0''.295 \pm 0''.003$. Almost contemporarily, Struve at Dorpat derived the parallax of Vega ($\pi = 0''.261 \pm 0''.025$, twice as large as the contemporary value $0''.134 \pm 0''.006$) with an excellent 9-inch reflector made again by Fraunhofer, while Henderson at the Cape measured that of the triple star $a$ Cen with a meridian quadrant. The value of Henderson too was decidedly larger that the contemporary value. The third star and faintest star of the trio, Proxima Cen, some 2° away from A and B, holds the record of being the closest star, with $\pi = 0''.7685 \pm 0''0.0002$. Proxima has at least one, and perhaps two, planets orbiting it, the nearest known extraterrestrial planets.

- The determination of stellar parallaxes was very slow, even when photographic plates became available. Of fundamental importance was the work of Frank Schlesinger at Yerkes, Allegheny and Yale Observatories. His *First General Catalogue of Parallaxes* of 1924 contained 1652 objects. The third edition by Jenkins in 1952 (*General Catalogue of Trigonometric Stellar Parallaxes*) contained 5822 stars, augmented to 6399 in the Supplement of 1966. These numbers suffice to illustrate the tremendous impact of Hipparcos and GAIA. Two more papers on nearby stars deserve to be quoted, namely, the one by Gliese (1969) and its extension by Gliese and Jahreiss (1979), see the References section. For a readable account of the historical developments, see the book by Hirshfeld (Bibliography section).

- For a discussion of asteroids trails in HST images, see the paper by Evans et al. (1998). Further discussions and images can be found in http://www.stsci.edu/communications-and-outreach searching for Evans 1998.

- Since the publication of the Hipparcos and Tycho Catalogs (1997, 2000, see the Web Sites section), doubts were raised about the reliability of part of the data, most noticeably those related to the distance determination of some open clusters, such as the Pleiades. See, for instance, Makarov (2002). An extensive discussion on the defects in the raw data and the provisions introduced in the new reduction can be found in Leeuwen (2005, 2007) and Leeuwen and Fantino (2005).

- On April 23, 2020, NASA's New Horizons conducted the first interplanetary parallax experiment. The spacecraft's distance from Earth was approximately $7 \times 10^9$ km, providing the largest baseline ever available to humankind. Two nearby stars were imaged by the on-board camera, namely, Wolf 359 and Proxima Centauri. Their position was distinctively different from that observed by ground telescopes. For details, see https://www.nasa.gov/feature/nasa-s-new-horizons-conducts-the-first-interstellar-parallax-experiment.

## EXERCISES

1. Describe the diurnal parallax in terms of Cartesian coordinates. In the geocentric reference system centered in C, let

$$\mathbf{r}\left(a_\oplus \sin\pi\cos\alpha\cos\delta,\ a_\oplus \sin\pi\sin\alpha\cos\delta,\ a_\oplus \sin\pi\sin\delta\right)$$

be the vector to the object P and $\mathbf{R}$ ($\rho a_\oplus\cos\phi'\cos CO,\ \rho a_\oplus\cos\phi'\sin CO,\ \rho a_\oplus\sin\phi'$) the vector to the observer O, at any given instant $ST$ of sidereal time. The topocentric vector position of P is $\mathbf{r}' = \mathbf{r} - \mathbf{R}$, with components

$$x' = r'\cos\alpha'\cos\delta' = a_\oplus \sin\pi\cos\alpha\cos\delta - \rho a_\oplus\cos\phi'\cos OC$$

and corresponding expressions for $y'$ and $z'$. The topocentric equatorial coordinates can be computed in the usual way:

$$\alpha' = \arctan\frac{y'}{x'}, \quad \delta' = \arctan\frac{z'}{\sqrt{x'^2 + y'^2}}$$

with due consideration of the quadrant of $\alpha'$ (add $12^h$ if $x' < 0$).

A worked example of the inverse procedure for a low Earth orbiting satellite can be found in the book by Green (Bibliography section).

2. Derive the effect of the diurnal parallax on the ecliptic coordinates. Determine the ecliptic coordinates of the geocentric zenith $(\lambda_Z', \beta_Z')$ at the instant of observation, and from Equation 8.9, compute

$$\begin{cases} \Delta\lambda' = -\rho\dfrac{\pi_\odot''}{d(\mathrm{AU})}\sec\beta\sin\left(\lambda_{Z'} - \lambda\right)\cos\beta_{Z'} \\[4mm] \Delta\beta'' = -\rho\dfrac{\pi_\odot''}{d(\mathrm{AU})}\left(\sin\beta_{Z'}\cos\beta - \cos\beta_{Z'}\sin\beta\cos\left(\lambda_{Z'} - \lambda\right)\right) \end{cases}$$

3. Compute the approximate position of the Solar System barycenter for the year 2000. Table 8.2 shows the heliocentric ecliptic position of Earth, Jupiter and Saturn with respect to the ecliptic and mean equinox at J2000.0 every 2 months since December 27, 1999 UT1 = $0^h$, until December 21, 2000. Table 8.2 shows that during that period, the longitude of Jupiter gradually reached and overtook that of Saturn. From the orbital elements of the two planets, we could notice that the distance of Saturn from the Sun was in that period decidedly smaller than the semi-major axis of its orbit. From these data, and from the masses of the planets, we can calculate the $(x, y, z)$ coordinates in the ecliptic heliocentric system and then by rotation the equatorial ones $(X, Y, Z)$.

4. Discuss the influence of the other planets. To verify your calculations, consider Table 8.3, which contains the accurate barycentric coordinates of the center of the Sun given by Multiyear Interactive Computer Almanac (MICA, US Naval Observatory, 1990–2005) taking into account all other planets. The table has been extended until January 2003 in order to show that $X$ changes sign with the passage of Jupiter in the second quadrant.

**TABLE 8.2**

**Heliocentric Ecliptic Positions of Earth, Jupiter and Saturn with Respect to the Mean Ecliptic and Equinox of J2000.0**

| Date (UT1 = 0$^h$) | Longitude | Latitude | Distance (UA) |
|---|---|---|---|
| | | **Earth** | |
| 1999 Dec 27 | 94°46′25″.4 | −0°00′00″.6 | 0.983441058 |
| 2000 Feb 25 | 155 41 26.6 | −0 00 00.5 | 0.989699191 |
| 2000 Apr 25 | 215 05 48.3 | +0 00 00.2 | 1.006041897 |
| 2000 Jun 24 | 272 47 36.7 | +0 00 00.9 | 1.016436859 |
| 2000 Aug 23 | 330 10 09.3 | +0 00 00.5 | 1.011210382 |
| 2000 Oct 22 | 28 53 16.4 | −0 00 00.6 | 0.995157692 |
| 2000 Dec 21 | 89 25 07.2 | −0 00 01.1 | 0.983751820 |
| | | **Jupiter** | |
| 1999 Dec 27 | 35°47′36″.0 | −1°10′46″.2 | 4.964672267 |
| 2000 Feb 25 | 41 15 16.0 | −1 07 15.9 | 4.973298467 |
| 2000 Apr 25 | 46 41 39.7 | −1 03 10.1 | 4.983834442 |
| 2000 Jun 24 | 52 06 32.9 | −0 58 31.5 | 4.996175289 |
| 2000 Aug 23 | 57 29 42.4 | −0 53 23.3 | 5.010203092 |
| 2000 Oct 22 | 62 50 56.6 | −0 47 49.0 | 5.025787816 |
| 2000 Dec 21 | 68 10 05.3 | −0 41 52.0 | 5.042780745 |
| | | **Saturn** | |
| 1999 Dec 27 | 45°31′24″.7 | −2°18′23″.1 | 9.185031429 |
| 2000 Feb 25 | 47 41 40.9 | −2 16 11.0 | 9.172285980 |
| 2000 Apr 25 | 49 52 18.1 | −2 13 46.7 | 9.159972260 |
| 2000 Jun 24 | 52 03 15.5 | −2 11 10.3 | 9.148103239 |
| 2000 Aug 23 | 54 14 32.5 | −2 08 22.2 | 9.136696298 |
| 2000 Oct 22 | 56 26 08.3 | −2 05 22.3 | 9.125772846 |
| 2000 Dec 21 | 58 38 02.3 | −2 02 10.9 | 9.115348720 |

**TABLE 8.3**

**Barycentric Coordinates (J2000.0) of the Center of the Sun from December 1999 to January 2003**

| Date (UT1 = 0$^h$) | X (UA) | Y (UA) | Z (UA) |
|---|---|---|---|
| 1999 Dec 27 | −0.007168681 | −0.002605028 | −0.000904252 |
| 2000 Feb 25 | −0.006833660 | −0.002999961 | −0.001082030 |
| 2000 Apr 25 | −0.006467923 | −0.003369512 | −0.001249870 |
| 2000 Jun 24 | −0.006069952 | −0.003709978 | −0.001406332 |
| 2000 Aug 23 | −0.005645329 | −0.004017280 | −0.001549221 |
| 2000 Oct 22 | −0.005198099 | −0.004292663 | −0.001678861 |
| 2000 Dec 21 | −0.004726104 | −0.004534023 | −0.001794670 |
| 2001 Jan 20 | −0.004481906 | −0.004639521 | −0.001846341 |
| 2002 Jun 14 | −0.000169340 | −0.004961843 | −0.002100710 |
| 2002 Jul 14 | +0.000069611 | −0.004897186 | −0.002079631 |
| 2002 Aug 13 | +0.000304080 | −0.004825351 | −0.002055347 |
| 2003 Jan 10 | +0.001422324 | −0.004353087 | −0.001884571 |

# 9 Radial Velocities and Proper Motions

The movements of the stars with respect to the observer reveal themselves in two different ways: as radial velocities $\mathbf{V_r}$ along the line of sight and as angular tangential motions on the celestial sphere. The radial velocity is detected and measured spectroscopically, thanks to the Doppler effect, directly in kilometer per second or other convenient units. The second one, named proper motion $\boldsymbol{\mu}$, is seen as a variation in time of the position of the star on the celestial sphere; it is expressed as angular velocity per unit time, for instance, milliarcsec per year or arcsec per century. It is much easier to measure proper motions than parallaxes and radial velocities because the displacements, and therefore the precision of the measurements, increase with the passage of time. Historically, it was E. Halley in 1718 who discovered the first proper motion, namely that of Arcturus ($\alpha$ Boo), by comparing his own coordinates with those given by Hipparchus two millennia earlier: after allowance for precession, the two determinations differed by more than 1°, too large to be attributed to errors.

Radial velocities could be measured only toward the end of the 19th century, after Christian Doppler discovered the effect that bears his name and spectroscopy entered the astronomical field (see Chapter 17). The intrinsic precision of the radial velocities does not depend on the distance of the star, but only on its apparent brightness, namely, on the photon flux collected by the telescope.

From a formal point of view, let $\mathbf{r}(t_0)$ be the heliocentric (or better yet barycentric) position vector of a star at a given initial epoch $t_0$, and $r_0 = 1/\sin \pi_0 \approx 1/\pi_0$ its distance in terms of the trigonometric parallax $\pi_0$. In matrix form,

$$
\mathbf{r}(t_0) = \begin{bmatrix} r_0 \cos \alpha_0 \cos \delta_0 \\ r_0 \sin \alpha_0 \cos \delta_0 \\ r_0 \sin \delta_0 \end{bmatrix}
$$

The instantaneous velocity vector $\mathbf{V}(t_0)$ is given by differencing the previous equation:

$$
\mathbf{V}(t_0) = \dot{\mathbf{r}}(t_0) = \begin{bmatrix} -\sin\alpha_0 \cos\delta_0 & -\cos\alpha_0 \sin\delta_0 & \cos\alpha_0 \cos\delta_0 \\ \cos\alpha_0 \cos\delta_0 & -\sin\alpha_0 \sin\delta_0 & \sin\alpha_0 \cos\delta_0 \\ 0 & \cos\delta_0 & \sin\delta_0 \end{bmatrix} \begin{bmatrix} r_0\, \mu_{\alpha 0} \\ r_0\, \mu_{\delta 0} \\ \dot{r}_0 \end{bmatrix}
$$

where $\dot{r}_0$ is the modulus of radial velocity, and $\mu_{\alpha 0}$ and $\mu_{\delta 0}$ are the proper motions in right ascension and declination at the initial epoch, respectively. In the practical application of the formula, care must be taken to the dimensions of the proper motion in RA: if the catalog gives the proper motion in RA in arc units, so that the tabulated value is $\mu_\alpha = 15\mu_\alpha \cos \delta$, the previous matrix becomes

$$
\dot{\mathbf{r}}(t_0) = \begin{bmatrix} -\sin\alpha_0 & -\cos\alpha_0 \sin\delta_0 & \cos\alpha_0 \\ \cos\alpha_0 & -\sin\alpha_0 \sin\delta_0 & \sin\alpha_0 \cos\delta_0 \\ 0 & \cos\delta_0 & \sin\delta_0 \end{bmatrix} \begin{bmatrix} r_0\, \mu_{\alpha 0} \\ r_0\, \mu_{\delta 0} \\ \dot{r}_0 \end{bmatrix}
$$

In the same manner, the velocity is sometimes required in kilometer per second, in other occasions in AU per day or in AU per century. For instance, the already quoted NOVAS software (see the Web Sites section) writes

$$
\begin{bmatrix}
15 s r_0 \, f \mu_{\alpha 0} \\
s r_0 \, f \mu_{\delta 0} \\
\dot{r}_0 \, f k
\end{bmatrix}
$$

where the proper motion in right ascension is in unit of time, $f$ is a relativistic correction (if required, in our assumption $= 1$), while $s$ and $k$ convert the units of velocity, namely, $s = 2\pi/(360 \times 3600 \times 36{,}525)$ from arcsec per century to radians per day, $k = 86{,}400/c\tau_A$ from km/s to AU/day.

The position of the star at a different date $T$, but referred to the same epoch, can then be derived from

$$
\mathbf{r}(T) = \mathbf{r}(t_0) + (T - t_0)\dot{\mathbf{r}}_0
$$

However, usually we know either the velocity or the proper motions, rarely both quantities, so that in the following, the radial velocities and the proper motions will be considered separately. It is also clear that the treatment of the system of proper motions and velocities is fundamentally linked to that of precession, a great complication indeed for transferring astrometric catalogs from one epoch to another, if the highest precision is to be maintained.

## 9.1  RADIAL VELOCITIES

The radial velocity is measured through the Doppler effect, namely, through the variation of the wavelength $\lambda$ (or frequency $\nu$) of the radiation, caused by the relative motion $\mathbf{V}_r$ along the line of sight. The velocities of the planets or of the stars of the Milky Way are usually small in comparison with the velocity of the light $c$. Therefore, no sensible error is made by using the pre-relativistic formula:

$$
\mathbf{V}_r = c \frac{\lambda_O - \lambda_S}{\lambda_S} = c \frac{\Delta\lambda}{\lambda} = c \cdot z, \qquad z = \frac{\Delta\lambda}{\lambda} = \frac{\mathbf{V}_r}{c}, \qquad \frac{\lambda_O}{\lambda_S} = 1 + \frac{\mathbf{V}_r}{c} = 1 + z \tag{9.1}
$$

where $\lambda_O$ is the wavelength measured by the observer, and $\lambda_S$ is that in the rest-frame of the source. The use of the letter $z$ to indicate $\Delta\lambda/\lambda$ is fairly widespread. Notice that the radial velocity can be positive ($z > 0$, namely, the wavelength is *red-shifted*) or negative ($z < 0$, namely, the wavelength is *blue-shifted*).

When the velocities exceed say $0.01c$, special relativity must be taken into account through the Lorentz transformations, as already seen in Chapter 7. In several cases, it is advantageous to work in terms of frequency $\nu$ instead of $\lambda$ (in vacuum, $\nu = c/\lambda$, $d\nu = -c d\lambda/\lambda^2$); the classical formula would be $1 + z = \nu_S/\nu_O$.

To derive the relativistic formula, let O be an observer in uniform motion with velocity $\mathbf{V}$ with respect to an inertial observer S, and let $\tau$ be the time measured by O (proper time) and $t$ the time measured by S. Between the two times, the following relation applies:

$$
d\tau = \sqrt{1 - \frac{\mathbf{V}^2}{c^2}} \, dt < dt \tag{9.2}
$$

This effect is also named *slowing down* of the moving clock. Therefore, the frequency of the light measured by O and the frequency of the light measured by S are related by

$$
\nu_O = \nu_S \sqrt{1 - \frac{\mathbf{V}^2}{c^2}} \left(1 - \frac{1}{c}\mathbf{V} \cdot \mathbf{n}\right)^{-1} \tag{9.3}
$$

where $n$ is the normal to the wavefront. Therefore, special relativity foresees a transverse Doppler effect (when **n** is perpendicular to **V**), which is not present in the pre-relativistic formula. In other words, the whole velocity vector, and not only its radial component, enters in the observed frequency displacement.

After some simple manipulation of Equation 9.3, we obtain the series expansion:

$$z = \frac{\Delta\lambda}{\lambda} = -1 + \sqrt{\frac{(1+\mathbf{V})/c}{(1-\mathbf{V})/c}} \approx \frac{\mathbf{V}}{c} + \frac{1}{2}\left(\frac{\mathbf{V}}{c}\right)^2 + \cdots \qquad (9.4)$$

which shows that actually the Doppler effect is not symmetric in the sign of **V**. See the "Exercises" section for a fuller discussion.

Inside the Solar System, velocities of few tens of km/s prevail, with the notable exception of comets and asteroids skimming the surface of the Sun; in this case, their heliocentric velocity can exceed 700 km/s. The classical formula is therefore adequate for most applications; however, the extreme precision (say ±1 mm/s) with which the velocity of a spacecraft, possibly orbiting a planet, can be measured, implies that general relativity must be taken into account in expressing the metric of space-time.

The radial velocities of the normal stars of the Galaxy rarely reach 500 km/s. Among the nearest stars, velocities higher than 50 km/s are seldom encountered. There are notable exceptions, such as Barnard's star, which has a radial velocity of −108 km/s with respect to the Sun. The group of nearby high velocity stars is extremely important for the comprehension of the overall kinematics and dynamics of the Milky Way (see, e.g., Binney and Merrifield (1998) in the Bibliography section). For the great majority of stars, a precision of at best ±100 m/s is reached. Only in favorable cases and with refined techniques, e.g., in searches for extra-solar planets by means of radial velocity variations, precisions better than ±1 m/s are achieved. The limiting factors are on one side the technical limitations and, on the other, the structure of the stellar spectral lines themselves. Therefore, the classical formula is usually adequate. There are cases, however, where special relativity must be applied, e.g., for the variable star SS433 or for the expanding gaseous envelopes of explosive variables (novae, supernovae) and velocities of tens of thousands kilometer per second are encountered.

For nearby galaxies, negative velocities can be found; for example, M31 in Andromeda has $\mathbf{V}_r = -182$ km/s.

For galaxies beyond a distance roughly coinciding with that of the cluster of galaxies in Virgo, whose velocity is approximately +1000 km/s, only positive velocities are encountered, with an amount $z_c$ increasing with the distance, as was discovered by Edwin Hubble around 1930 (see the Bibliography section). This observational effect, at the basis of cosmology, is referred to as the *expansion of the Universe*. According to Hubble's measurements, the radial velocities of distant galaxies increase proportionally to their distances; the proportionality constant (Hubble's constant) is indicated with $H_0$. Its value, from the original determination around 500 km/s·Mpc, was revised several times and now is estimated around 70 km/s·Mpc. The inverse of $H_0$, being dimensionally a time, is referred to as the *age of the Universe*; its current value is approximately 13.8 Gyr. In 2019, the IAU deliberated to rename $H_0$ as Hubble–Lemaître constant, to acknowledge the great contributions of the Belgian priest and mathematician Georges Lemaître to cosmology. The spectral lines of the distant galaxies and other objects of cosmological significance, e.g., the Quasi Stellar Objects (QSOs, also named *quasars* from the original discoveries as quasi-stellar radio sources), are redshifted by great amounts. Values of $z_c$ as high as 10 have already been measured, with an immense displacement of the wavelength of the spectral lines. For instance, at $z_c = 6$ the ultraviolet spectral line Lyman-$\alpha$ of hydrogen, having $\lambda = 1216$ Å in the laboratory, is observed in the near infrared at $\lambda = 8512$ Å. To derive the radial velocity from such high redshifts, the relativistic formula must be employed; however, in cosmology the simple connection between $z_c$ and velocity breaks down, because the expansion of the universe is an expansion of the metric itself and it does not reflect the motion of the source in a fixed coordinate frame. Nor it is possible to reason in terms of three spatial coordinates and one time coordinate. For these reasons, the radial velocities derived from Equation 9.4 are better called *indicative* velocities.

General relativity predicts another redshift for the light emitted by atoms in a gravitational field, e.g., on the surface of a star. From a spherical source of mass $M$ and radius $R$, this gravitational redshift is expressed by

$$\frac{\lambda}{\lambda_0} = \sqrt{1 - 2\frac{GM}{c^2 R}} = \sqrt{1 - \frac{r_S}{R}}$$

where $G$ is the gravitational constant and $r_S$ the already introduced Schwarzschild's radius (see Chapter 7). If the radius $R \gg r_S$, then

$$z_g = -1 + \frac{1}{\sqrt{1 - \frac{2GM}{c^2 R}}} \approx \frac{GM}{c^2 R} = \frac{r_S}{2R} \tag{9.5}$$

On the Sun, the effect amounts to 0.64 km/s. On the white dwarf Sirius B, having a mass approximately equal to the Sun's, but radius only 80% of that of the Earth, the effect is correspondingly much larger. The difficulty is that the gravitational effect cannot be separated from the radial velocity, unless a large number of stars well distributed over the celestial sphere can be used to average out the radial velocity component. For instance, Falcon et al. (2010), using a number of nearby white dwarfs, found an average value for $z_g$ corresponding to 32.6 ± 1.2 km/s, with an average mass of 0.65 solar masses, and indicative radii around 0.01 solar radii. Therefore, the measurement of the gravitational redshifts from white dwarfs constitutes another test of the correctness of general relativity in the weak gravitational field approximation.

In the limit of $R$ shrinking to the Schwarzschild radius $r_s$ (namely, at the surface of a black hole), the light will be infinitely reddened.

The closest massive black hole (MBH) is at the center of the Milky Way, at a distance of about 8.1 kpc and a mass $4 \times 10^6$ solar masses. Its Schwarzschild radius is about 0.08 AU, corresponding to 20 microarcsec apparent diameter. It is coincident with a very compact, variable X-ray, infrared and radio source called SgrA*, which is surrounded by a very dense cluster of orbiting stars. The MBH profoundly affects the dynamics of the surrounding stars. One in particular, named S2, was observed for radial velocity and proper motion over 26 years, including the pericenter date, which occurred in 2018. The orbital velocity of the star is approximately 7650 km/s. The data have permitted to measure the combined gravitational redshift and relativistic transverse Doppler effect of $c \cdot z \approx 200$ km/s. Moreover, it can be concluded that the S2 data are inconsistent with pure Newtonian dynamics. The results quoted here for SgrA* have been published by the GRAVITY collaboration (see the "Notes" section for references and further details.

If an object possesses kinematical, cosmological and gravitational redshifts, the combination of the different effects is given by

$$1 + z_{\text{tot}} = (1 + z_V)(1 + z_c)(1 + z_g), \qquad z_{\text{tot}} \approx z_V + z_c + z_g \tag{9.6}$$

where the last approximate equality holds true only for small redshifts.

When measuring radial velocities, the observations must be corrected for the annual and diurnal motions of the observer. The necessary formulae are easily determined if the wanted precision is around 100 m/s. The heliocentric velocity of the Earth $V_\oplus$ varies between 29.3 km/s at aphelion and 30.3 km/s at perihelion; its projection toward a direction of ecliptic coordinates $(\lambda, \beta)$ is

$$V_\oplus = -Kc\cos\beta\left[\sin(\lambda - \lambda_\odot) + e\,\sin(\lambda_\Pi - \lambda)\right] \tag{9.7}$$

(see Chapter 7), where $Kc = 29.79$ km/s, $\lambda_\odot$ is the longitude of the Sun in that particular date, $\lambda_\Pi$ the longitude of the perigee and $e$ the eccentricity of Earth's orbit. Notice that the term in eccentricity, amounting at most to 0.50 km/s, is essentially constant for any given star, with a secular variation due to the very small changes in $\lambda_\Pi$ and $e$.

Another way of expressing this correction is by making use of equatorial coordinates of the line of sight and of the Cartesian components of the Earth's velocity, given in AU per day for each day by the following low precision formula (see the *Astronomical Almanac, Table E7*):

$$V_\oplus = 1731.5\left(\dot{X}\cos\alpha\cos\delta + \dot{Y}\sin\alpha\cos\delta + \dot{Z}\sin\delta\right)\text{km/s} \tag{9.8}$$

For instance, for JD2458480.5 (December 28, 2018) the values of $(\dot{X}, \dot{Y}, \dot{Z})$ were (−0.01739, −0.0017, −0.00017); for JD 2458720.5 (August 28, 2019), the values were (+0.01632, −0.00420, −0.00018) AU/day, respectively.

Regarding the diurnal rotation, the velocity of the observer at the equator is approximately 0.465 km/s: therefore, for a generic geocentric latitude $\phi'$, the projection of this velocity on the line of sight to a star of declination $\delta$ and hour angle $HA$ will be

$$V_{\text{rot}} = 0.465\cos\phi'\cos\delta\sin HA\,\text{km/s} \tag{9.9}$$

with the complication that $V_{\text{rot}}$ can appreciably change on very long exposures. Furthermore, for better precision, one should take into account the elevation of the observer. This is particularly true for the HST, for which a complete knowledge of the geocentric velocity vector is required. Finally, the wanted heliocentric radial velocity is obtained by

$$\mathbf{V} = \mathbf{V}_r - \mathbf{V}_\oplus - \mathbf{V}_{\text{rot}}$$

Should one need a precision better than 100 m/s, more accurate formulae must be employed, e.g., for exoplanet search by the radial velocity method, as already mentioned. At this level of precision, one should refer the velocities to the Earth–Moon and Solar System barycentric observers, the relative velocities with respect to the geocentric and the heliocentric observers being both of approximately 13 m/s.

## 9.2 PROPER MOTIONS

Let us consider first the proper motion of the nearer stars. Be S the heliocentric observer and X a generic star at a distance $d$ with heliocentric velocity **V**, at a certain date (see Figure 9.1).

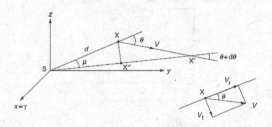

**FIGURE 9.1** Heliocentric velocity and proper motion of star X. Notice that after 1 year, when the star reaches X′, the angle between the line of sight and V will change by $d\theta = -\mu$.

Owing to the enormous distances, for many decades or centuries the velocities can be considered as rectilinear and uniform, apart from exceptions such as Barnard's star. Expressing the modules in kilometer per second and indicating with $n$ the number of seconds in 1 year, after 1 year the star will be seen in X', having traveled a course of $V n$ km along a rectilinear path forming an unknown angle $\theta$ with axis SX. On the plane tangent to the celestial sphere, the star will appear to have moved by the small angle:

$$\mu = n \frac{V}{d} \sin\theta \text{ (radians)} \tag{9.10}$$

The component of $\mathbf{V}$ perpendicular to the line of sight, $\mathbf{V}_t = \mathbf{V} \sin\theta$, is called transverse or *tangential* velocity. The corresponding apparent angular velocity $\mu$ is said to be *proper motion* of the star X; it is commonly measured in arcsec per year, or arcsec per century or other convenient units. By using the parallax $\pi$ in arcsec instead of the distance $d$ in kilometers and taking into account the appropriate conversion factors (namely, 1 km/s = 0.21095 AU/year, 1 AU/year = 4.74045 km/s), we derive the following values of the modules:

$$V_t = V \sin\theta = 4.740 \frac{\mu}{\pi} \text{ km/s}, \qquad \mu = \frac{\pi}{4.740} V_t \text{ arcsec/year}$$

Barnard's star has a proper motion as high as 10″/year; very few stars have $\mu$ larger than 2″/year. The nearer stars have, in general, the larger proper motions, but the vice versa is not true; many nearby stars have small motions because of the orientation of their velocity vectors. For the nearer stars, the annual ellipse of parallax is trailed by the proper motion, as evidenced in Figure 9.2.

If the radial velocity of the star X can be measured from the Doppler effect, then

$$V_r = V \cos\theta = c \frac{\Delta\lambda}{\lambda} = 4.740 \frac{\mu}{\pi} \cos\theta \text{ km/s} \tag{9.11}$$

(the nonrelativistic formula being certainly valid) and then the full velocity vector of the star can be reconstructed.

It is convenient to consider the proper motion as a vector $\boldsymbol{\mu}(\mu, q)$ on the plane tangent to the celestial sphere, with modulus $\mu$ expressed in angular units (i.e., arcsec per year) along the great circle XX″ and direction expressed by a position angle $q$ measured from the north through the East ($0° \leq q < 360°$). Alternatively, the two equatorial components ($\mu_\alpha$, $\mu_\delta$) can be given; attention must

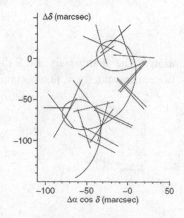

**FIGURE 9.2**   The proper motion of $\mu = 0″.03$/year trails the annual parallax ellipse of a nearby star having $\pi = 0″.08$: the segments give the instantaneous positions and associated errors; the continuous line is the best fitting path. (Adapted from Hipparcos satellite data.)

be paid to the units of $\mu_\alpha$, because the angular distance between two successive positions of the star must be measured along the great circle passing through XX''; often though, $\mu_\alpha$ is derived by the difference between two successive right ascensions and expressed in seconds of time.

Therefore,

$$\mu_\alpha(s/y) = \frac{1}{15}\mu(''/y)\sin q \sec\delta, \qquad \mu_\delta(''/y) = \mu(''/y)\cos q \qquad (9.12)$$

$$\mu_\alpha(''/y) = 15\mu(s/y)\cos\delta, \qquad \tan q = \frac{\mu_\alpha(''/y)}{\mu_\delta(''/y)}$$

while the tangential velocity components are expressed as

$$V_{t\alpha} = 4.740\frac{\mu_\alpha(''/y)}{\pi}, \qquad V_{t\delta} = 4.740\frac{\mu_\delta(''/y)}{\pi} \qquad (\text{km/s})$$

From the previous relationships, we also derive

$$\mu = \sqrt{\mu_\alpha^2 + \mu_\delta^2} = \frac{\pi}{4.740}\sqrt{t_\alpha^2 + t_\delta^2}, \qquad \tan q = \frac{\mu_\alpha}{\mu_\delta} = \frac{t_\alpha}{t_\delta}$$

Namely, the position angle of the proper motions is the same as that of the tangential velocity and is independent of the parallax of the star. This property, which applies also to the distribution function of the proper motions in a given small area of the sky, has been used to study the stellar kinematics in several investigations where only the position angles were known. We will discuss this argument further in the following sections. Again, with reference to Figure 9.3, from the spherical triangle XPX'' we notice that

$$\cos\delta \cdot \sin q = \cos\delta'' \cdot \sin q''$$

In other words, the quantity $\cos\delta \sin q$ is conserved during the movement of the star, implying that

$$\frac{d}{dt}(\cos\delta \cdot \sin q) = 0, \qquad \dot{q} = \tan q \cdot \tan\delta \cdot \dot{\delta} \qquad (9.13)$$

a property that will be utilized in the following.

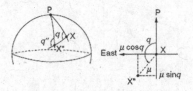

**FIGURE 9.3** The proper motion resolved in two equatorial components.

## 9.3 VARIATION OF THE EQUATORIAL COORDINATES

In order to find the effect of the proper motion on the celestial coordinates of a given star X, a first-order approximation will be sufficient for short time intervals; $\Delta t$ years after the epoch $t_0$, the mean coordinates will be

$$\alpha(t_0 + \Delta t) = \alpha(t_0) + \mu_\alpha \Delta t + (m + n \sin\alpha \tan\delta)\Delta t, \tag{9.14}$$

$$\delta(t_0 + \Delta t) = \delta(t_0) + \mu_\delta \Delta t + (n \cos\alpha)\Delta t$$

The formula includes the general precession (do not confuse in this section the constant of precession $n$ with the number of seconds in 1 year), with due attention to the units (the sign of $\Delta t$ can obviously be reversed). However, to be rigorous we must note that $(\mu_\alpha, \mu_\delta)$ do vary in time even if the velocity is rectilinear and constant. This happens for two distinct reasons: on the one hand, the reference system rotates because of precession; on the other, the changing perspective alters the apparent length of equal arcs, as shown (with great exaggeration) in Figure 9.4.

Therefore, consider the successive terms in Equation 9.14:

$$\alpha(t_0 + \Delta t) = \alpha(t_0) + \mu_\alpha \Delta t + (m + n \sin\alpha \tan\delta)\Delta t + \frac{1}{2}\dot\mu_\alpha (\Delta t)^2 \tag{9.15}$$

$$\delta(t_0 + \Delta t) = \delta(t_0) + \mu_\delta \Delta t + (n \cos\alpha)\Delta t + \frac{1}{2}\dot\mu_\delta (\Delta t)^2 \tag{9.16}$$

The time derivatives of $(\mu_\alpha, \mu_\delta)$ are composed of two terms: the first term, due to precession and independent from distance and radial velocity, is the dominating one. The second term, due to the variation of the projection of the velocity on the line of sight, can be written down explicitly only when the parallax and the radial velocity are both known.

The components due to precession, which clearly modify only the direction of $\mathbf{\mu}$ but not its modulus ($d\mu/dt = 0$), are found in a somewhat elaborate manner which is not given here (see Zagar in the Bibliography section and also the "Exercises" section of Chapter 6):

$$\dot\mu_{\alpha,p} = \frac{n}{R''}\left(\mu_\alpha \cos\alpha \tan\delta + \mu_\delta \sin\alpha \sec^2\delta\right) \; (\text{arcsec/year}) \tag{9.17}$$

$$\dot\mu_{\delta,\,p} = -\frac{n}{R''}\mu_\alpha \sin\alpha \; (\text{arcsec/year})$$

where the general precession in declination $n$, $\mu_\alpha$ and $\mu_\delta$ are in arcsec/year.

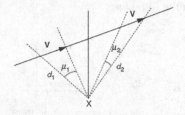

**FIGURE 9.4** Perspective acceleration of a star X in uniform motion. The proper motion changes with time even if vector **V** stays constant. The perspective effect is obviously present also in the radial velocity.

Regarding the second intrinsic term, let us take the time derivatives of Equation 9.12, with the caution of remembering that all derivatives must be in circular units (for instance, $\dot{\delta} = \mu_\delta / R''$) Then, recall Equation 9.13 and that

$$\dot{\mu} = \frac{V}{4.740}[\dot{\pi}\sin q + \pi \cos q \dot{q}], \qquad \dot{\pi} = -\frac{\pi^2 V_{\rm r}}{4.740} \qquad (9.18)$$

After some manipulation, it can be shown that these intrinsic components are

$$\dot{\mu}_{\alpha,i} = \frac{1}{R''}\left(2\mu_\alpha \mu_\delta \tan\delta - 0.422 V_{\rm r}\pi\mu_\alpha\right),$$

$$\dot{\mu}_{\delta,i} = -\frac{1}{R''}\left(\mu_\alpha^2 \sin\delta \cos\delta + 0.422 V_{\rm r}\pi\mu_\delta\right)$$

As before $\mu_\alpha$ and $\mu_\delta$ are in arcsec/year, $\pi$ in arcsec and $V_{\rm r}$ in km/s.

For long time intervals, we must include in Equations 9.15 and 9.16 the terms $1/2\left(\dot{\mu}_{\alpha,p} + \dot{\mu}_{\alpha,i}\right)T^2$ and $1/2\left(\dot{\mu}_{\delta,p} + \dot{\mu}_{\delta,i}\right)T^2$. Recourse to even more complicated procedures (for instance, considering another term in the series expansion) is seldom necessary, unless very long time spans are considered, or utmost precision is sought.

The discussion made in Chapters 7–9 of precession, nutation, parallax, gravitational light deflection, radial velocities and proper motions has given the possibility to transfer the stellar coordinates from an initial epoch to a wanted date, although in an approximate manner, by superimposing one after the other the several effects. Such approximation is sufficient in many cases; however, if utmost precision is required, a more elaborate procedure must be followed, as briefly discussed in Chapter 10. For Solar System objects, further elements will be provided in Chapter 13.

## 9.4  INTERPLAY BETWEEN PROPER MOTIONS AND PRECESSION CONSTANTS

The proper motion of a star can be derived from the difference of its equatorial coordinates at two successive dates, provided the effects of precession and nutation are first removed; namely, the two coordinate sets must be reported to the same equinox before taking the difference. Therefore, the uncertainties in the precession constants will enter into the uncertainty of the proper motion. Such uncertainty is not so important for a single star, but instead for the system of proper motions. For instance, the FK5 could be affected by a spurious rotation at the level of about $0''.15$/century. This seemingly small systematic error enters into the knowledge of the overall field of motions and forces of the Milky Way, as will be discussed in a later paragraph. From the other point of view, a consistent model of the distribution of proper motions can lead to an improved determination of the precession constants. It is worth, therefore, looking into the connection between proper motions and precession with more insight.

For simplicity, let us limit the discussion to the first-order formulae 9.14, which can be rewritten as

$$\begin{cases} \mu_\alpha = \dot{\alpha} - (m + n \cdot \sin\alpha \cdot \tan\delta) \\[2mm] \mu_\alpha = \dot{\delta} - n \cdot \cos\alpha \end{cases} \qquad (9.19)$$

where (see Chapter 5) $m = \psi \cos\varepsilon - g$, $n = \psi \sin\varepsilon$, $\psi$ is the lunisolar and $g$ the planetary precession, containing also the very small geodesic precession term $\psi_{\rm g}$, $\psi = \psi_0 + \psi_{\rm g}$ mentioned in Chapter 5. The geodesic precession derives from general relativity, which foresees a very slow rotation of the frame connected with the Earth orbiting around the barycenter of the Solar System, as discussed by de Sitter and Brouwer (1938). According to Lieske et al. (1977), its value is

$$\psi_g = \frac{3}{2}K^2(1-e)n \approx -1''.92/\text{century}$$

where $K$ is the aberration constant in radians, $e$ the eccentricity of the Earth's orbit and $n$ its mean motion in arcsec/century. Note the negative sign; the geodesic precession is a motion of $\gamma$ on the ecliptic opposite to the lunisolar one.

Equation 9.19 shows that the determination of the proper motions of a large number of stars, well distributed over the celestial sphere and in the volume containing the Sun, can lead to the determination of the precessional constants and how the errors on the former propagate on them. The method to determine the precessional constants via the stellar kinematics was discussed, for instance, by Fricke (1977). As an example of how intricate the problem is, one can recall that transforming the proper motions from the pre- to the post-1984 reference frames, namely, from FK4 to FK5, was a truly challenging task, as explained in detail by Hohenkerk et al. (1992).

The following paragraphs will make clear that the system of proper motions contains, in addition to any spurious undetected systematic error, two major components. The first is the reflex of the motion of the Sun with respect to the nearer stars (the so-called Local Standard of Rest, LSR, see also Section 9.7); the second is the overall galactic rotation and its variation with distance and direction (namely, the presence of the $A$ and $B$ constants in Oort's and Lindblad's theory of galactic rotation).

Therefore, all effort must be made to obtain a system of proper motions as precise as possible and free from systematic effects. Recalling that by definition, the reference system ICRF is devoid of rotation, proper motions can be derived with respect to the background of the distant quasars. See, for instance, the paper by Charlot et al. (1995), who found a correction of $-3.00 \pm 0.20$ mas/year to the 1997 IAU lunisolar constant, by combining data provided by VLBI and lunar laser ranging.

The satellite Hipparcos could not produce major improvements to the proper motion system, because it did not reach the wanted orbit, so that its operational life was too short. GAIA is actively pursuing this important task.

## 9.5  ASTROMETRIC RADIAL VELOCITIES

Astrometric satellites such as Hipparcos and a fortiori GAIA allow the determination of radial velocities without the intermediary of the spectroscopy. The advantage of astrometry is that the velocities are not affected by possible systematic errors of spectroscopy (e.g., convective motions in the star's atmosphere, transverse Doppler effect, or gravitational redshift). Seeliger in 1900 and Schlesinger in 1917 had already put forward the idea, but only recently the method has been brought to fruition, as further discussed in the "Notes" section. With reference to Figure 9.4, we notice that the distance to the star changes with the passage of time, so that the radial velocity can be expressed as

$$V_r = -a_\odot \frac{\dot{\pi}}{\pi^2}$$

and as

$$V_r = -a_\odot \frac{\dot{\mu}}{2\pi\mu}$$

where $a_\odot$ is the astronomical unit. The difficulty is to have a reliable determination of $\dot{\pi}$ and $\dot{\mu}$. For instance, for Barnard's star $\dot{\pi}$ amounts to only +34 microarcsec/year.

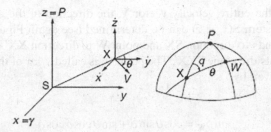

**FIGURE 9.5**   On the celestial sphere, the direction of $V$ corresponds to point W (apex of the stellar motion); the great circle XW corresponds to the plane passing through SX and containing the vector **V**.

Another effect of the passage of time is visible in the relative distance between two stars having parallel and equal velocities but slightly different positions, e.g., two stars belonging to the same cluster; the cluster will appear to contract or to expand with the same amount of the changing parallax.

Although the astrometric method to determine radial velocities is attractive and promising, it also has some systematic biases that must be accurately checked. In particular, care must be taken to remove gravitational perturbations due to nearby stars (it is well known that the majority of stars are members of binary or multiple systems) or to the overall gravitational field of the Milky Way.

## 9.6   APEX OF STELLAR MOTIONS AND GROUP PARALLAXES

Let us consider, in Figure 9.5, the heliocentric equatorial reference system. Let $(\dot{x}, \dot{y}, \dot{z})$ be the velocity components of star X.

These Cartesian components can be expressed in terms of $(V_{t\alpha}, V_{t\delta}, V_r)$ by a suitable rotation of coordinates. Table 9.1 shows the rotation matrix that transforms $(V_{t\alpha}, V_{t\delta}, V_r)$ into $(\dot{x}, \dot{y}, \dot{z})$ and vice versa, for instance,

$$\dot{x} = -V_{t\alpha} \sin\alpha - V_{t\delta} \cos\alpha \sin\delta + V_r \cos\alpha \cos\delta,$$

$$V_{t\alpha} = -\dot{x}\sin\alpha + \dot{y}\cos\alpha$$

and similar for the other components.

Recalling the expression of the tangential velocity $V_t$, we derive the Cartesian equatorial components by the observable quantities:

$$\dot{x} = \frac{4.740}{\pi}\left(\mu_\alpha \sin\alpha + \mu_\delta \cos\alpha \sin\delta\right) + V_r \cos\alpha \cos\delta$$

$$\dot{y} = +\frac{4.740}{\pi}\left(\mu_\alpha \cos\alpha - \mu_\delta \sin\alpha \sin\delta\right) + V_r \sin\alpha \cos\delta \qquad (9.20)$$

$$\dot{z} = +\frac{4.740}{\pi}\mu_\delta \cos\delta + V_r \sin\delta$$

---

**TABLE 9.1**
**Rotation Matrix between $(V_{t\alpha}, V_{t\delta}, V_r)$ and $(\dot{x}, \dot{y}, \dot{z})$**

|  | $V_{t\alpha}$ | $V_{t\delta}$ | $V_r$ |
|---|---|---|---|
| $\dot{x}$ | $-\sin\alpha$ | $-\cos\alpha \sin\delta$ | $\cos\alpha \cos\delta$ |
| $\dot{y}$ | $\cos\alpha$ | $-\sin\alpha \sin\delta$ | $\sin\alpha \cos\delta$ |
| $\dot{z}$ | $0$ | $\cos\delta$ | $\sin\delta$ |

---

From the knowledge of the entire velocity vector **V**, the direction of the motion of the star in the heliocentric reference system S($x$, $y$, $z$) can be determined (see again Figure 9.5). On the celestial sphere, point X corresponds to direction SX and point W to direction XX′ of Figure 9.1, so that the great circle XW represents the plane SXX′. The point W is called *apex* of the stellar motion and has equatorial coordinates given by

$$\left\{ \begin{array}{l} \sin\delta_W = \cos\theta\sin\delta + \sin\theta\cos\delta\cos q \\ \cos(\alpha_W - \alpha)\cos\delta_W = \cos\theta\cos\delta - \sin\theta\sin\delta\cos q \\ \sin(\alpha_W - \alpha)\cos\delta_W = \sin\theta\sin q \end{array} \right. \qquad (9.21)$$

where ($\alpha$, $\delta$) are the initial coordinates of the star. The angular distance between X and W is given by

$$\cos\theta = \sin\delta_W\sin\delta + \cos\delta_W\cos\delta\cos(\alpha_W - \alpha)$$

so that the equatorial Cartesian components of vector **V** can be expressed in a second way:

$$\left\{ \begin{array}{l} \dot{x} = \mathbf{V}\cos XW = \mathbf{V}\cos\delta_W\cos\alpha_W \\ \dot{y} = \mathbf{V}\cos YW = \mathbf{V}\cos\delta_W\sin\alpha_W \\ \dot{z} = \mathbf{V}\cos ZW = \mathbf{V}\sin\delta_W \end{array} \right. \qquad (9.22)$$

In general, however, the parallax is unknown, so that the observations do not provide the length of the arc $\theta$. Nevertheless, the statistical study of the stellar motions in certain areas of the sky has shown the existence of stars having motions converging toward the same apex. Therefore, these stars seem to belong to a group having a common motion, *as if* their space velocities were parallel. We could therefore speak of a *co-moving group* or even of a *stellar current*. However, having convergent motions is not *per se* a proof of parallel velocities, unless the parallaxes are known to be identical or there are reasons to believe that the stars are at the same distance. In such instances (a strong case is when the stars are in the same open cluster, Hyades, Pleiades, etc.), we can obtain the parallax of the group by the determination of the convergent point and of the radial velocities of the stars; a hidden assumption is that the diameter of the cluster does not vary in time (see Hanson, 1975). We have indeed

$$V\cos\theta = V\left[\sin\delta_W\sin\delta + \cos\delta_W\cos\delta\cos(\alpha_W - \alpha)\right] = V_r, \qquad \pi = \frac{4.740\mu}{V_r\tan\theta}$$

Let us consider in particular the very near open cluster of the Hyades, whose distance is around 45 pc and whose extension in space is approximately 10 pc. The convergence of the proper motions of the stars to a common apex was ascertained long ago, as shown in Figure 9.6, reproduced here for its historical interest.

The data obtained by the Hipparcos satellite has provided a much improved knowledge (see Perryman et al., 1998). This is one of the few cases where the tridimensional structure can be obtained directly from the observations. Therefore, the Hyades play a central role in the calibration of several relationships, e.g., the Hertzsprung–Russell diagram (see, e.g., de Bruijne et al., 2001). Chapter 16 will give recent GAIA results.

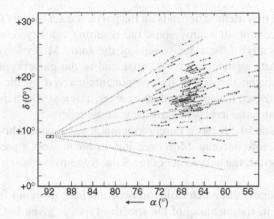

**FIGURE 9.6**   The convergence of the proper motions of the Hyades stars. Notice the great angular distance between the cluster and its convergent point. (Adapted from data by Pearce, 1995).

## 9.7   THE PECULIAR MOTION OF THE SUN AND THE LOCAL STANDARD OF REST

In this section, we complete the discussion of the secular parallaxes initiated in Chapter 8. The observed proper motions contain the reflex of the motion of the Sun with respect to the nearer stars. Therefore, from the ensemble of the motions of these stars we can construct a velocity reference system (not a spatial one!) to which the motion of the Sun can be referred. In 1783, W. Herschel had already laid down the foundations of this method, utilizing the position angles $q$ of 12 stars. Indeed, $q$ is independent from the parallax and coincides with the position angle of the tangential velocity. Herschel's method can be easily visualized: draw on the celestial sphere the great circles defined by the proper motions, and consider the semicircle oriented as the motion itself. All these semicircles will intersect, within the errors, in a point (more realistically, in a small area) which is the antapex of the solar motion; the modulus of the velocity of the Sun will remain undetermined by this method.

Let us now examine the case of a group of $N$ nearby stars, for which the four quantities ($\pi$, $\mu$, $q$, $V_r$) and therefore also the heliocentric velocity vectors $\mathbf{V}(\dot{x}_i, \dot{y}_i, \dot{z}_i)$, ($i = 1, ..., N$) are known. Let us calculate the average value, change its sign and define this quantity as the *peculiar motion* of the Sun with respect to the given ensemble:

$$\dot{x}_\odot = -\langle \dot{x} \rangle = -\frac{1}{N}\sum_1^N \dot{x}_i, \qquad \dot{y}_\odot = -\langle \dot{y} \rangle = -\frac{1}{N}\sum_1^N \dot{y}_i,$$

$$\dot{z}_\odot = -\langle \dot{z} \rangle = -\frac{1}{N}\sum_1^N \dot{z}_i$$

The modulus $s_\odot$ and apex ($\alpha_\odot$, $\delta_\odot$) are derived in the usual manner:

$$s_\odot = \sqrt{\dot{x}_\odot^2 + \dot{y}_\odot^2 + \dot{z}_\odot^2}, \qquad \tan\alpha_\odot = \frac{\dot{y}_\odot}{\dot{x}_\odot}, \qquad \sin\delta_\odot = -\frac{\dot{z}_\odot}{s_\odot}$$

The early observations provided a velocity of approximately 20 km/s, in direction $\mathbf{W}_\odot$ ($\alpha_\odot$, $\delta_\odot$) $\approx$ (18$^h$ + 30°), not far from Vega.

The value provided by this method depends on the particular set of stars used to define it. Ideally, if one could take into account all nearby stars, the resulting velocity reference system would be of great interest for the study of the velocity field of the entire Milky Way. Therefore, it is worth to transform the coordinate system from the equatorial to the galactic one (see Chapter 3). Let us denote with $(u_i, v_i, w_i)$ the three velocity components derived by the appropriate rotation of $(\dot{x}_i, \dot{y}_i, \dot{z}_i)$, with axis $\mathbf{u}$ directed toward the galactic center, axis $\mathbf{v}$ at 90° in the galactic plane (toward the constellation of Cygnus) and axis $\mathbf{w}$ toward the galactic pole.

Notice that the so-defined LSR has a purely kinematic significance, because the masses of the stars have not been taken into account. Moreover, it does not possess a precise origin in space; we assume that the Sun (or better, the barycenter of the Solar System) is passing through the LSR origin at the present time.

Tables 9.2 and 9.3 contain three determinations of the LSR deriving from three different sets of stars chosen to define it; the meaning of the spectral types is given in Chapters 16 and 17. The designation of *typical* is for the value determined using all available stars in the catalogs, and it is the value to be used in the determination of secular parallaxes; *basic* refers to the nearer stars and *fundamental* with respect to the hypothetical LSR in circular orbit around the center of the Galaxy, with zero velocity dispersion. The noticeable differences in the various columns have a profound significance in the frame of all theories on the origin and evolution of the Milky Way.

We mention only that after the initial hypothesis put forward by Kapteyn of the evidence for two star currents, Karl Schwarzschild in 1907 and Carl Charlier in 1915 proposed that the distribution function of the velocities is a three-dimensional Gaussian function having different dispersions along the three axes. Moreover, the dispersion depends on the particular class of stars. The tables show that the dispersion is always smaller in the direction perpendicular to the plane and larger toward the galactic center (the dispersion ellipsoid indeed points to the galactic center). Other relationships in the tables were noted long ago, for instance, Bengt Strömgren pointed out a linear augmentation of $\sigma^2$ with $v$, a dependence called *asymmetric drift,* so that one could extrapolate the solar motion to the hypothetical population having $\sigma^2 = 0$.

### TABLE 9.2
### Solar Motion in Galactic Coordinates, Resulting from Three Different Choices of the Defining Stars

|  | $u_\odot$ | $v_\odot$ | $s_\odot$ | $w_\odot$ | $t_\odot$ | $b_\odot$ | $\alpha_\odot$ | $\delta_\odot$ |
|---|---|---|---|---|---|---|---|---|
| Typical | −10.4 | +14.8 | +7.3 | 20.0 | 56° | 23° | 18$^h$0 | +30° |
| Basic | −9.0 | +11.0 | +6.0 | 15.0 | 51° | 23° | 17$^h$9 | +26° |
| Fundamental | −9.0 | +12.0 | +7.0 | 16.5 | 51° | 23° | 17$^h$8 | +23° |

### TABLE 9.3
### Solar Motion Referred to Different Classes of Stars

| Spectral Types | $u_\odot$ | $v_\odot$ | $w_\odot$ | $\sigma_u$ | $\sigma_v$ | $\sigma_w$ | $\sigma^2$ |
|---|---|---|---|---|---|---|---|
| Supergiants O-B5 | −9.0 | +13.4 | +3.7 | 12 | 11 | 9 | 364 |
| A Giants | −13.4 | +11.6 | +10.3 | 22 | 13 | 9 | 734 |
| M Giants | −4.5 | +18.3 | +6.2 | 31 | 23 | 16 | 786 |
| Subgiants | −8.0 | +28.0 | +8.0 | 48 | 23 | 16 | 3875 |
| G0-V | −14.5 | +21.1 | +6.4 | 26 | 18 | 20 | 1400 |
| White dwarfs | −6.0 | +37.0 | +8.0 | 50 | 33 | 25 | 5614 |
| Planetary nebulae | −8.0 | +29.0 | +8.0 | 45 | 35 | 20 | 3650 |

For a pre-GAIA treatment of this rich and evolving field, reference is made to several texts on the dynamics of the Milky Way. See, for instance, Binney and Merrifield (1998) cited in the Bibliography section. Great advances are expected in the near future from GAIA's data.

## 9.8   SECULAR AND STATISTICAL PARALLAXES

The astrometric satellites Hipparcos and GAIA have extended the trigonometric parallaxes to a point where those two statistical indicators are mostly of historical significance, but worth mentioning to appreciate the developments in the field. The knowledge of the solar motion $(s_\odot, \mathbf{W}_\odot)$ allows the definition of two distance indicators whose statistical validity can be extended to several hundred parsecs, namely, the *secular* and the *statistical parallaxes*. Suppose we have a set of stars all at the same distance from the Sun and well distributed over a large area of sky (ideally, over the entire celestial sphere); furthermore, let their velocities and their intrinsic proper motions be randomly distributed. Following a method first indicated by Jacobus Kapteyn, the individual proper motions can be resolved in two orthogonal components: one along the great circle passing for the star and the solar apex, called component *upsilon* $(v, \mu_v)$, and one perpendicular to it, called component *tau* $(\tau, \mu_\tau)$:

$$\mu_v = 15\mu_\alpha \cos\delta \sin\psi - \mu_\delta \cos\psi, \qquad \mu_\tau = 15\mu_\alpha \cos\delta \cos\psi - \mu_\delta \sin\psi$$

where both components are in arcsec/year, and $\psi$ is the position angle of the solar apex $\mathbf{W}_\odot$ with respect to star X. The same decomposition can be made for the velocity vectors.

Obviously, the solar motion is present only in component $v$, being easily demonstrated that for each star:

$$\mu_{v,i} - \frac{\pi s_\odot \sin\psi_i}{4.740} = \frac{\pi V_{v,i}}{4.740}$$

By averaging over all stars, the right side will vanish; however, before averaging it is advisable to multiply both sides for $\sin\psi_i$, in order to properly apply the least squares method. Finally, the secular parallax $H$ is obtained as

$$H = \langle\pi\rangle = \frac{4.740}{s_\odot \sum \sin^2\psi_i} \sum \mu_{v,i} \sin\psi_i$$

The analysis of component $\tau$ can be made by adding the knowledge of the radial velocities, which contain the reflex of the solar velocity, with the additional hypothesis that the individual stars have a small random isotropic velocity component superimposed to that of the entire group. Since the average of the peculiar $\tau$ velocities and proper motions is zero, it is necessary to consider their absolute values before averaging. Then, the statistical parallax is

$$\langle\pi\rangle = \frac{4.740 \sum |\mu_{\tau,i}|}{\sum |V_{r,i} + s_\odot \cos\psi_i|}$$

The methods of secular and statistical parallaxes have been applied to stars as distant as 500 pc, but with great uncertainties. The apparently simple process of averaging is indeed affected by the selection of the stars and other subtle choices; even the assumptions of the method can be questioned.

## 9.9   DIFFERENTIAL ROTATION OF THE GALAXY AND OORT'S CONSTANTS

When the distance from the Sun increases beyond a few hundred parsecs, it is not legitimate to ignore the effects of the differential rotation of the Milky Way on radial and tangential velocities. Let us assume, in a very simplified model, that the stellar system is a thin disk and that each star, including the Sun, performs a coplanar circular orbit with angular velocity $\omega(R)$ around the galactic center GC distant $R$:

$$\mathbf{V}(R) = \omega(R)\mathbf{R}$$

Within these assumptions peculiar motions are not allowed: the stars perform circular orbits around the galactic center GC with an angular velocity $\omega(R)$, which is an unknown function of the distance. If $\omega(R) = const.$, the rotation would be a *rigid* one. Observations of sufficiently distant stars can reveal the behavior of $\omega(R)$. In other words, we should be able to detect the differential effects of the rotation both on the radial velocities and on the proper motions. We have already highlighted in Chapter 3 that the sense of rotation of the Milky Way is clockwise if seen from the north pole of the heliocentric galactic coordinates system, so that the galactic longitudes $l$ increase contrary to the rotation. Furthermore, the distance of the Sun from the GC is approximately $R_\odot = 8.1$ kpc (see also the "Notes" section). In the process of analyzing radial velocities and proper motions, we are led, following the ideas of Jan Oort, to introduce the two constants $A$ and $B$, thus defined

$$A = -\frac{1}{2}R_\odot\left(\frac{d\omega}{dR}\right)_\odot, \qquad B = A - \omega_\odot$$

which have dimensions (time)$^{-1}$, in practical units km/s·kpc.

Let us calculate the heliocentric radial velocity of a star having galactic coordinates $(l, b)$ and distant $R$ from C, $r$ from the Sun. Since only circular velocities are allowed, the value will be

$$V_r = R_\odot(\omega - \omega_\odot)\cos b \sin l \approx R_\odot\left(\frac{d\varpi}{dR}\right)(R - R_\odot)\cos b \sin l$$

Observations in the visible band are limited to $r \ll R$, because of the strong obscuration by the dust in the galactic plane (this limitation is not present in radio or infrared observations), so that from the triangle Sun–star–GC, we obtain

$$V_r \approx Ar\cos^2 b \sin 2l$$

Therefore, Oort's constant $A$ measures the local shear, i.e., the radial velocity gradient of the orbital motions of stars in the solar neighborhood. Current values of $A$ are approximately $15.3 \pm 0.4$ km/s·kpc (as already mentioned, if the Milky Way rotated as a rigid disk, then $\omega = const.$, and $A = 0$).

Regarding the proper motions, let us decompose each vector $\boldsymbol{\mu}$ in its components along $l$ and along $b$. After some passages, we obtain

$$4.740\mu_l = A\cos 2l + B, \qquad 4.740\mu_b = -A\sin 2l \cos b \sin b$$

As the quantity $VR = \omega R^2$ is the specific angular momentum of the material in the galactic disk, Oort's constant $B$ measures the angular momentum gradient. Notice that the effect of the galactic rotation on the proper motions is independent from the distance of the star. Averaging the first relation over all longitudes, an overall rotation of the entire celestial sphere of $(B/4.740)$ mas/year is predicted. Since the current values of $B$ are around $-11.9 \pm 0.4$ km/s·kpc, this systematic rotation is at the level of 0″.2/century, i.e., the same amount of the uncertainty of the rotation of the FK5. These considerations are a further proof of the importance of all systematic effects in the proper motions and of the superiority of any reference system based on very distant, nonrotating objects, such as the ICRF provided by the VLBI.

From the values of $A$ and $B$, we can derive the speed and the period of rotation of the Milky Way at the distance of the Sun:

$$V(R_\odot) = \omega(R_\odot)R_\odot = (A - B)R_\odot \approx 220\,\text{km/s},$$

$$P(R_\odot) = 2\pi/\omega(R_\odot)\,2.26 \times 10^8 \text{ year}$$

## NOTES

- A classical treatise on proper motions is given by Luyten (1963), see the Bibliography section.
- Stumpff (1979, 1985 in the References section and in the Bibliography section) called attention to the rigorous definition of the heliocentric motions of the stars in a truly inertial system. These papers deal also with relativistic velocities and superluminal expansions. The differences with the loose definitions given here do not amount to more than a few mas after several decades.
- Ives and Stilwell (1938, 1941) were among the first to determine with precision the validity of Equation 9.3. These first experiments have been repeated several times with improved precision and different technical equipment. A discussion of the transverse Doppler effect in the cosmological situation is given, for instance, by Dishon and Weber (1977).
- The results quoted here for SgrA* have been published by the GRAVITY collaboration in (2018a). In a subsequent paper, the same GRAVITY collaboration (2018b) refined the determination to the distance of the Milky Way center, finding the value $R = 8178$ with an estimated error not larger than about 30 pc.
- The astrometric radial velocities are discussed in detail by Dravins et al. (1999), Lindegren et al. (2000) and Madsen et al. (2002). These papers contain many references to previous literature, in particular to the 1917 paper by Schlesinger. Eichhorn (1981) proposed the opposite method, of deriving parallaxes from radial velocities and proper motions extended over 100 years. The GAIA results in connection with astrometric radial velocity have been discussed by Reino et al. (2018) and by Leão et al. (2019).
- For the moving cluster method, see Hanson (1975) and also the volume *Star Clusters,* Hesser ed., in the Bibliography section.
- For Oort's constants $A$ and $B$, see the classic paper by J. E. Oort in the Bibliography section. The values of the constants here adopted derive from GAIA measurements (Bovy, 2017). This paper discusses two additional constants, C and K, both = 0 in the classical treatment. The four constants $A$, $B$, $C$ and $K$ are the four first-order coefficients in a Taylor series expansion of the two-dimensional, in-plane mean velocity field of a given stellar population with respect to distance from the Sun. $C$ is present as an additional term $(-C \sin 2l \sin b)$ in the proper motions in longitude and $(-C \cos 2l)$ in the proper motions in latitude, while $K$ is present as $(-K \sin b \cos b)$ in the proper motion in latitude. The GAIA data provide the values $C = -3.2 \pm 0.4$ km/s kpc and $K = -3.3 \pm 0.6$ km/s kpc. Furthermore, the paper points out that in reality, the "constants" vary in time and depend on the particular stellar population.
- The rotation curve of the Milky Way to the distance of 20 kpc from the galactic center has been measured by the OGLE collaboration (Mróz et al., 2019) using classical Cepheids. The rotation speed at the Sun turns out to be $233.6 \pm 2.8$ km/s using the already quoted

distance of 8.1 kpc determined by the GRAVITY collaboration. The same research group constructed a 3D model of the Milky Way constraining the warped shape of its disk (Skowron et al., 2019).

## EXERCISES

1. The relativistic Doppler effect has been given in Equation 9.3 as

$$\nu_O = \nu_S \frac{\sqrt{1 - \frac{V^2}{c^2}}}{1 - \frac{V \cos\theta}{c}}$$

where $\nu_O$, $\nu_S$ are the frequencies of the light in the reference frame of the observer O and of the source S, $V$ is the relative velocity and $\theta$ is the angle between the line of sight and the normal to the wavefront. Using wavelengths, letting $\beta = V/c$ and $\alpha = \pi - \theta$, the angle between the line of sight and the velocity vector of S we have

$$\frac{\lambda_O}{\lambda_S} = -1 + \frac{1 + \beta\cos\alpha}{\sqrt{1 - \beta^2}}$$

or else

$$1 + z = 1 + \frac{\lambda_O - \lambda_S}{\lambda_S} = \frac{1 + \beta\cos\alpha}{\sqrt{1 - \beta^2}} \qquad (9.23)$$

which is the expression we wish to discuss. Let $V$ be all along the line of sight, $V = V_r$. If $\alpha = 0°$ (source receding):

$$1 + z = 1 + \frac{\lambda_O - \lambda_S}{\lambda_S} = \frac{1 + \beta}{\sqrt{1 - \beta^2}} = \sqrt{\frac{1 + \beta}{1 - \beta}} \approx 1 + \beta + \frac{1}{2}\beta^2 + \cdots$$

If $\alpha = 180°$ (source approaching),

$$1 + z = 1 + \frac{\lambda_O - \lambda}{\lambda} = \frac{1 - \beta}{\sqrt{1 - \beta^2}} = \sqrt{\frac{1 - \beta}{1 + \beta}} \approx 1 - \beta + \frac{1}{2}\beta^2 + \cdots$$

To the first order, the classical Doppler effect is

$$1 + z = 1 + \frac{\lambda_O - \lambda_S}{\lambda_S} \approx 1 \pm \beta$$

The second term, however, shows that the relativistic expression is not symmetric in the sign of $V$. The redshift of a receding source can become arbitrarily high (galaxies and quasars have been observed with $z \approx 10$), but if the source approaches the observer, its blueshift tends to the limit $z = -1$, as can be seen by discussing the cubic equation obtained with $\cos \alpha = 1$ and taking $\beta$ with the appropriate sign:

$$(\beta - 1)z^2 + 2(\beta - 1)z + 2\beta = 0$$

For approaching velocities, the Doppler effect is always numerically very close to the classical one.

Let $V$ now be all transverse, $V_r = 0$; the shift is always positive and approximately equal to

$$1 + z \approx 1 + \frac{1}{2}\beta^2$$

(see Chapter 1), vanishingly small for small velocities.

In the general case, the calculations can be carried out for all values of $\alpha$ between 0 and 180°, maintaining $\beta$ as a parameter, e.g., up to $\beta = 0.999999$. Notice that for angles greater than 90°, redshifts can be observed even with approaching sources! Indeed, the observed redshift depends on a combination of the radial and transverse ones, and the latter only produces positive values for $z$.

2. Determine and compare the Oort's constants in three rotational models, namely, rigid body, flat rotation curve and Keplerian curve. Show that the flat rotation is closer to the measured values, although the velocity at the solar distance must actually decrease, because $-A-B > 0$.

# 10 The Astronomical Times, the Atomic Time and the Earth Rotation Angle

In previous chapters, time was found necessary to describe the movements of the celestial sphere with respect to the meridian. Time also enters into the Newtonian dynamical theory, as the fundamental independent variable in differential equations. This chapter gives several operative definitions of time, together with the transformations among them, in order to complete the notions expounded in Chapters 2 and 4 by taking into account precession, nutation, aberration and the Earth rotation angle (ERA) and its variations. We will consider four different timescales: sidereal, solar, dynamic and atomic, the first three scales being associated with astronomical observations. Furthermore, when general relativity matters, it will be necessary to distinguish between proper time and coordinate time, and time will become a component of the overall space-time geometry.

## 10.1  THE SIDEREAL TIME *ST*

The present and the next paragraphs are based on the traditional (pre-2003) treatment making use of the equinox. Section 10.6 will introduce the notions and algorithms based on the ERA.

The sidereal time *ST* was defined in Chapters 2 and 4 as the hour angle of the vernal point $\gamma$, which is in diurnal motion with respect to the meridian because of the Earth's rotation. Correspondingly, the variation of $HA(\gamma)$ can be utilized to define a timescale in which the elementary unit of time is the interval between two successive passages of $\gamma$ through the meridian. Such time unit is called *sidereal day*, which is divided into $24^h$ of $3600^s$ of sidereal time. However, even disregarding the irregularities of the Earth's rotation, the sidereal time is affected by the secular motion of precession and by nutation, composed by the superposition of many different periodic terms, in particular by the one depending on the longitude of the node of the lunar orbit. Therefore, one has to distinguish between *apparent* and *mean* sidereal time. The difference $EE = \Delta\psi \cos \varepsilon$ (in the sense apparent *ST* minus mean *ST*), due to the nutation in longitude of the equinox, is called *equation of the equinoxes* (before 1960, *EE* was also called nutation in right ascension). The amount of EE is always between $\pm 1^s.179$, with a main periodicity of 18.61 year. For instance, in 2001 *EE* was $-0^s.94$ at the beginning and $-1^s.02$ at the end of the year. Recent values are provided in Table 10.1.

*EE* became well measurable around 1930, when the precision of clocks became better than 1 ms/day. The mean sidereal time is more uniform than the apparent one, but it is the latter that enters in telescopic observations.

The sidereal day should not be confused with the interval of time between two culminations on the upper meridian of an ideal equatorial star devoid of parallax and proper motion. Although very similar, this interval does not coincide with the sidereal day: first, because of the precession, which generates a difference of about $0^s.0084$/day (the stellar day being longer than the sidereal one), and second, because of the diurnal variation of the annual aberration. Therefore, the stellar day would provide a better measurement of the Earth's rotation period, but it is not used in astronomy. The ratio between the mean sidereal day and the Earth's rotation period at the present epoch is 0.99999990; it varies very slowly, because of the varying precessional constants, by about one part over $6 \times 10^{13}$ each century. See Section 10.6 for precise values.

We open at this point a parenthesis about the many irregularities of the Earth's rotation, some having a short-term behavior and others a secular one. These irregularities appear in exactly the same manner in the observations of sidereal and stellar times. Therefore, they cannot be detected by meridian observations, unless very precise clocks, independent of astronomical observations, are available, such as the ephemerides of planets (the dynamical *Ephemeris Time*), or unless reference is made to phenomena independent of the Earth's rotation (a laboratory time, *Atomic Time*). Consequently, the timescales based on the Earth's rotation are not fully adequate for dynamical purposes. This matter will be treated in more detail in the following sections.

## 10.2   THE SOLAR TIME $T_\odot$

A second timescale was defined in Chapter 4 by means of the hour angle of the Sun. The *solar day* is the interval of time between two consecutive upper transits of the true Sun (we use in this context the adjective *true* to indicate the apparent Sun; we have already commented about the slight difference between transit and culmination times). The true solar time $T_\odot$ is therefore the hour angle of the apparent Sun, augmented by $12^h$ in order to comply with the contemporary convention that the day starts at midnight, not at noon: $T_\odot = HA_\odot + 12^h$. Furthermore, we recall that the largest non-uniformities of $T_\odot$ are eliminated through the introduction of the mean Sun $M_\odot$, which is in uniform motion *on the equator* and whose hour angle defines the local mean time. In particular, the universal time is the hour angle of $M_\odot$ with respect to the meridian of Greenwich, plus $12^h$:

$$UT = HA_{\text{Greenwich}}\left(M_\odot\right) + 12^h$$

Some authors call "Universal Sun" the mean Sun moving with respect to the Greenwich meridian.

Actually, the mean time is not a fully satisfactory concept, because it presupposes the availability of a strictly uniform timescale (we shall introduce the Ephemeris Time in Section 10.4) and the absence of interpolation errors.

The instant of the passage of the true Sun in meridian is affected by nutation, by the two terms of the annual aberration (constant plus elliptical) and by the diurnal aberration. The *duration* of the solar day instead does not depend on the constant term of the annual aberration nor on the diurnal one. Therefore, by definition *ST* and UT, although different in rhythm and origin, have the same degree of uniformity, namely, that of the terrestrial rotation.

Let us discuss the origins of the two times. According to Newcomb, the mean longitude of $F_\odot$ (the non-aberrated fictitious Sun on the ecliptic) at $12^h$ UT (noon) of January 1, 1900 had the value:

$$\lambda\left(F_\odot\right) = 280°40'56''.37 = 18^h42^m42^s.391$$

At the same instant, that was also the *ST* at Greenwich. Notice that the non-aberrated Sun, not the apparent one, which is $20''.45$ behind it on the ecliptic, appears in this definition.

Using the IAU pre-2000 values of the constants, the GMST at *midnight* of January 1, 2000 was

$$GMST = ST_{\text{Greenwich}}\left(0^h UT, JD2451544.5\right) = 6^h41^m50^s.54841$$

For a successive date, counting the time $T$ in Julian centuries since January 1, 2000 at $12^h$ UT (*noon*, namely, the number of days since JD 2451545.0 divided by 36,525; remember also that year 2000 was *not* a leap year, according to the Gregorian reform of the calendar), the expression, accurate to about 1 ms for few decennia, is

$$GMST\left(0^h UT, T\right) = 6^h41^m50^s.54841 + 8640184^s.81287\ T + 0^s.09310\ T^2 \tag{10.1}$$

where the last term derives from the variation of the precessional constants.

**TABLE 10.1**

**Values of GMST and *EE* at January 1, 12ʰ UT, from 2015 to 2022 (GAST = GMST + *EE*)**

| Year | h | m | s | *EE* (ˢ) |
|------|-----|-----|---------|---------|
| 2015 | 18 | 43 | 17.4095 | +0.2999 |
| 2016[a] | 18 | 42 | 20.0653 | −0.0536 |
| 2017 | 18 | 45 | 19.3837 | −0.3939 |
| 2018 | 18 | 44 | 22.0931 | −0.7020 |
| 2019 | 18 | 43 | 24.8025 | −0.9221 |
| 2020[a] | 18 | 42 | 27.5120 | −1.0101 |
| 2021 | 18 | 45 | 26.7768 | −0.9852 |
| 2022 | 18 | 44 | 29.4863 | −0.8709 |

[a] Leap year

The equivalent expression, given by the *Astronomical Almanac* with the post-2003 values, rounded to eight digits and ignoring terms in $T^3$ and higher, is the following:

$$\text{GMST}(D_U, T) = 86{,}400^s \left(0.77905727 + 0.00273781 D_U + D_U \bmod 1\right) + 0^s.00096707$$

$$+ 307^s.47710227 T + 0^s.09277211 T^2$$

where $D_U$ is the interval in days elapsed since the epoch J2000 (2000 Jan 1 12ʰUT, JD2451545.0), $D_U$ mod1 is the fraction of the UT day remaining after removing the integer and $T$ is measured in Julian centuries from J2000. Rigorously, UT should be UT1 and $T$ in the TT scale (see later), but the distinction is irrelevant at this level of precision.

Table 10.1 provides values accurate to 0.1 ms for *GMST* and *EE* from 2015 to 2022. The apparent sidereal time GAST derives from GMST by adding the nutational term *EE*.

It is worth mentioning that while *ST* originates from the diurnal rotation, the mean solar time has a more hybrid nature, because the yearly revolution appears in its definition in addition to the rotation. To remove this intrinsic source of ambiguity, the *Ephemeris Time* ET, based entirely on dynamical considerations, was introduced in 1960 and used until 1984. Moreover, since the IAU 2000 resolutions, the ERA is preferred, as already indicated in Chapter 2. The dismissal of ET and the introduction of ERA will be treated in Sections 10.4–10.6.

## 10.3　THE YEAR

The yearly revolution of the Earth provides a new timescale and of a new unit of time, namely, the year, which can be defined in several different ways as detailed in the flowing paragraphs.

### 10.3.1　TROPICAL YEAR

The tropical year is the interval of time between two consecutive passages of the Sun through the vernal point γ, or else the time needed for the right ascension of the Sun to increase by 360°. Its value, expressed in mean solar days (j), can be obtained with great precision simply by counting the number of days between two widely separated equinoxes. From a discussion of all the available data since antiquity, Simon Newcomb derived the expression:

$$1\,\text{tropical year} = 365^{\text{j}}.24219879 - 0^{\text{j}}.00000614\,T \tag{10.2}$$

where $T$ is the number of Julian centuries since mean *noon* at Greenwich on January 1, 1900.

Taking into account that the duration of the mean solar day has a minute change over the centuries, we should add to this definition that j is that measured at the same initial date. The second term in Equation 10.2 is due to the variation of the precessional constant in right ascension; it amounts to a diminution of the duration of the tropical year of approximately $0^{\text{s}}.53$/century. Correspondingly, the mean motion of the Sun, which refers to the moving equinox, will also increase. The mean longitude of $F_{\odot}$, the fictitious Sun on the ecliptic at any date $T$, is expressed by

$$\lambda\left(F_{\odot}\right) = 280°40'56''.37 + 129602768''.13\,T + 1''.089T^2 - 20''.45 \tag{10.3}$$

where the last term distinguishes between the geometric and apparent (aberrated) Sun.

A more refined expression for the average duration of the tropical year, based on Laskar's orbital elements (1985, 1986), is

$$1\,\text{tropical year} = 365^{\text{j}}.2421896698 - 0^{\text{j}}.0000065359\,T - 7.29\times10^{-10}\,T^2 \tag{10.4}$$

where $T$ is now measured in centuries from the new fundamental epoch J2000.0, $T = (JD - 2451545.0)/36{,}525$.

### 10.3.2 Besselian Year B or Annus Fictus

As already stated in Chapter 4, the tropical year does not have a definite origin. Following Bessel, the origin of the year was defined as the instant when the *apparent* longitude of the *aberrated* fictitious Sun $\lambda(F_{\odot})$, referred to the mean equinox of date, is exactly 280°. The year having this origin is called the Besselian year and is indicated by the letter B before the numeral. The initial instant, designated by apposing .0 after the numeral (e.g., B1950.0), is always within 1 day of the midnight of December 31. Any other instant in the course of the year is designated by an appropriate number of decimal digits, e.g., B1986.12345.

The next Besselian year will start at the next passage of the Sun through the same longitude. The duration of the Besselian year is therefore essentially the same as the duration of the tropical year, apart from a minute secular acceleration coming from the different definitions of $\lambda(F_{\odot})$ and $\alpha(M_{\odot})$:

$$\text{Duration Besselian year} = \text{Duration tropical year} - 0^{\text{s}}.148T' \tag{10.5}$$

being $T'$ in tropical centuries since 1900.0. This different duration means an increasing difference in the origins; for instance, in year 2000 the Besselian year started about $7^{\text{s}}.4$ before the Julian one, a shift so small to be irrelevant in almost all instances, e.g., in precessional calculations.

### 10.3.3 Sidereal Year

The sidereal year is the interval of time between two passages of the Sun over an ideal *ecliptic* star devoid of proper motion. Therefore, the sidereal year is longer than the tropical one by the amount of the precession of $\gamma$ along the ecliptic, namely, by the factor $1296000''/(1296000'' - 50''.26) \approx 1.000039$, corresponding to $365^{\text{j}}.256363$. The difference amounts to $20^{\text{m}}24^{\text{s}}$, or else to $36{,}720\,\text{km}$ along the orbit of the Earth. Notice that the value of the sidereal year is not a measured one, but is derived from the length of the tropical year. From it, we also get the mean sidereal solar motion:

$$n = 1296000''/365^{j}.256363 = 3548''.1928/j \qquad (10.6)$$

whose value is not affected by the secular variation of the precessional constant and therefore has the same uniformity of the diurnal rotation.

### 10.3.4   ANOMALISTIC YEAR

The anomalistic year is the interval of time between two consecutive passages of the Sun through the perigee. We have already underlined in Chapter 4 the influence of the Moon on the date of the passage through the perigee. In addition, the direction of the major axis of Earth's orbit (in other terms, the *line of the apses*) is not fixed in the inertial space: it slowly precesses in the same direction of the yearly motion, by an amount of $11''.63$/year determined by the gravitational perturbations of the other planets. Therefore, the longitude of the perigee increases by approximately $11''.63 + 50''.26 = 61''.89$/year. The anomalistic year is thus longer than the previous ones, its average duration being approximately $365^{j}.259636$, with a secular acceleration of $0^{s}.263$/century. It is easily seen that perigee and equinox coincide every 21,000 years; because the duration of the seasons depends on the distance between equinox and perigee, the consequence is their appreciable variation of duration, at a level of $1^{h}$/century (see Table 4.1).

### 10.3.5   DRACONITIC (OR ECLIPSE) AND GAUSSIAN YEARS

We quote 2 more years: the draconitic (or draconic or eclipse) and the Gaussian year.

The draconitic year is the interval of time between two passages of the Sun through the ascending node of the lunar orbit. Therefore, it is connected to the occurrence of eclipses and named also *eclipse* year. Owing to the retrograde motion of the lunar nodes on the ecliptic, this year is the shortest, its average duration being $346^{j}.62008$.

The Gaussian year derives from Kepler's third law: it is the period of revolution of a massless body in circular orbit around the Sun at the distance of 1 AU. Its value is $365^{j}.25690$. We shall discuss the Gaussian year again in Chapter 12.

## 10.4   THE DYNAMICAL EPHEMERIS TIME ET

We discuss now the non-uniformities of the sidereal and universal times caused by the non-uniformities of the diurnal rotation. Let us set aside the acceleration due to the secular variation of the precessional constant, in order to concentrate our attention on the rotation itself. Kepler had already alluded to the possibility of a non-uniform rotation of the Earth. Immanuel Kant in 1754 put forward the idea of a progressive slowing down due to the tides. A sound experimental evidence of secular effects came around 1870, when Newcomb compared the available observations with the very precise tables of lunar positions calculated by Peter Hansen. The agreement in longitude was very good in the second part of the 18th century and in the first part of the 19th; it gradually deteriorated before 1750 (the data being the instants of occultations measured in Paris) and after 1850. However, Newcomb himself had serious doubts that the slowing down of the rotation was the only reason for the disagreement. More data coming from different sources, such as the instants of ancient eclipses and better clocks available after 1930, were needed to solve the problem. We can nowadays distinguish three types of irregularities (see Figure 10.1):

- a secular slowing down of the rotation, amounting to a variation of the mean solar day by approximately 2 ms/century, partly but not totally explained by tidal dissipation of the rotational energy. Since this increase accumulates over the ages, the effect on phenomena that occurred millennia ago can amount to several hours. The records of eclipses in the distant past, which are available from 4000 B.C., are of considerable help in establishing

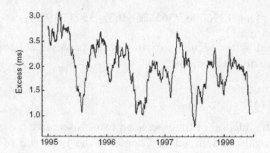

**FIGURE 10.1**    Fluctuation of the duration of the day, expressed as excess to 86,400 SI, from 1995 to 1998. (Adapted from the IERS site).

    this variation of the length of the day. For the utilization of historical eclipses to determine the slowing down of the Earth rotation, see, e.g. Stephenson and Morrison (1984) and Morrison and Stephenson (1998);

- seasonal variations due to meteorological causes, of periodic nature and amplitude of few milliseconds;
- irregular fluctuations of geophysical origin, implying a transfer of angular momentum between core and mantle.

Table 10.2 provides a summary of the largest long-term variations. The mathematical expressions given there are simply convenient interpolation formulae, with no physical basis. Different authors give different formulae (see later for the meaning of TT and UT1).

    These variations, however small and difficult to measure, must appear in the ephemerides of the planets and of the Moon (as further discussed in Chapter 14). A second effect is expected for the Moon: the tides dissipate a fraction of the rotational energy of the Earth; therefore, the Earth–Moon distance has to increase to keep constant the total angular momentum of the Earth–Moon system. By virtue of Kepler's third law, the mean motion of the Moon must decrease. This secular acceleration has opposite sign to that due to the lengthening of the day. This is the reason for the many debates on the lunar accelerations during the 19th and 20th centuries. Today, laser and radar echoes continuously monitor the Earth–Moon distance, with a precision at a level of a few millimeters per year that allows ascertaining and measuring even minute variation (see, for instance, the lunar laser ranging Apollo project in the Web Sites section).

    Following the previous considerations, we can discuss with greater insight the universal time, arriving in successive steps to four realizations of it: UT0, UT1, UT2 and then UTC, discussed in Section 10.5.

---

## TABLE 10.2
### Long-Term Variations of the Duration of the Day

| | |
|---|---|
| From 390 B.C. to 948 A.D. | $\Delta j = 2.4$ ms/century |
| | $\Delta(TT-UT1) = 1360^s + 320T + 44.3T^2$ |
| From 148 A.D. to 1600 A.D. | $\Delta j = 1.4$ ms/century |
| | $\Delta(TT-UT1) = 25.5T^2$ |
| From 1600 A.D. to date | $\Delta j = (7^s.286 \pm 0^s.170) \times 10^{-6}(year - 1819)$ |
| | $\approx 0.7$ ms/century |
| | $\Delta(TT-UT1) = 5^s.156 \pm 0^s.404 + (13.3066 \pm 0.3264)(T - 0.19 \pm 0.01)^2$ |

$T$ in Julian Centuries after 1800.

---

We shall call UT0 the apparent mean time at Greenwich. UT0 cannot be used in high-precision works because of the polar motion. By measuring the local sidereal time and knowing the longitude and latitude of the observing station, the effects of the polar motion can be removed by a correction of the type:

$$UT1 = UT0 - (u_x \sin \Lambda + u_y \cos \Lambda) \tan \phi' \qquad (10.7)$$

where $(u_x, u_y)$ are the coordinates of the pole in time units and $(\Lambda, \phi')$ the geocentric longitude and latitude. Therefore, UT1 is the observatory-independent time, the one that should enter in all previous formulae referring to UT. UT1, determined by the a-posteriori analysis of the data coming from many stations, is distributed by the IERS. For the pre-2000 definition of universal time, see, for instance, Aoki et al. (1982); the 2001 IAU implementation of UT1 is described by Capitaine et al. (2003a).

Although much more uniform than UT0, UT1 is still affected by the non-uniformities of the Earth's rotation.

By removing the periodic components of these non-uniformities, the third-level UT2 is derived, which is however not used in astronomy. Indeed, UT2 is not entirely satisfactory for dynamical purposes because of the remaining irregularities. To realize a truly uniform time, one could resort to the mean sun originally defined by Newcomb, both as origin and rhythm. According to the definition, the geometric mean longitude of the Sun is

$$\lambda = 279°41'48''.04 + 129602768''.13\,T + 1''.089\,T^2 \qquad (10.8)$$

$T$ being expressed in Julian centuries after 1900, Jan.0, $12^h$ UT.

The Equation 10.8 could be used to indicate a uniform time, whose rhythm is given by the coefficient of $T$ and which has a small precessional acceleration. This time was called *Ephemeris Time* ET. The second of ET is then defined by the number $N$ of seconds in the tropical year 1900:

$$N = \frac{1,296,000 \times 35,525 \times 86,400}{129,602,768.13} = 31,556,925.9747$$

In other words, $1^s$ ET is the fraction $1/N$ of the length of the tropical year 1900. Indeed, at this stage we should redefine the initial epoch as $12^h$ ET, not UT. Notice that ET is a *dynamical solar time* defined by the revolution of the Earth and apparently independent from the rotation.

The body of knowledge accumulated since the original definition has proved that two different mean suns have to be distinguished: one whose right ascension increases uniformly with UT and one whose right ascension increases uniformly with ET; only if the rotation of the Earth were strictly uniform, would the two suns coincide in a single body.

In other words, in order to free ET, whose origin is purely the annual revolution, from the rotational slowing down of the Earth, we must introduce a mobile Greenwich meridian (*ephemeris meridian*) in very slow eastward motion with respect to the conventional one. The hour angle of $\gamma$ with respect to the ephemeris meridian is called ephemeris sidereal time. By convention, the two meridians coincided in 1902, at the present epoch they are at about $2''$ from each other.

While UT can be obtained from meridian observations, ET must be derived from the longitude of the Sun, which however moves too slowly along the ecliptic to provide a good clock. The Moon is much better for this purpose. Thus, ET was identified with the dynamical time that appears in the ephemerides of the Moon.

A first difficulty of ET is that its knowledge implies the reduction of a great amount of data, so that the difference UT−ET is known only a posteriori; moreover, it can be affected by residual errors in the lunar ephemerides. Finally, and perhaps more decisively, ET is still a pre-relativistic concept; therefore, its utilization in the almanacs, introduced in 1960, was discontinued in 1984. Nevertheless, ET retains some usefulness, as we will discuss in Section 10.5.

## 10.5   THE ATOMIC TIME

In the previous sections, the time was defined, or mathematically derived, from the movements of heavenly bodies. Since 1955, a different physical time of very high regularity became available, namely, the international atomic time (TAI), which was officially adopted as standard time in 1972.

TAI is defined by the radiation coming from two hyperfine levels of the fundamental energy level of cesium, when the atom is far from magnetic fields and at sea level. The frequency of this resonant transition is 9,192,631,770 Hz, with a stability of about $2 \times 10^{-13}$. It defines the international second SI, which, always by definition, is equal to the second of ET. The duration of the mean solar day is then 86,400 SI. In practice, some 200 stations well distributed over the Earth keep the atomic time to within 1 ns/day and distribute it via radio and navigational systems (e.g., GLONASS, GPS, Galileo GNSS).

Regarding the origin of TAI, by international agreement its zero point was at epoch January 1, 1958 at $0^h$ UT2. This decision implied an offset between ET and TAI: $ET = TAI + 32^s.184$.

Adopting now TAI as the fundamental timescale, the quantity $TAI + 32^s.184$ was named terrestrial dynamic time, TDT. From 1986 to 2000, TDT was the tabular argument of the ephemerides. TDT maintains the continuity with ET, but its realization no longer depends on observations of the Sun or the Moon, but on laboratory clocks. From 2001 onwards, by IAU decision, TDT has been renamed Terrestrial Time TT.

TAI (and consequently TT) is certainly a very uniform time; nevertheless, according to general relativity, the frequency of any clock varies with the varying gravitational potential in which it is immersed. Therefore, TAI must be interpreted as a *proper time*. In the differential equations of mechanics, it must be transformed to *coordinate time*, according to the position and velocity of the observer with respect to the barycenter of the Solar System. This ideal time of the inertial observer is called dynamical time of the barycenter (TDB). The TT–TDB difference is expressed by purely periodic terms; if a precision of 1 ms is sufficient, the following expression can be used:

$$TDB = TT + 1.658 \times 10^{-3}(\sin E + 0.0368) + 2.03 \times 10^{-6} \cos\phi'(\sin(UT + \Lambda) - \sin\Lambda) + \cdots \quad (10.9)$$

where $E$ is the eccentric anomaly of the Sun, and $\phi'$, $\Lambda$ are latitude and longitude of the clock. The difference TDB–TT never exceeds $0^s.0017$, so at this level of precision we could adopt a generic dynamical time TD in its different realizations.

In the end, however, what matters for astronomy and navigation is the true angle of rotation of the Earth, namely, UT in its various realizations. Therefore, by international agreement, a time is broadcasted having the rhythm of TAI, but the origin always coincident, within 900 ms, with that of UT1. This hybrid time is called UTC (coordinated universal time). Since UT is not uniform, UTC cannot be a continuous function; according to the need, a leap second is added (or in theory subtracted, but since 1972 this never happened) at the beginning or at the midpoint of each particular year. The introduction is decided by the IERS about 6 months in advance, and its necessity is fairly erratic. For instance, no necessity for the leap second occurred from January 1, 1999 to June 30, 2005, nor in 2007 and 2008, nor in 2017 and 2018. It never was necessary to introduce 2 leap seconds in 1 year, except in 1972. UTC is therefore extremely practical, inexpensive and sufficient for many astronomical purposes; however, at least in principle, it is not correct to measure

**TABLE 10.3**

**Values of $\Delta T = ET - UT$ from 1890 to 1983, $\Delta T = TDT - UT$ from 1894 to 2001, $\Delta T = TT - UT$ from 2001**

| Year | $\Delta T$ (s) | Year | $\Delta T$ (s) |
|------|-----------|------|-----------|
| 1620 | +124 | 1975 | +45.48 |
| 1700 | +9 | 1985 | +54.34 |
| 1750 | +13 | 1995 | +60.78 |
| 1800 | +13.7 | 2000 | +63.83 |
| 1850 | +7.1 | 2005 | +64.69 |
| 1900 | −2.72 | 2010 | +66.07 |
| 1902 | −0.02 | 2015 | +67.64 |
| 1950 | +29.15 | 2020 | (+70?) |

the duration of an event by differencing UTC, because of its discontinuities. The broadcasted radio signals actually contain also the UTC-UT1 difference.

By means of UTC and then UT1, it is possible to redefine precisely the longitude of the ephemeris meridian with respect to Greenwich:

$$\Lambda(ME) = 1.002738(TT - UT1)$$

The adoption of the dynamical time suggests redefining the Julian day (JD) using TD and not UT. This JD is often indicated by JED or JDE according to others.

The *Astronomical Almanac* in Table K8 gives the TD–UT difference since 1620. Before 1984, TD was ET. The numbers are extremely uncertain before 1890, being based on an adopted value of −26″/(century)$^2$ for the tidal term in the mean motion of the Moon. The difference instead is accurately known for the last decades. Some values of $\Delta T$ are given in Table 10.3; at the time of writing (2019), the value for 2020 is still an extrapolation.

## 10.6  THE EARTH ROTATION ANGLE (ERA)

The ERA measures the rotation of the Earth from a nonrotating origin on the celestial equator, namely, from the so-called Celestial Intermediate Origin (CIO), which has no instantaneous motion along the equator. The introduction of the ERA in 2003 implied a redefinition of *ST*, in particular of GMST and more fundamentally (because if its connection with the observations) of GAST (see Table 10.1).

Let us recall first some basic quantities. According to the IERS resolutions, the Earth true rotation period has the value 86,164.098903691 UT1 seconds ($23^h56^m04^s.098903691$, i.e., 0.99726966323716 mean solar days). The corresponding angular velocity is approximately 7.292 radians per SI second, i.e., 360°.986 per solar day. Using the WGS84 value of 6378.137 m for the equatorial radius, the equatorial speed is 465.10 m/s.

The rotation period relative to the precessing mean equinox, i.e., the mean sidereal day, is 86,164.090530832 88 UT1 seconds ($23^h56^m04^s.09053083288$, i.e., 0.99726956632908 mean solar days). As already said, the sidereal day is shorter than the rotation period by about 8 ms; the ratio between the length of the mean solar day to that of the sidereal day is 1.002737909350795.

Therefore, while ERA is measured from a fixed origin, GAST is measured from the true equinox, whose motion is due to the movements of the equator and the ecliptic.

ERA, measured in radians, is related to UT1 by the expression:

$$\theta(t_U) = 2\pi\left(0.7790572732640 + 1.002737909350795 t_U\right)$$

where $t_U$ is the Julian UT1 date $-2451545.0$.

With ERA, the new definition of GMST and GAST is

$$\text{GMST}(t_U, t) = \theta(t_U) + E_p(t)$$
$$\text{GAST}(t_U, t) = \theta(t_U) - E_0(t)$$

where $E_p$ is the accumulated precession and $E_0$ is equation of the origins, which represents accumulated precession and nutation.

As stated before, the ERA measures the rotation of the Earth from the CIO, which is fixed along the equator. The *geocentric* equatorial coordinate system having as fundamental plane the true equator of date, origin of RA on that plane the CIO, whose pole is the Celestial Intermediate Pole (CIP), is named Celestial Intermediate Reference System (ICRS). To be more precise, the CIP is the reference pole of the IAU2006/2000 precession and nutation theories. The ICRS is the best realization of an inertial, time-independent, kinematical and barycentric system. The axes of the ICRS are those of the ICRF, based on the VLBI positions of distant quasars and tied to a subset of optical objects. The true coordinates are calculated taking into account the nutation.

Let us recall, for a better comparison the traditional reference systems introduced in the previous chapters, the J2000.0 system, whose fundamental plane is the *mean* equator and whose origin on that plane is the *mean* equinox at epoch J2000.0. The pole of the fundamental plane is the mean pole at the same epoch. The origin is the *barycenter* of the Solar System. The *true* equatorial coordinates are referred to the true equator of date and to the true equinox of date; they are calculated taking into account nutation.

In the limit of our treatment, J2000.0 coincides with the ICRS. Not discussed in this text are the Barycentric Celestial Reference System (BCRS) for Solar System objects and the Geocentric Celestial Reference System (GCRS), obtained via relativistic transformations from the ICRS according to IAU resolutions.

At any rate, the axes of J2000.0, ICRS, BCRS and GCRS have the same orientation to better than 100 mas.

In parallel to the new definitions of celestial systems, there are new definitions of the terrestrial ones.

The International Terrestrial Reference System (ITRS) is the geocentric non-inertial system rotating with the Earth. It is the recommended system to express geodetic longitudes and latitudes on Earth, being consistent with the WGS84 to about 1 cm. The International Terrestrial Reference Frame (ITRF) is the realization of the ITRS through a number of fiducial stations with well-known coordinates and velocities. Its initial orientation was the 1984 position of the Greenwich meridian (Terrestrial Intermediate Origin, TIO).

The fundamental plane of the geocentric Terrestrial Intermediate Reference System is the true equator of date; the pole is the CIP. The origin of the GHA on the fundamental plane is the nonrotating TIO.

The 2019 Report of the IAU/IAG Joint Working Group on Theory of Earth Rotation and Validation by Ferrándiz and collaborators summarizes the uncertainties affecting the current theories of Earth rotation, precession and nutation. See the Web Sites section for additional information.

## NOTES

- After the great work of Huygens at the end of the 17th century, the precision of clocks reached a few tens of seconds per day; it gradually improved through several mechanical innovations, mostly in England (Harrison, Kendall) and France (Bertoud), until the masterly work of Shortt in 1920 (a combination of electrical and mechanical devices). The Shortt bi-pendulum could claim a precision of about 10 ms /day and provided decisive evidence of the non-uniformity of the Earth's rotation. Then, in 1925, piezoelectric technology became available and finally atomic technology. The Web Sites section quotes places where to find useful information on timescales and on historical and modern clocks.
- In 1967, Jocelyn Bell Burnell and Anthony Hewish discovered with the Mullard Radio Telescope the first source emitting short pulses of radiation at very regular intervals (Hewish et al., 1968). These pulsating radio sources, renamed *pulsars*, provide a natural high-frequency, high-stability clock. They have been identified with fast rotating neutron stars. For the utilization of the pulsars as high-precision clocks, see: Backer et al. (Bibliography section). Germanà et al. (2012, see in particular their Note 6) discuss the difficulties of properly reducing the arrival times of radio and optical pulses. For example, rotational period of he pulsar in the Crab nebula (M1) is slightly shorter in the TDB system than the one in the Barycentric Coordinate Time (TCB) system.
- The web site https://www.iers.org/IERS/EN/Science/EarthRotation/LODsince1623.html gives the excess of the duration of the day to $86,400^s$, and angular velocity of the Earth's rotation, since 1623. As already stated in Chapter 6, in addition to the IERS web site (https://www.iers.org/), the reader is advised to check on the US Naval Observatory site (http://www.usno.navy.mil/)

## EXERCISE

Table 10.4 provides the dates of the perigee of the Sun from 2008 to 2016, as taken from Section A1 of the *Astronomical Almanac*. The second row gives the corresponding approximate JD. The third row gives the different durations of the passages from the previous year. Why only in 2011 and 2015 are the durations reasonably close to the standard value of the anomalistic year ($365^j.259636$)?

Solution: Notice that 2008, 2012 and 2016 were leap years. The average over 8 years of the values of the third row comes out 265.25, namely, the expected value within the precision. Then, there is a matter of definition. In Section A1 (Phenomena), the times are defined by the instant when the Earth–Sun distance (approximately 0.983 AU) is minimum, namely, when $\dot{R}(t) = 0$. These times do not agree with those corresponding to the passage through the perigee in the mean elliptical orbit, because of perturbations by the planets. The same reasoning applies to the instances of the apogee.

---

**TABLE 10.4**

**Dates of the Perigee of the Sun and Duration of the Passage with Respect to Previous One from 2008 to 2016**

| (UT1) | 2008 | 2009 | 2010 | 2011 | 2012 | 2013 | 2014 | 2015 | 2016 |
|---|---|---|---|---|---|---|---|---|---|
| January (d h m) | 2 23 51 | 4 15 30 | 3 00 09 | 3 18 32 | 5 00 32 | 2 04 38 | 4 11 59 | 4 06 36 | 2 22 49 |
| JD 245 + | 4468.5 | 4836.1 | 5199.5 | 5565.3 | 5931.5 | 6294.7 | 6662.0 | 7026.8 | 7390.5 |
| Difference ($\Delta$–JD) | | 367.6 | 363.4 | 365.8 | 366.2 | 363.2 | 367.3 | 364.8 | 363.7 |

---

# 11 The Terrestrial Atmosphere

This chapter is devoted to the examination of the influence of the Earth's atmosphere on the apparent coordinates of the stars and on the shape of their images; the discussion will be limited essentially to the visual band. The effects of the atmosphere on astronomical photometry and spectrophotometry will be expounded in Chapters 16 and 17.

Tycho Brahe had already recognized the importance of the atmospheric refraction. Great contributions came later, thanks to Cassini, Laplace, Bouguer and others. We will treat here only the case of sources at infinite distance from the observer; much more complex is the case of nearby sources embedded in the same refracting medium, which often produce quite spectacular phenomena such as the mirage and the *fata Morgana*; see, for instance, the books of Danjon (1980) and Minnaert (1993) quoted in the References section.

## 11.1   THE VERTICAL STRUCTURE OF THE ATMOSPHERE

Figure 11.1 provides a schematic indicative representation of the complicated vertical temperature profile of the atmosphere, subdivided into its main regions. The density profile can be represented by an exponential decrease of six orders of magnitude from the ground up to about 1000 km, as discussed in Section 11.6.

The *troposphere*, where approximately 90% of the total mass of the atmosphere is contained, extends from the ground to approximately 15 km of height, with noticeable latitude and seasonal dependence (several km higher at the equator than at the poles, in summer than in winter). The structure of the troposphere is controlled essentially by its temperature, which has an average gradient of −6 K/km (we will see later on the dependence on humidity of such gradient). The air does not significantly absorb solar radiation; instead, it is warmed by contact with the ground. The warm air expands as it warms, becomes less dense than surrounding cooler air and rises as buoyant and turbulent bubbles. Therefore, *convection* is the main process by which the troposphere mixes and heats. Although convection mixes the troposphere insuring an essentially uniform chemical composition, the higher it is the colder and less dense it becomes. The decrease in temperature, and so

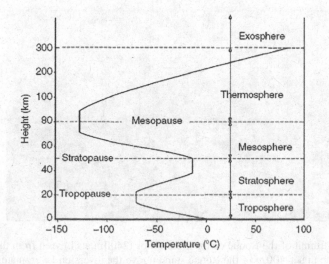

**FIGURE 11.1**   The indicative vertical structure of the temperature of Earth's atmosphere.

the convection, stops at the tropopause, as explained later on. Notice that in the stratosphere and the thermosphere, convection cannot develop; therefore, these regions have an intrinsic stability and consequent lack of turbulence.

Actually, the temperature profile in the troposphere can be more complicated than shown in Figure 11.1, because inside a critical layer situated in the first few kilometers (in some cases extending down to the ground), the temperature gradient can be inverted. The two most common ways to produce an inversion are

1. in winter time, when a cold and heavy air from polar regions moves southward (or northward, depending on which hemisphere) and slips under a warmer air mass;
2. during the summer months, when an upper-level high-pressure area causes air in the mid-levels to sink and warm, creating a capping inversion.

Telescopes on the ground are immersed in the troposphere, which affects astrometric, photometric and spectroscopic observations (see also Chapters 16 and 17). The temperature inversion has beneficial effects on astronomical observations performed from observatories situated above the inversion boundary, thanks to the minimal turbulence of the wind, which does not interact with the ground and thus acquire a laminar flow. This is the case, for instance, of the Observatory of the Roque de los Muchachos (Canary Islands, height 2400 m above sea level, a.s.l.), where the inversion layer is usually a few hundred meters below the telescopes at the top of the mountain (see Figure 11.2). Notice the wavy but essentially flat top of the clouds layer: clouds do not spill above the inversion boundary. Added beneficial effects of that such thick layer of clouds are to greatly dim the artificial lights and impede pollution from the cities at sea level to raise to the observatory level.

The chemical composition of the unpolluted troposphere is mostly molecular nitrogen $N_2$ (78%), molecular oxygen $O_2$ (21%), argon Ar (1%) and traces of water vapor $H_2O$ (the water vapor concentration may be as high as 3% at the equator and decreases toward the poles), carbon dioxide $CO_2$ and methane $CH_4$. Minute quantities of molecular hydrogen $H_2$ and helium are also present. Water vapor, $CO_2$ and $CH_4$, often called *greenhouse gases*, have great importance to establish the temperature of the surface of the Earth, essentially by blocking the re-radiation to space in the infrared (IR) of the ground heat. The nitrogen is normally inert, except under the action of lightening and in some biochemical processes; moreover, it can be enriched by human activity in the form of $NO_2$. In the same manner, the amount of $CO_2$ can fluctuate for anthropic causes.

**FIGURE 11.2**   The summit of the Roque de los Muchachos (2400 m a.s.l.) seen from the Teide mountain in Tenerife at sunset. The upper 400 m of the Roque stand above the inversion layer, materialized by the thick cloud layer at 2000 m height. (Photo by CB).

The height of the *tropopause* (a layer of almost constant temperature) from the ground ranges from 8 km at high latitudes to 18 km above the equator. As already said, it is also highest in summer and lowest in winter. Above the tropopause, at higher heights in the *stratosphere*, the temperature rises considerably, thanks to the absorption of solar UV radiation by the ozone molecule ($O_3$), with the process: UV photon + $O_3$ = $O_2$ + O + heat.

The *mesosphere* ranges from about 50 to 80 km; in this region, concentrations of $O_3$ and $H_2O$ vapor are negligible; hence, the temperature is lower than in the stratosphere. The chemical composition of the air becomes strongly height-dependent, with heavier gases stratified in the lower layers. Meteors entering the atmosphere start to warm up and mostly burn up as they enter the mesosphere. For that reason, special care must be taken to avoid excessive heating of re-entering spacecraft.

Coherent movements inside the atmosphere, known as internal thermosphere *gravity waves*, are common features in the mesosphere. They are frequently visible in the night-time airglow emissions, such as from OH near 90 km. Figure 11.3 provides an example, taken with the ASIAGO all-sky multispectral imager of the Center for Space Physics of the Boston University (Baumgardner et al., 2013; Smith et al., 2017).

Following the smooth decrease in the mesosphere, the temperature rises again in the *thermosphere*, because the solar UV and X-rays and the cosmic rays (mostly particles from the solar wind plus a small percentage of very energetic particles of extragalactic origin) partly ionize its very thin gases. This weakly ionized region, which conducts electricity and reflects back to the ground the radio waves with frequencies less than 30 kHz, is called *ionosphere*. It is divided into the regions *D* (60–90 km), *E* (90–140 km) and *F* (140–1000 km), according to the respective electron density profiles. A layer of neutral sodium Na between say 95 and 105 km is very important for laser-aided *adaptive optics* devices, as explained in Section 11.6.

Finally, above 1000 km, the gas composition is dominated by atomic hydrogen escaping the Earth's gravity, which is seen by satellites as a bright *geocorona* in the resonance line Ly-α at $\lambda = 1216$ Å.

Several interesting phenomena take place from 80 to 1000 km of height, where, for instance, the International Space Station at 400 km and the HST at 600 km navigate. These phenomena, such as noctilucent clouds, aurorae and stable red arcs (SAR arcs), and airglow, are controlled largely by the solar activity and terrestrial magnetic field. Their study is a most powerful source of information

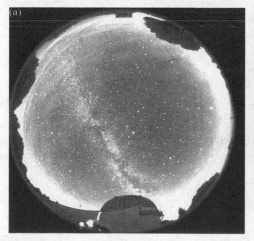
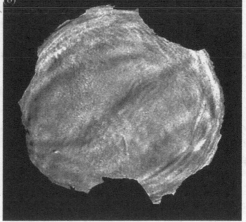

**FIGURE 11.3**   (a) Normal all-sky image taken at Cima Ekar Asiago Observatory with the BU ASIAGO multispectral. The ASIAGO multispectral atmospheric imager. The Milky Way is well visible. Bottom: silhouette of the Copernicus telescopes. Upper center: silhouette of the Schmidt telescopes and trees. (b) A still image of the fast-moving OH bands at a height around 90 km.

about the very complex *Sun–Earth interactions*, where light, particles, magnetic fields and inter-planetary plasma all play important roles (see, for instance, Mendillo et al., 2012 in the References section and Mendillo et al., in the Bibliography section). For centuries, the only means to study the solar activity were aurorae and sunspots. Today, there is a variety of devices on ground and in Space providing a continuous global surveillance. See the Web Sites section for more information on the Sun–Earth interactions.

## 11.2    THE REFRACTION

It is well known that the light propagates in a straight line in any medium of constant refraction index $n$, with a phase velocity given by $V = c/n = 1/(\varepsilon\mu)^{1/2}$, where $\varepsilon$ is the dielectric constant, and $\mu$ is the magnetic permeability of the medium, both quantities being wavelength dependent. The group velocity is $U = V - \lambda dV/d\lambda$. At the separation surface between two media of different refraction index (e.g., vacuum/air), the ray changes direction, so that the observer immersed in the second medium sees the light coming from an apparent direction different from the true one (see Figure 11.4). The propagation in a medium of continuously varying index of refraction can be treated by successive recourse to Huygens' principle, but in the following sections, a simpler treatment based on geometrical optics will suffice.

Consider again Figure 11.4 and suppose that the atmosphere can be treated as a succession of parallel planes (hypothesis of *plane-parallel stratification*), by virtue of its small vertical extent with respect to the Earth's radius. According to Snell's laws, when the ray coming from the region of refraction index $n_0$ encounters the separation surface with a medium of refraction index $n_1 > n_0$, part of the energy will be reflected back to the same hemispace (to the left in the figure) with the same angle $r_0$ with respect to the normal. This reflected fraction will not be considered here; it only implies a dimming of the source, which will be examined in Chapter 16. The remaining part will be refracted, in the same plane defined by the incident ray and the normal, to an angle $r_1 < r_0$. In reality, a clear atmosphere without clouds does not have sharp separation surfaces. The refraction index gradually increases from 1 to a final value $n_f$ near the ground, with typical scale lengths for significant changes of $n$ much greater than the wavelength of light (as already mentioned, we limit our treatment to the visual band). Therefore, the continuously varying direction can be considered as a series of finite steps in the plane passing through the vertical and the direction to the star (see Figure 11.5):

$$n_0 \sin r_0 = n_1 \sin r_1, \ldots, n_i \sin r_i = n_{i+1} \sin r_{i+1}, \ldots, n_{f-1} \sin r_{f-1} = n_f \sin r_f$$

where $n_{i+1} > n_i$, and $r_{i+1} < r_i$. By equating each term,

$$n_0 \sin r_0 = n_f \sin r_f \qquad (n_0 = 1) \tag{11.1}$$

Hence, the important result: in a plane-parallel atmosphere, the total angular deviation only depends on the refraction index close to the ground, independent of the exact law with which it varies along the path. The net effect is shown in Figure 11.6.

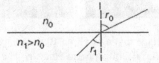

**FIGURE 11.4**    Vacuum–air refraction.

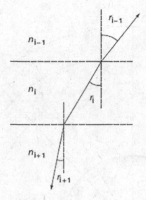

**FIGURE 11.5** The successive refraction angles in a plane-parallel stratified atmosphere.

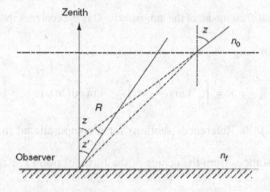

**FIGURE 11.6** The total atmospheric refraction.

With respect to the vertical, the star is seen in a direction $z'$ smaller than the true direction $z$, namely, closer to the local Zenith, by an amount $R$ that expresses the atmospheric refraction: $R = z - z'$. By virtue of Equation 11.1, and for small $R$ (in practice, if $z < 45°$),

$$n_f \sin z' = \sin z = \sin(z' + R) = \sin z' \cos R + \cos z' \sin R$$

$$\approx \sin z' + R \cos z'$$

(11.2)

or else

$$R = (n_f - 1) \tan z'$$

(11.3)

In the visual band, for average values of temperature and pressure ($T = 273$ K, $P = 760$ mmHg), $n_f \approx 1.00029$, so that in round numbers $R(15°) \approx 16''$, $R(45°) \approx 60''$, a very large effect indeed. For Zenith distances larger than 45°, the path of the ray inside the atmosphere is so long that the curvature of the Earth cannot be ignored. The mathematical treatment becomes more intricate, even restricting it to successive refractions in the same plane and with $n$ decreasing outwards with continuity.

With reference to Figure 11.7, let $dn$ be an infinitesimal variation of the refraction index in passing through the surface AA′ separating layer M from layer M′:

$$n \sin r = n' \sin r' = (n + dn)\sin(r - dr), \qquad dr = dR = \frac{dn}{n} \tan r$$

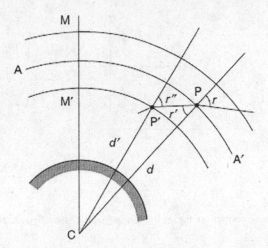

**FIGURE 11.7**   The spherical shell model of the atmosphere. C is the center of the Earth.

Integrating from $n = 1$ to $n = n_f$,

$$R = \int_1^{n_f} \tan r \, \frac{dn}{n} = \int_0^{\ln n_f} \tan r \, d(\ln n) \tag{11.4}$$

See Auer and Standish (2000, References section) for a computational method to solve Equation 11.4 for all Zenith angles.

Let us call $d$, $d'$ the distances from the center of the Earth of point P, P'; then, $d' \sin r'' = d \sin r'$, and therefore,

$$d' \sin r'' = d \sin r', \quad dnr = d'n' \sin r'' = \text{const} \tag{11.5}$$

In other words, the product of the distance of a particular point P from the center of the Earth times the refraction index at that point times the sin of the incidence angle is constant along the path of the ray. The radius of the Earth being $a_\oplus$, by the invariant relation Equation 11.5, we have on the ground

$$dn \sin r = a_\oplus n_f \sin z', \qquad \tan r = \frac{a_\oplus \, n_f \sin z'}{\sqrt{d^2 n^2 - a_\oplus n_f^2 \sin^2 z'}}$$

$$R = a_\oplus n_f \, \sin z' \int_1^{n_f} \frac{dn}{n\sqrt{d^2 n^2 - a_\oplus^2 n_f^2 \, \sin^2 z'}} \tag{11.6}$$

This integral has a fundamental importance in several studies of ray propagation inside an atmosphere. Here, we simplify the problem arriving to an expression, independent of the exact relationship between the height and density of the air, valid down to 75° of Zenith distance. Let us introduce a new parameter $s$ by letting $d = a_\oplus + H = a_\oplus (1 + s)$. The parameter $s$ is in any case a small quantity. Ignoring $s^2$ and higher powers, the integral in Equation 11.6 can be split into two parts, one independent and the other dependent on $s$:

$$R = \int_1^{n_f} \frac{n_f \sin z' dn}{n\sqrt{n^2 - n_f^2 \sin^2 z'}} - n_f \sin z' \int_1^{n_f} \frac{sn \, dn}{\sqrt{\left(n^2 - n_f^2 \sin^2 z'\right)^3}} = R_1 + R_2$$

The primitive of the integrand function in $R_1$ is $-\arcsin(n_f \sin z'/n)$, so that after several passages, the first integral provides

$$\tan \frac{R_1}{2} = \frac{\left(n_f - 1\right)}{\left(n_f + 1\right)} \tan\left(z' + \frac{R_1}{2}\right) = \frac{2n_f\left(n_f - 1\right)}{\left(n_f + 1\right)^2 - \left(n_f^2 - 1\right)\tan^2 z'} \tan z'$$

For angles $z'$ smaller than approximately 75°, the term in $\left(n_f^2 - 1\right)\tan^2 z'$ can be ignored; furthermore, $n_f$ is approximately equals to 1, so that $R_1$ is

$$\frac{R_1}{2} \approx \frac{2n_f\left(n_f - 1\right)}{\left(n_f + 1\right)^2} \tan z' \approx \frac{\left(n_f - 1\right)}{2} \tan z'$$

i.e., the same as in the elementary theory.

For the second term $R_2$, again down to $z' = 75°$, we can approximately put $n = n_f = 1$, so that

$$R_2 = \frac{\sin z'}{\cos^3 z'} \int_1^{n_f} s\, dn = \left(\tan z' + \tan^3 z'\right) \int_1^{n_f} s\, dn$$

$$= \left(\tan z' + \tan^3 z'\right) \frac{1}{a_\oplus} \int_1^{n_f} h\, dn \tag{11.7}$$

The last integral can be solved by parts:

$$\int_1^{n_f} s\, dn = \frac{1}{a_\oplus} \int_1^{n_f} H\, dn = \frac{1}{a_\oplus}\left[Hn\right]_1^{n_f} - \frac{1}{a_\oplus} \int_{H_1}^0 n\, dH$$

where $H_1$ is the value of $H$ for which $n = 1$. Recalling that $n = n_f$ corresponds to $H = 0$, we conclude that

$$R_2 = \left(\tan z' + \tan^3 z'\right) \int_{0^-}^{H_1} (n - 1)\, dH$$

As shown in Section 11.5, the term $(n-1)$ is nearly proportional to the density of the air. Therefore, the above integral is simply proportional to the total mass of the air column from the ground to the summit of the atmosphere, quite independent of the exact law between height and density. We can conclude that

$$R_2 = \left(\tan z' + \tan^3 z'\right) l \rho_0$$

where $l$ is the length of a hypothetical column of air at constant density having the same temperature and pressure observed on the ground. Collating the two results, we have finally

$$R = A \tan z' + B \tan^3 z' = \left(n_f - 1\right)\left[\left(1 - \frac{l}{a_\oplus}\right)\tan z' - \frac{l}{a_\oplus}\tan^3 z'\right] \tag{11.8}$$

where

$$A = \left(n_f - 1\right) + B, \qquad B = -\frac{l}{a_\oplus}\left(n_f - 1\right)$$

Typical values of $l$ at sea level are around 8 km; in addition to the temperature, the actual value depends on the elevation and latitude (namely on the gravity) of the site; typical values for $B$ are therefore around $-0''.07$.

## 11.3   EFFECTS OF REFRACTION ON THE APPARENT COORDINATES

We have seen that the main effect of the refraction is to move the star closer to the Zenith in the vertical plane, thus raising its elevation $h$ but leaving essentially unchanged its azimuth $A$. From Figure 11.8, we get

$$\begin{cases} -\Delta HA \cos\delta = (\alpha' - \alpha)\cos\delta = R\sin q \\ \Delta\delta = \delta' - \delta = R\cos q \end{cases} \tag{11.9}$$

where the parallactic angle $q$ can be found by one of the two formulae (see Chapter 3):

$$\cos\delta\cos q = \sin\varphi\cos h + \cos\varphi\sin h\cos A$$

$$\sin A\sin h = \cos HA\sin q + \sin HA\cos q\sin\delta$$

In meridian, the refraction is all in declination and this is true also for the Sun at true noon.

For Zenith distances not greater than approximately 45°, after several passages we finally get

$$\begin{cases} \Delta\alpha = (n_f - 1)\dfrac{\sec^2\delta\sin HA}{\cos HA + \tan\varphi\tan\delta} \\ \Delta\delta = (n_f - 1)\dfrac{\tan\varphi - \tan\delta\cos HA}{\cos HA + \tan\varphi\tan\delta} \end{cases} \tag{11.10}$$

Formulae 11.10 give the corrections to the true coordinates needed to calculate the apparent ones, according to the topocentric position of the observer. Obviously, no such correction is necessary for a telescope in outer space.

It is worth mentioning at this point the effects of the atmosphere on astrometry at radio frequencies. We have underlined how radio astronomy became more and more important for fundamental astrometric work, as discussed in detail in the context of the VLBI and of the ICRF. The variable electron content of the ionosphere and the variable humidity of the troposphere affect the propagation of the radio waves. The *ionosphere* will introduce a delay on the arrival time of the wave, given by

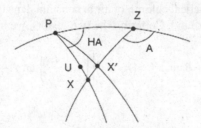

**FIGURE 11.8**   The effect of the refraction on the equatorial coordinates. $XX' = R = \Delta h$, $PXX' = PXZ = q$, $ZX = z$, $PX = 90 - \delta$, $XU = \Delta\delta$.

$$\Delta T = \frac{40.3}{c\nu^2} \int_I N_e \, ds \text{ seconds}$$

where $\nu$ is the frequency of the wave, $N_e$ is the electron content (which varies with the night and day cycle, with the season and also with the solar cycle) and $I$ is the total path inside the ionosphere along the line of sight.

The *tropospheric delay* can be divided into two components, a dry one and a humid one. The dry component amounts to about 7 ns at the Zenith and varies with a modified "cosec $z$" relationship:

$$\Delta T = 7\left( \cos z + \frac{0.0014}{0.0045 + \cot z} \right) \text{ns}$$

The wet component is about 10% of the dry one, but it depends strongly on the content of water vapor, and varies rapidly and is not foreseeable.

These fluctuations can be compensated with dual-frequency transmitters and receivers and appropriate corrections.

We only mention that two other media affect the propagation of the radio waves, namely, the solar corona and the ionized interstellar medium (see the "Notes" section).

## 11.4 THE CHROMATIC REFRACTION OF THE ATMOSPHERE

The refraction index $n$ depends on the wavelength, diminishing from blue to red; the same is true for the refraction angle $R$. Consequently, the image on the ground of the star is a succession of monochromatic points aligned along the vertical circle. The blue ray is below the red one, so that the blue star appears above the red one (see Figure 11.9). Therefore, the atmosphere behaves as a prism producing a short spectrum in the vertical plane, whose length increases with the Zenith distance, reaching several arc seconds at low elevations.

The relationships $n(\lambda)$ can be expressed by the so-called Cauchy's formula:

$$n(\lambda) = 0.00028\left( 1 + \frac{0.00566}{\lambda^2} + \frac{0.000047}{\lambda^4} \right) \quad (\lambda \text{ in } \mu m) \tag{11.11}$$

corresponding to a variation of approximately 2% over the visible range. Table 11.1 gives representative values of the refraction at a site such as the Roque (at 2400 m a.s.l) in several photometric bands from the blue to the near IR.

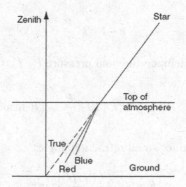

**FIGURE 11.9** The atmospheric chromatic refraction.

**TABLE 11.1**

**Indicative Chromatic Refraction at 45° Elevation for an Observatory at 2400 m a.s.l.**

| Band | Spectral Range (nm) | Refraction (milliarcsec) |
|------|---------------------|--------------------------|
| B    | 391–489             | 730                      |
| V    | 505–595             | 290                      |
| R    | 590–810             | 350                      |
| I    | 780–1020            | 190                      |

As already said, at low elevations of the star above the horizon, the spectrum can become several arcsec long. Therefore, systematic effects due to chromatic dispersion of the atmosphere must be feared on broad-band *astrometry*, if the color of the object under investigation is different from that of the reference stars (e.g., when measuring the coordinates of a red asteroid against a net of blue stars or of a very blue quasar against redder stars). The influence of refraction in astrometry and geodesy is discussed, for instance, by Saastamoinen (Bibliography section). In addition, the chromatic refraction introduces severe complications in accurate *spectrophotometric* works, in particular if the light of star enters the spectrograph through a slit not aligned in the vertical plane. Indeed, it is always advisable to align the slit of the spectrograph in the vertical plane, as we will also discuss in Chapters 16 and 17. To minimize these adverse effects, resort can be made to appropriate (and complex) optical devices called atmospheric dispersion correctors.

## 11.5   RELATIONSHIPS BETWEEN REFRACTION INDEX, PRESSURE AND TEMPERATURE

At any given wavelength $\lambda$, the refraction index depends on the density $\rho$ according to Gladstone–Dale's law:

$$n - 1 = k\rho \tag{11.12}$$

With the hypothesis that the atmosphere behaves as a perfect gas of pressure $P$, temperature $T$ and molecular weight $\mu$, the density is given by

$$\rho = \frac{\mu P}{RT} \tag{11.13}$$

where $R$ is the gas universal constant, so that

$$n - 1 = k' \frac{P}{T} \tag{11.14}$$

By referring to normal values of temperature and pressure $P_0$, $T_0$ (i.e., 760 mmHg, 273 K) one has

$$\frac{n-1}{n_0-1} = \frac{P}{P_0}\frac{T_0}{T}, \qquad n-1 \approx 78.7 \times 10^{-6}\frac{P}{T} \quad \left(P \text{ in mmHg}, \quad T \text{ in K}\right)$$

Therefore, in the visible band and for small refraction angles,

$$R \approx 60''.4\frac{(P/760)}{(T/273)}\tan z$$

Calling $h$ the height over the ground, from Equation 11.14 we have

$$dn = k'\frac{1}{T}\left(dP - \frac{P}{T}dT\right), \quad \frac{dn}{dh} = k'\left(\frac{1}{T}\frac{dP}{dh} - \frac{P}{T^2}\frac{dT}{dh}\right) = k'\frac{P}{T^2}\left(\frac{T}{P}\frac{dP}{dh} - \frac{dT}{dh}\right)$$

The variation of pressure with the height is equal to the weight of the air in the elementary volume having unitary base and height $dh$, $dP = -g\rho(h)$ dh (which is the condition of *hydrostatic equilibrium*), so that

$$\frac{dn}{dh} = k'\mu\frac{P}{T^2}\left(\frac{-g}{R} - \frac{dT}{\mu dh}\right) \tag{11.15}$$

where $g$ is the (assumed constant) gravity, and the constant $g/R \approx 3.4$ K/km is called the *adiabatic lapse*. From Equation 11.15, we reach the conclusion that the variations of the refraction index depend on the vertical gradients of the temperature.

From the condition of hydrostatic equilibrium and the further hypothesis of thermal equilibrium (*isotheral Boltzmann's atmosphere*), we derive an exponential decrease of density and pressure from the ground, with a typical scale height $H = kT/mg$, where $m$ is the mass of the elementary volume:

$$\rho(h) = \rho_0 e^{-mgh/kT}, \quad P(h) = P_0 e^{-mgh/kT}$$

At $h = H$, density and pressure are reduced to $1/e$ the value at ground level. Typical values of $H$ are around 7–8 km. It is easy to demonstrate that approximately 63% of the mass of the atmosphere is below $H$.

Although the isothermal Boltzmann's model in not correct, it provides a useful reference model. In non-isothermal conditions, with $T = T(h)$ but always at constant gravity and molecular weight, one must distinguish the scale height for density, $H_\rho$, from the previous one, which is more properly the scale height for pressure:

$$\frac{1}{H_\rho(h)} = \frac{1}{T(h)}\frac{dT(h)}{dh} + \frac{gm}{kT(h)}$$

On closer examination, the scale height should be molecular mass dependent; for instance, the scale height of $CO_2$ is 44 times lower than that of $H_2$. Thus, the Boltzmann's law would predict a *stratification* of the troposphere, which however does not take place thanks to *convection*.

The discussion of the convection should be done more properly, taking into account the humidity of the air, namely, the ratio $\gamma$ between its specific heats at constant volume and constant pressure. We limit here to quote that for dry air, whose average molecular mass is $m = 28.7m_p$, (where $m_p \approx 1.66 \times 10^{-24}$ g is the proton mass) and the adiabatic index is $\gamma = 7/5$, the adiabatic lapse becomes −9.8 K/km; for very humid air, it reduces to −5 K/km.

We postpone the discussion of the balance between gravitational and thermal energies and Maxwell's equation to Chapter 12. We simply quote here that

- the most probable velocities of the $N_2$ ($m = 28m_p$) and $O_2$ ($m = 32m_p$) air molecules near the ground and at 0C (273 K) are around $V_t \approx 500-450$ m/s, while for $H_2$ ($m = 2m_p$), $V_t \approx 1700$ m/s. These values, compared with the escape velocity from the ground ($\approx 11$ km/s) and the long tail of Maxwell equation, explain on one side the capability of the Earth to retain a stable dense atmosphere, and on the other, why the lighter elements Hydrogen and Helium are essentially absent in the lower layers;

- the mean free path between two collisions is around $10^{-5}$cm; dividing this distance by the average speed, we see that each molecule experiences about $10^9$ collisions per second, with an efficient equipartition of energies and consequent local thermal equilibrium.

## 11.6  SCINTILLATION AND SEEING

The unavoidable thermal turbulence of the atmosphere above the observatory produces unpredictable (although small) variations of the direction of arrival of the rays from a given star. From a different point of view, the turbulence is responsible of erratic distortions of the wavefront. To the unaided human eye, the most impressive phenomena are the scintillation and the color variation of the brightest stars (in popular terms, the stars "twinkle"). Indeed, the small aperture of the pupil accepts only a small portion of the wavefront, so that at a certain moment, all rays converge and the star will appear brighter; at another instant, all rays diverge and the star will dim. The colors will also change, for the same reason of small spatial sampling of the wavefront. If we look at a planet of non-negligible angular size (e.g., Jupiter) instead of a point-like star, all points of the planetary disk will scintillate with different phases, so that the total effect will essentially vanish (planets do not twinkle). Scintillation is present not only in the eye vision, but it can be detected and measured also on telescopes of small or moderate aperture; see the three papers by Dravins and collaborators (1997a, b, c) quoted in the References section.

On the focal plane of a telescope, the position and shape of the stellar image will rapidly vary, with characteristics that depend on the size of the aperture. These phenomena, which originate from erratic wavefront curvatures and tilts, produce angular displacements and affect the structure of the stellar images; they are called collectively stellar *seeing*. A complete discussion should take into account, in addition to the size and shape of the telescope aperture (usually but not necessarily circular), also the wavelength and the temporal scale over which the seeing is sampled. A crucial parameter is the ratio between the typical timescale of the passage of a wavefront disuniformity over the aperture (which ranges from say a few thousandths to a few seconds) and the response time of the particular detector. While the human eye has response times of few tenths of a second and does not integrate the signal, detectors such as the photographic plate and several solid-state detectors integrate the faint stellar light for seconds, minutes and even hours. As a consequence of its fast response time, the eye can sometimes discern details invisible on long exposures (typical is the case of visual binary stars). Instead, at the end of a long exposure, the light from the star will fill an approximately circular region whose diameter is called the *seeing diameter*. Inside this circle, the light distribution can be approximated by a Gaussian, or by the sum of two Gaussians of different widths, or by a Lorentz function or even by an empirically determined numerical expression. Therefore, a convenient parameter to characterize the curve is its full width half maximum (FWHM), namely, the width of the curve (usually in arcsec) at half its maximum value. Figure 11.10 provides an example of seeing FWHM obtained in the visual band from years of measurements at the European Southern Observatory (ESO) of Cerro Paranal, in the Chilean Atacama desert. The fraction of the stellar energy contained in any circular radius from the center at a given seeing FWHM depends on the actual shape of the curve. Figure 11.11 provides an indicative estimate.

Special detectors are available with response time of the order of 1 ms. They can sample the instantaneous structure of the light distribution over the focal plane. An image with such short exposure time, in an almost monochromatic band, reveals that the light is distributed in a large number of so-called *speckles*, whose dimension is essentially the diffraction figure of the telescope at that wavelength, namely, essentially $\lambda/D$ radians, where D is the diameter of the telescope. A discussion of the image formation and structure, tailored to the astrometric needs from ground and space, is provided for instance by Kovalevsky (Bibliography section).

The complex statistical theory of the air turbulence and instantaneous image structure is associated with names such as Kolmogorov and Tatarski, who provided the theoretical foundations

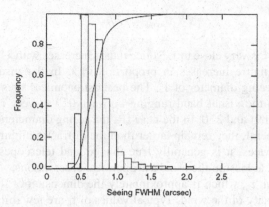

FIGURE 11.10   The seeing distribution at Cerro Paranal (2400 m a.s.l) in the visual band. The ordinate is the frequency of a given seeing bin in the histogram. The curve is the cumulative distribution function. The median FWHM is 0″.67. (Courtesy of ESO).

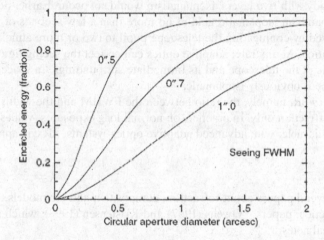

FIGURE 11.11   Indicative value of the fraction of the stellar energy contained inside a circle of a given radius on the focal plane, according to three different values of the seeing FWHM.

of methods employed to contrast the detrimental effects of the seeing on the image quality. The theoretical complexity is reflected by the very sophisticated technologies implemented on all large modern telescopes. These techniques are known collectively as *adaptive optics* devices. See the pioneering paper by W.H. Babcock (1953, References section). In advanced systems, the devices on the focal plane of the telescope are complemented by powerful lasers, which define an artificial yellow star by exciting the already mentioned Na layer around 100 km. Such artificial star provides the wavefront reference. Some advanced adaptive optics systems like SPHERE at the ESO VLT (see the Web Sites section) are equipped with coronagraphic masks meant to suppress the light of the star, thus permitting to obtain images of their accompanying planets, which are typically only a fraction of arcsec away.

We quote some of the figures of merit characterizing these techniques:

- the *coherence length* on the plane perpendicular to the propagation vector, namely, the distance over the wavefront necessary to change the phase by more than 1 radian. Such distance is known as Fried's parameter $r_0$, which is also, roughly speaking, the diameter of a telescope whose diffraction figure has an angular extension equal to the seeing:

$$\theta_{\text{seeing}} \approx k\lambda/r_0 \tag{11.16}$$

where the constant $k$ is very close to 1. Notice that $r_0$ increases with $\lambda^{6/5}$, while the extension of the diffraction figure increases in proportion to $\lambda$. In the visual band, $r_0 = 10\,\text{cm}$ corresponds to a seeing diameter of $1''$. The best astronomical sites are characterized by seeing FWHMs in the visual band ranging between $0''.3$ (a very rare figure indeed, as shown by Figure 11.9) and $2''.0$. In the near IR, the seeing diameters are correspondingly smaller, but, in general, they remain larger than the diffraction figure. Therefore, without adaptive optics devices, it is generally true that ground telescopes are seeing, and not diffraction limited, at least in the usual astronomical observations;

- the *correlation time* $\tau_0$, which is approximately the dimension of the average air bubble divided by the velocity of the wind. Typical values of $\tau_0$ are few milliseconds. As $r_0$, also $\tau_0$ increases with $\lambda^{6/5}$;
- the *isoplanatic angle* $\theta_0$: if the atmosphere were a single thin layer of air, any ray would suffer the same deflection, independently of the arrival direction; the distortion would therefore be the same all over the field of view (5A*isoplanatism*). However, this is not the case, and already with two layers, isoplanatism would not occur. Earlier devices provided, in the visual band, an isoplanatic field of no more than a few seconds of arc. This figure can be improved by conjugating the telescope pupil to two or more atmospheric layers of different elevation. At any rate, adaptive optics can correct the seeing figure of the star on the optical axis of the telescope and its immediate surroundings. In space, the entire field of the telescope is obviously isoplanatic;
- The *Strehl ratio* SR, namely, the ratio between the FWHM and the width expected in the presence of diffraction only. In practice, on normal long exposures, values of SR = 0.1 are very rarely achievable; with advanced adaptive optics systems, SR can approach 1.

## NOTES

- Among the several papers about refraction, composition and air models, we quote in the References section: papers – Owens (1967) and Kristensen (1998), which also gives interesting historical notes.
- The refraction alters the apparent position of the pole and the apparent speed of rotation of the celestial sphere. These effects cannot be ignored in telescopes with a large field of view (FoV), such as the Schmidt telescopes. For equatorial telescopes, it is possible to alter the inclination of the polar axis with respect to the horizontal plane according to the elevation of the FoV. Nevertheless, the differential refraction over a FoV of several degrees can be noticeable.
- Another differential effect of the refraction is the contraction of apparent diameters, essentially all in the vertical plane. If $s$ is the true diameter of say a planet, from 11.3 we get

$$s' = s - dR = s - (n_f - 1)s/\cos^2 z'$$

Even at $z \approx 80°$ the effect is small, but it grows rapidly; at the horizon, it becomes about $6'$ for the Sun and the Moon. This shortening of the vertical axis is therefore well measurable. The subjective judgment can however be deceiving: we tend to judge the Moon (or the Sun) much larger and flattened at the horizon than at the Zenith. Several explanations have been put forward for this so-called Moon's illusion; see Minnaert (Bibliography section). Images taken on board the International Space Station provide a spectacular visualization of differential refraction. From there, the length of the ray path is practically double than for ground observers. See, for instance, the series of photos of the setting full moon taken by astronaut Don Pettit on April 16, 2003 aboard the International Space Station. These images are shown and discussed in several places, e.g., http://fer3.com/arc/m2.aspx/Odd-star-brightness-seen-from-ISS-night-orbits-FrankReed-mar-2018-g41599

- The refraction also alters the instant of the eclipses and of the occultations. Owing to refraction, the observer aligned with the Sun and Moon at a certain instant $t$ is not the ground observer in P, but one slightly higher in P′. Chauvenet gave a formula, valid down to $z \approx 70°$, for the distance $H$ of P′ from P:

$$1 + h/a = n\sin(z - R)/\sin z$$

where $a$ is the distance of P from the center of the Earth. Typical values of $H$ are around 20 m.

- In addition to our atmosphere, all transparent media along the light path can affect the propagation of the light, however distant they might be from the observer. An interesting discussion is given by Moniez (2003).
- Regarding adaptive optics, among the many papers we quote in the References section Ragazzoni (1996), Ragazzoni et al. (2000); in the Bibliography section: Vernet et al. (2002) and two useful booklets by SPIE: Andrews (2004), Tyson and Frazier (2004).
- In addition to devices with deformable elements and laser guide stars, two more techniques are applied to remove the atmospheric distortions:

  *Speckle interferometry*, where a large number of rapid narrow band images are deconvolved via Fourier analysis, see Weigelt G. (1991, Bibliography section);

  *Lucky imaging and beyond.* Lucky imaging consists in selecting images of good quality in a large collection of short exposure ones. See Law et al. (2006). In successive developments, lucky imaging has been combined with low-order adaptive optics deformable mirrors. See Mackay et al. (2014).
- Although the discussion of these effects is outside the scope of this book, we underline that dispersion of the radio waves in the interstellar medium complicates the comparison with the arrival time of optical pulses from pulsars. See, for instance, Zampieri et al. (2014) for a comparison of the arrival times of pulses detected with the photon counter Iqueye at the 3.5m NTT ESO telescope and the radio pulses detected at Jodrell Bank from the pulsar in the Crab nebula.
- In addition to adaptive optics, the resolution in the extended visible band can be greatly augmented by interferometric devices, like CHARA and VLTI; a peculiar arrangement is possible by combining the two 8.4 mirrors of the Large Binocular Telescope. See the Web Sites section. The atmospheric turbulence and dispersion causes great difficulties and require sophisticated technological solution, such as vacuum pipes and accurate thermal control of the light path from the focus of the telescope to the interferometric station. The best results achieved to date are in the near IR, as expected from the dependence of Fried's radius, coherence time and other parameters from wavelength. Therefore, red stars are the preferred targets for those techniques. The theories behind these interferometers go back to Fizeau and Michelson. There is a different type of interferometry, still affected by atmospheric refraction and chromatism, but essentially independent from the seeing, namely, the Hanbury Brown and Twiss Intensity Interferometry (HBTII), experimented at Narrabri (Australia) around 1970 (see Hanbury Brown, 1974, Bibliography section). The interferometer is realized exploiting the correlation of the arrival times of photons on the two apertures; resolution times of the order of few nanoseconds and large photon fluxes are necessary. No optical link connecting the telescopes is required, only a common high-stability, high-precision time reference system. The efficiency is better in the blue than in IR; blue stars are the best targets, thus complementing the data from phase interferometry (notice

that in some texts, the Narrabri interferometer is mistaken for a radio one). Modern detectors promise to resurrect the almost forgotten technique. Cavazzani et al. (2011) have calculated the dispersion of the arrival times of optical photons, accurate at the picosecond level, required by very high time resolution astrophysical applications, such as the comparison of radio and optical data of Giant Radio Bursts from optical pulsars and by HBTII. Their calculations were based on the Marini–Murray model (Marini & Murray, 1973) correction to the optical path of photons in air. See Dravins (2019, Bibliography section) for a review of intensity interferometry.

# 12 The Two-Body Problem

In this chapter, we consider the dynamics of two bodies under their mutual gravitational attraction, a fundamental problem in many astronomical situations, from the Earth–Moon and Sun–planet systems to exoplanets, binary stars and pairs of galaxies. The two bodies will be considered of very small extent with respect to their mutual distance, in order to regard them as point-like masses (particles). The present approach is based on the Newtonian dynamics, with the underlying assumptions of Galilean relativity (space is homogeneous and isotropic, time is uniform, all inertial system are equivalent with respect to the dynamical properties). No mention will be made of chaotic dynamics (see, for instance, the book by Celletti quoted in the Bibliography section). Occasionally, we will introduce some modifications due to general relativity. One of the basic disagreements with the Newtonian dynamics originated indeed from astronomical evidence accumulated at the end of the 19th century, namely, by the secular advance of Mercury's perihelion. Once allowance was made for all known planetary perturbations, a fraction of the observed value, namely, 43″/century out of the total 550″/century, could not be explained. The correct value was derived from general relativity, as described in Chapter 14.

As is well known, some of the results obtained for the gravitational interaction can be applied to the electrostatic force between two charged particles, although the electrostatic force can also be repulsive. The reason for such commonality is that they both are central forces, with same line of action and dependence of the intensity from the inverse square of the inter-particle distance.

## 12.1 THE BARYCENTRIC TREATMENT

Let $P_1$ and $P_2$ be two particles of masses $m_1$ and $m_2$, respectively. Given an arbitrary but inertial frame of reference O(x, y), let $\mathbf{r}_1$, $\mathbf{r}_2$, $\mathbf{V}_1$, $\mathbf{V}_2$ be the position and velocity vectors of the two particles in it. The gravitational force between them is directed along the joining line, with an intensity depending only on their relative distance:

$$\mathbf{F} = F(r)\frac{\mathbf{r}}{r}, \quad \mathbf{r} = \mathbf{r}_2 - \mathbf{r}_1 \tag{12.1}$$

Furthermore, the force can be derived from a potential function $-U(r) = -U(|\mathbf{r}_2 - \mathbf{r}_1|)$:

$$\mathbf{F} = -\frac{dU}{dr}\frac{\mathbf{r}}{r}, \quad F_x = -\frac{\partial U}{\partial x}, \quad F_y = -\frac{\partial U}{\partial y}, \quad F_z = -\frac{\partial U}{\partial z} \tag{12.2}$$

The gravitational force is then a particular case of central forces; the potential belongs to the class of *conservative* fields. Notice the minus sign in the definition of the potential; the function $U(r)$ is then the potential *energy* of the system. While $\mathbf{F}$ is a vector, the potential $-U(r)$ is a scalar quantity.

The movements $\mathbf{r}_1(t)$, $\mathbf{r}_2(t)$ of the two particles can be derived from the fundamental Newtonian equations:

$$m_1\frac{d^2\mathbf{r}_1}{dt^2} = m_1\ddot{\mathbf{r}}_1 = \mathbf{F}_1 = -\mathbf{F}, \quad m_2\frac{d^2\mathbf{r}_2}{dt^2} = m_2\ddot{\mathbf{r}}_2 = \mathbf{F}_2 = \mathbf{F} \tag{12.3}$$

It is convenient to translate the origin of the inertial system in the barycenter B, a point always on the line joining the two particles. With respect to B, the distances of the two particles satisfy the relationships:

$$m_1\mathbf{r}_1 + m_2\mathbf{r}_2 = 0, \; \frac{r_1}{r_2} = \frac{m_2}{m_1}, \quad \mathbf{r}_1 = -\mathbf{r}\frac{m_2}{m_1 + m_2}, \quad \mathbf{r}_2 = \mathbf{r}\frac{m_1}{m_1 + m_2} \tag{12.4}$$

Therefore, the barycenter is closer to the heavier of the two masses.

The Lagrange function $L$ of the system, namely, the difference between the kinetic energy $T$ and the potential energy $U$, is

$$L = T - U = \frac{1}{2}m_1\dot{\mathbf{r}}_1^2 + \frac{1}{2}m_2\dot{\mathbf{r}}_1^2 - U(r)$$

As is well known, the Lagrange function $L$ satisfies the conditions:

$$\frac{\mathrm{d}}{\mathrm{d}t}\frac{\partial L}{\partial \mathbf{V}} = \frac{\partial L}{\partial \mathbf{r}} = \frac{\mathrm{d}U}{\mathrm{d}r}, \quad \frac{\mathrm{d}}{\mathrm{d}t}\frac{\partial L}{\partial \dot{r}_i} - \frac{\partial L}{\partial r_i} = 0 \quad (i = 1, 2, 3)$$

In the barycentric system, from Equation 12.4 we derive

$$\dot{\mathbf{r}} = \mathbf{V} = \dot{\mathbf{r}}_2 - \dot{\mathbf{r}}_1, \quad T = \frac{1}{2}\left(m_1 V_1^2 + m_2 V_2^2\right) = \frac{1}{2}\frac{m_1 m_2}{m_1 + m_2}V^2 = \frac{1}{2}mV^2$$

where

$$m = \frac{m_1 m_2}{m_1 + m_2} = \frac{m_2}{1 + m_2/m_1}, \quad \frac{1}{m} = \frac{1}{m_1} + \frac{1}{m_2}$$

is the so-called *reduced mass* of the two particles. If $m_1$ is the heavier of the two,

$$\frac{1}{2}m_2 \leq m \leq m_2$$

Therefore, in the barycentric system the Lagrange function formally coincides with that of a single particle of mass $m$ in movement in an external field $-U(r)$, which is symmetric with respect to the fixed origin of coordinates. If we determine the movement $\mathbf{r}(t)$ of this fictitious particle, then by Equation 12.4 we can obtain separately $\mathbf{r}_1(t)$ and $\mathbf{r}_2(t)$.

The equations of movement are three second-order differential equations (or six first-order equations in the Hamiltonian notation). Therefore, six initial arbitrary constants are needed in order to specify the movement of the particles, for instance, position and velocity at a given time. In astronomical applications, those initial constants are the six orbital elements, which will be introduced in Chapter 13. Furthermore, the masses of the two particles (or of one of them) are usually unknown, at least initially.

In addition to the reduced mass, it is convenient to consider the total mass $M$ of the two particles: $M = m_1 + m_2$. If $m_1$ is the heavier of the two, then

$$M = m_1 + m_2 = m_1\left(1 + \frac{m_2}{m_1}\right), \quad m_1 \leq M \leq 2m_1$$

In particular, for the case Sun–planet:

$$m_1(= M_\odot) \gg m_2(= m_{\text{planet}}), \quad m \approx m_2(= m_{\text{planet}}), \quad M \approx m_1(= M_\odot) \tag{12.5}$$

The largest deviation from Equation 12.5 is for Jupiter, because $M_\odot \approx 1048 \; m_{\text{Jupiter}}$. The same result applies to the proton–electron pair, because the proton is 1836 times heavier than the electron.

Owing to the properties of the force field (which is central and conservative), the isotropy of space imposes the conservation of the angular momentum **K** of the fictitious particle, namely, of the vector product:

$$\mathbf{K} = r \times mV = K\frac{r}{r} = Kk \tag{12.6}$$

Therefore, the vector **K** must be constant in both direction and modulus. The consequence of the constant direction is immediate: being the angular momentum perpendicular to the plane of **r** and **V**, the movement must be confined at all times in a plane perpendicular to **K**. The central movement takes place always on such plane (the orbital plane), whatever the mathematical expression of $-U(r)$, even for a repulsive force (e.g., proton–proton).

To examine the consequences of the constancy of the modulus of **K**, let us define a polar system $(r, \varphi)$ in the orbital plane, with origin in the barycenter B and arbitrary orientation of the axes. The vectors velocity **V** and acceleration **a** can be decomposed in two components: one along the radial direction and one along the perpendicular to the radius vector. If **i**, **j** are the unitary vectors along those directions, then

$$V = \dot{r}\mathbf{i} + r\dot{\varphi}\mathbf{j}, \quad a = \left(\ddot{r} - r\dot{\varphi}^2\right)\mathbf{i} + \frac{1}{r}\frac{d}{dt}\left(r^2\dot{\varphi}\right)\mathbf{j}$$

The modulus of the angular momentum and the kinetic energy are

$$K = mr^2\dot{\varphi}, \quad T = \frac{1}{2}m\left(\dot{r}^2 + r^2\dot{\varphi}^2\right) \tag{12.7}$$

The variable $\varphi$ does not appear explicitly in $K$, $T$ and $L$; it is then called a *cyclic* variable. The associated generalized momentum $k = mr^2\dot{\varphi}$ constitutes a so-called *integral of motion*. This result too is entirely independent of the actual mathematical expression of $-U(r)$. There is a very useful geometrical interpretation of such constancy of $K$; let

$$dA = \frac{1}{2}r \cdot r \, d\varphi \tag{12.8}$$

be the elementary area swept by the radius vector in the time $dt$; the areal velocity $dA/dt$ is then constant, a result known as Kepler's second law in the gravitational case, but which is valid in any central and conservative force field. In the astronomical notation, it is customary to introduce the area constant $C/2$ defined as in Chapter 4 by

$$\frac{C}{2} = \frac{dA}{dt} = \frac{1}{2}r \cdot r\dot{\varphi} = \frac{K}{2m}, \quad K = Cm \tag{12.9}$$

Then,

$$\dot{r} = \frac{dr}{d\varphi}\dot{\varphi} = \frac{C}{r^2}\frac{dr}{d\varphi} = -C\frac{d(1/r)}{d\varphi}, \quad \ddot{r} = -\frac{C^2}{r^2}\frac{d^2(1/r)}{d\varphi^2}$$

The transverse acceleration is always nil, as it should in a central field, while the radial one is given by Binet's formula:

$$a_r = -\frac{C^2}{r^2}\left[\frac{1}{r} + \frac{d^2(1/r)}{d\varphi^2}\right]$$

(12.10)

The kinetic energy is given by

$$T = \frac{1}{2}m\dot{r}^2 + \frac{K^2}{2mr^2}$$

(12.11)

The total energy (another integral of motion, deeply connected to the uniformity of time) is

$$E = T + U = \frac{1}{2}m\left[\dot{r}^2 + r^2\dot{\varphi}^2\right] + U(r)$$

(12.12)

Or else

$$dt = \pm\sqrt{\frac{2(E-U)}{m} - \frac{K^2}{m^2r^2}}\,dt, \quad d\varphi = \frac{K}{mr^2}dt$$

(12.13)

Relation (12.13) contains two differential equations in the variables ($r$, $\varphi$, $t$). By eliminating $t$, the equation of the *trajectory* $r(\varphi)$ in its plane is obtained; by eliminating $\varphi$, we obtain the equation of the *movement* $r(t)$; finally, by eliminating $r$, the equation of the *anomaly* $\varphi(t)$ can be derived. From the constancy of the sign of $d\varphi/dt$, we conclude that the function $\varphi(t)$ is a monotonic one: the movement cannot reverse direction, whatever the potential $-U(r)$. Furthermore, the radial movement takes place as if the effective potential was the sum of that due to the external force plus a centrifuge potential:

$$-U_{\text{eff}} = -U(r) - \frac{K^2}{2mr^2}$$

If $K \neq 0$, the particle cannot reach the center (the two particles cannot collide), unless $-U(r)$ approaches zero faster than $1/r^2$, or as $U_0/r^2$ provided $U_0 > K/2m$.

For those values of $r$ for which the effective potential energy $U_{\text{eff}}(r) = E$, it is also $dr/dt = 0$; namely, the distance inverts its behavior in the points of minimum or maximum $r(t)$. However, even in those points $d\varphi/dt$ is different to zero and maintains the same sign.

Notice that, in general, there is no insurance that $r$ varies between a minimum and a maximum value ($r_{\min}$ and $r_{\max}$): there is no *a priori* reason why the orbit should be confined to a ring in its plane. There are cases, indeed, when $r_{\max}$ can become arbitrarily large: the two particles approach to a minimum distance, and then separate again to infinity.

Furthermore, even a confined orbit will not necessarily be a closed one. The general case is a succession of arches going from $r_{\min}$ to $r_{\max}$ and *vice versa*, for instance, in a "rosette" configuration as shown in Figure 12.1.

After an infinitely long time, the particle will pass through every point of the ring. The figure also shows that the line joining $r_{\min}$ to $r_{\max}$ (the *line of apses*) rotates in the plane in the same direction of the movement.

The orbit is necessarily a closed one in two important cases, namely, when

$$U(r) = U_0 r^2, \quad U(r) = -\frac{U_0}{r}$$

The first case, not considered further, is the so-called *spatial oscillator*. The second case covers the attractive gravitational force.

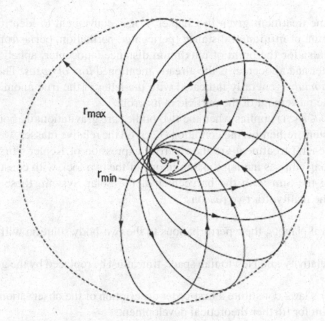

**FIGURE 12.1** A so-called "rosette" movement.

## 12.2 THE GRAVITATIONAL ATTRACTION

Let us now restrict the discussion to the gravitational attraction:

$$U(r) = -\frac{U_0}{r}, \quad F(r) = U_0/r^2, \quad U_0 = Gm_1m_2$$

where $G$ is Newton's gravitational constant. From Equation 12.13, we obtain

$$\varphi = \varphi_0 + \arccos\frac{K/r - mU_0/K}{\sqrt{2mE + m^2U_0^2/K^2}} \tag{12.14}$$

By introducing the two positive quantities $p$ and $e$ defined by

$$p = \frac{K^2}{mU_0} \ (\text{parameter}), \quad e = \sqrt{1 + \frac{2EK^2}{mU_0^2}} \ (\text{eccentricity})$$

we finally obtain the trajectory $r(\varphi)$ as

$$\frac{1}{r} = \frac{1}{p}\left[1 + e\cos(\varphi - \varphi_0)\right] \tag{12.15}$$

The generic orbit is therefore a conic curve having its focus in the center of the coordinates. The sign of the total energy $E$ determines the value of the eccentricity $e$:

if $E < 0$ then $0 \le e < 1$ (the orbit is an ellipse)
if $E = 0$ then $e = 1$ (a limiting case, the orbit is a parabola)
if $E > 0$ then $e > 1$ (the orbit is an hyperbole)

Maintaining the same treatment given in Chapter 4, it is convenient to identify the origin of the angles $\varphi$ with the point of minimum distance (pericenter, perihelion, periastron), which is located on the line passing also for the point of maximum distance (apocenter, aphelion, apoastron). The line joining pericenter and apocenter is the already mentioned *line of apses*. The angle $\varphi$ from that origin is the *true anomaly*, generally indicated with the letter $v$; the true anomaly increases in the same direction as the movement of the particle in its orbit.

The case $E < 0$ ($0 \leq e < 1$) applies when the two bodies are gravitationally bound together. In the case of the Solar System, remembering condition (12.5) on the relative masses, we see that the results obtained with the general treatment are the rigorous expression of Kepler's first and second laws: the planets move along ellipses having the Sun in one of the two foci, with constant areal velocity.

Even substituting the Sun with the barycenter of the Solar System, these two laws are only approximations to the reality, for two reasons:

1. there are several planets; their perturbations to the two-body solution will be examined in Chapter 14.
2. the Galilean relativity and Newtonian space-time must be replaced by the general relativity.

Nevertheless, Kepler's laws constitute a first-order description of the observations and an extremely useful departure point for further theoretical developments.

Let us examine again the eccentricity $e$ as function of the energy $E$: the ellipse becomes a circle ($e = 0$) if

$$\frac{2EK^2}{mU_0^2} = -1, \quad E = -\frac{mU_0^2}{2K^2}$$

which is the minimum possible value for $E$. For larger values of $E$, the semi-major and semi-minor axes of the ellipse are given, respectively, by

$$a = \frac{p}{1-e^2} = \frac{U_0}{2|E|}, \quad b = \frac{p}{\sqrt{1-e^2}} = \frac{K}{\sqrt{2m|E|}}$$

Therefore, the value of the semi-major axis depends on the total energy but not on the angular momentum of the particle, as instead the semi-minor axis does.

The apocenter is given by $r_{max} = p/(1 - e) = a(1 + e)$: the pericenter by $r_{min} = p/(1 + e) = a(1 - e)$.

Let us derive the period of revolution $P$ on the ellipse, whose area is $A = \pi ab$. From Kepler's second law, $2mA = PK$, and therefore,

$$P = 2\pi a^{3/2} \sqrt{\frac{m}{U_0}} = \pi U_0 \sqrt{\frac{m}{2|E|^3}} = 2\pi \frac{a^{3/2}}{\sqrt{GM}} \tag{12.16}$$

In conclusion, the period depends on the total energy, but not on the eccentricity. Its square is proportional to the third power of the semi-major axis and is inversely proportional to the sum of the masses of the two bodies. Therefore, the usual statement of Kepler's third law for Solar System planets, that the square of the periods are proportional to the cubes of the semi-major axes, is incorrect: the constant of proportionality actually depends on the mass of the particular planet (as already said, the largest deviation is for Jupiter). This remark must be taken into full account when discussing binary stars or pairs of galaxies. Notice also that the measure of $P$ and $a$ produces the product $GM$, not separately $G$ and $M$. For instance, for the Sun the product $GM_\odot$ is equal to $1.327124 \times 10^{20}$ m$^3$/s$^2$; for the Earth, $GM_\oplus$ is equal to $3.986004 \times 10^{14}$ m$^3$/s$^2$, omitting several more digits and the dependence of the last of them on the precise definition of the second, as given in the *Astronomical Almanac* Section K6. In other words, the high precision of the astronomical data does not transfer to the precise determination of $G$. Indeed, the value of $G$ ($\approx 6.67408 \times 10^{-11}$ m³/kg·s²) is the least precise of all fundamental constants, the fifth decimal digit already being uncertain.

We do not derive explicitly the equations for $E \geq 0$; if the two bodies are not gravitationally bound, the fictitious body approaches the center of force coming from and returning to infinity, with a velocity at infinity greater than zero (or equal to zero for the parabolic orbit). The minimum distance from the focus is $r_{min} = p/(1 + e)$, which becomes $= p/2$ for the parabola.

Recently, two comets have been discovered, coming through the Solar System on hyperbolic orbits. The first one (see Bailer-Jones et al., 2018) has been formally designated 1I/2017 U1 by the IAU, where 1I means *first interstellar*. The object has been nicknamed, in the Hawaiian language, Oumuamua, meaning a messenger coming from afar. The second object has been named 2I/Borisov (see www.iau.org/news/pressreleases/detail/iau1910/), from the name of the astronomer who discovered it with his 65-cm telescope in Crimea.

## 12.3 THE RELATIVE MOVEMENT

According to the general results obtained in the previous sections, the movements of the two particles in the barycentric system are obtained by solving the differential equation:

$$m \frac{d^2 \mathbf{r}}{dt^2} = m\ddot{\mathbf{r}} = \mathbf{F} = Gm \frac{\mathbf{r}}{r^3}$$

Then, separately

$$\ddot{\mathbf{r}}_1 = -Gm_2 \frac{\mathbf{r}}{r^3} = Gm_2 \frac{m_1^3}{(m_1 + m_2)^3} \frac{\mathbf{r}_1}{r_1^3}, \quad \ddot{\mathbf{r}}_2 = Gm_1 \frac{\mathbf{r}}{r^3} = -Gm_1 \frac{m_2^3}{(m_1 + m_2)^3} \frac{\mathbf{r}_2}{r_2^3}$$

It is like having in B a particle of fictitious mass $m_f$:

$$m_f = \frac{m_1^3}{(m_1 + m_2)^3}$$

attracting a particle of mass $m_2$ at distance $r_2$, and *vice versa* inverting the role of the two particles.

In astronomical applications, the barycentric system is not always convenient. In the Solar System, the barycenter closely coincides with the center of the Sun, but in more general cases, we have no *a priori* knowledge of the two masses, nor we can determine the position of the center of mass by the observations. Let us consider the specific example of a visual binary star. From the observational point of view, it is more immediate to refer the positions not to the invisible barycenter B, but to a nonrotating reference system centered in the brighter object $P_1$ (in the case of binary stars, the brighter star is usually also the heavier one) and describe the relative motion of $P_2$ with respect to $P_1$. To be sure, $P_1$ itself will be accelerated toward $P_2$, and therefore, this reference system will simply be parallel to the inertial barycentric one. Thus, this translation of origin introduces a fictitious radial acceleration, but not a Coriolis acceleration. Therefore, the movement of $P_2$ relative to $P_1$ is described by the equation:

$$\ddot{\mathbf{r}} = -G(m_1 + m_2) - \frac{\mathbf{r}}{r^3} = -GM \frac{\mathbf{r}}{r^3} \tag{12.17}$$

Equation 12.17 is the same as in the barycentric case, provided the reduced mass $m$ is substituted by the sum of the two masses $M$. The difference between the relative and barycentric description is shown in Figure 12.2.

Therefore, we shall proceed with the relative description and with the astronomical notation, using the previous area constant C/2 and the new constant $h$ expressing the energy in astronomical terms (in Latin, the *vis viva* constant):

$$h = \frac{1}{2}V^2 - \frac{GM}{r} \tag{12.18}$$

Notice that $2h$ is the square of the velocity at infinity.

The equation of the relative orbit can be derived by a second method. If $(x, y)$ is a Cartesian system and $(r, \varphi)$ a polar one, in the plane of the orbit and centered in $P_1$ the Newtonian differential equations are

$$\ddot{x} = -GM\frac{x}{r^3}, \quad \ddot{y} = -GM\frac{y}{r^3}, \quad x^2 + y^2 = r^2$$

Recalling Binet's expression 12.10 for the radial acceleration, after simple passages the following equation can be derived:

$$-\frac{GM}{r^2} = -\frac{C}{r^2}\left[\frac{1}{r} + \frac{d^2(1/r)}{d\varphi^2}\right] \tag{12.19}$$

namely, a linear nonhomogeneous differential equation with constant coefficients, in the unknown $1/r$ with respect to the variable $\varphi$. The solution is the conic curve of focus $P_1$ given by

$$\frac{1}{r} = \frac{GM}{C^2}\left[1 + \sqrt{1 + 2h\left(\frac{C}{GM}\right)^2}\cos(\varphi - \varphi_0)\right] \tag{12.20}$$

where the constant term is a particular integral of the equation. In the following, we will indicate the constant with $p$ (parameter of the conic) and eliminate the arbitrary constant $\varphi_0$ by counting the angles from the pericenter, thus using the true anomaly $v$.

A second differential equation is obtained by again considering the radial acceleration and taking into account the area integral in Equation 12.8:

$$-GM\frac{r}{r^2} = \ddot{r}r + \dot{r}r\dot{\varphi}^2 + r^2\dot{\varphi}\ddot{\varphi}$$

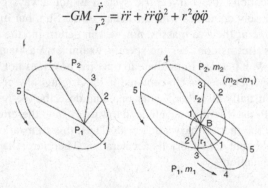

**FIGURE 12.2**   The relative and barycentric movements of two bodies, respectively, on the left and on the right panels (not to scale). The numbers refer to the same instants of time.

whose integral is given by the *vis viva* integral:

$$V^2 = \dot{r}^2 + r^2\dot{\varphi}^2 = 2\frac{GM}{r} + 2h$$

Summarizing these results,

$$p = \frac{C^2}{GM}, \quad e = \sqrt{1 + \frac{2hp}{GM}}, \quad r = \frac{p}{1 + e\cos v} = \frac{a(1-e^2)}{1 + e\cos v} \tag{12.21}$$

$$V^2 = GM\left[\frac{2}{r} - \frac{1-e^2}{p}\right] = GM\left[\frac{2}{r} - \frac{2h}{GM}\right], \quad 2h = \frac{GM}{p}(e^2 - 1) \tag{12.22}$$

Consider the instant of time when the distance of $P_2$ from $P_1$ is $r$ and its velocity is $V$, and compare the measured value of $V$ with the limit (parabolic) value $V_p$ given by

$$V_p = \sqrt{\frac{2GM}{r}} \tag{12.23}$$

$V_p$ is the velocity in $r$ of a body which moves toward $P_1$ starting from infinity with zero velocity and therefore describing a parabolic orbit (because $e(h = 0) = 1$). The orbit of $P_2$ will be an ellipse, a parabola or a hyperbola according to the ratio $V/V_p$ being less than 1, equal to 1 or greater than 1.

Let us discuss again in more detail the elliptical case (the energy constant $h$ is negative), identifying $P_1$ with the Sun. From the area integral, and from Equations 12.21 and 12.22 we get

$$\pi ab = \pi a^2\sqrt{1-e^2} = \frac{1}{2}\int_0^P r^2 \dot{v}\, dt = \frac{1}{2}CP,$$

$$C = r^2\dot{v} = \frac{2\pi}{P}\sqrt{1-e^2}\, a^2 = n\sqrt{1-e^2}\, a^2,$$

$$C^2 = pGM = aGM(1-e^2), \quad a = \frac{GM}{\dfrac{2GM}{r} - V^2}, \quad 2h = -\frac{GM}{a}$$

Finally,

$$P = 2\pi GM\left(\frac{2GM}{r} - V^2\right)^{-3/2}, \quad 4\pi^2\frac{a^3}{P^2} = n^2 a^3 \tag{12.24}$$

$$= GM = G(m_1 + m_2)$$

which again is Kepler's third law.

The quantity $n = 2\pi p/P$, which has the dimension of an angular velocity, is called the *mean motion*; from $n$, another useful quantity is derived, namely, the *mean anomaly* $M = n(t - t_0)$, which increases linearly with time. In several instances, it will be useful to substitute the time derivative, $d/dt$, with the derivative with respect to the mean motion, $n\,d/dM$.

Notice that the semi-major axis $a$ depends on $r$ and $V$, while the period $P$ depends on $a$ but not on $e$. Therefore, if two bodies $P_2$, $P_3$ pass through the same position $r$ with identical $V$ but different eccentricity, they will have the same $a$. After a time equal to the period $P$, they will be found again in the same place; therefore, if a planet disaggregates in several fragments, after the time $P$ all fragments will be found together again in the fragmentation point, quite independent of their individual orbits. The focusing properties were used by Encke, after the discovery of the first few asteroids in a small region of the Solar System, to put forward the theory of the fragmentation of a progenitor planet for their origin. A similar focusing effect can be found in an association of stars orbiting around the Milky Way center, after a full galactic rotation.

The square of the elliptical velocity is

$$V^2 = GM\left(\frac{2}{r} - \frac{1}{a}\right) = n^2 a^2 \frac{1 + 2e\cos v + e^2}{1 - e^2}$$

from which we obtain

$$V = na\sqrt{\frac{2a - r}{r}}, \quad V_{\text{aph}}^2 = \frac{GM}{a}\frac{1 - e}{1 + e}, \quad V_{\text{per}}^2 = \frac{GM}{a}\frac{1 + e}{1 - e},$$

$$V_{\text{aph}}/V_{\text{per}} = \frac{1 - e}{1 + e}$$

where $V_{\text{aph}}$ and $V_{\text{per}}$ are the velocities at the aphelion and perihelion, respectively. The velocity is equal to $na$ on the semi-minor axis.

We have already discussed (see Chapter 7) the convenience of decomposing the vector $\mathbf{V}$ along two non-orthogonal directions, namely, perpendicular to the radial direction and perpendicular to the major axis. Those two components will maintain a constant modulus, $na/\sqrt{1 - e^2}$,    $nae/\sqrt{1 - e^2}$, respectively, and the second also a constant direction.

For the angular velocity,

$$\dot{v} = na\frac{1 + e\cos v}{\sqrt{r(1 - e^2)}} = n\frac{a^2}{r^2}\sqrt{1 - e^2}, \quad \dot{v}_{\text{aph}}/\dot{v}_{\text{per}} = \left(\frac{1 - e}{1 + e}\right)^2$$

For the radial velocity,

$$\dot{r} = \frac{nae\sin v}{\sqrt{(1 - e^2)}}$$

To close this section, it will be useful to remember that in order to calculate the kinetic and total energy from the knowledge of the relative velocity $V$ of $P_2$ with respect to $P_1$, we have to return to the inertial system. Namely, we must concentrate in $P_2$ the reduced mass of the system:

$$T = \frac{1}{2}mV^2, \quad E = T + U = -Gm_1 m_2 \frac{1 - e^2}{2p}$$

## 12.4   PLANETARY MASSES FROM KEPLER'S THIRD LAW

In the previous sections, we have derived the three laws that Kepler empirically found analyzing the careful observations of Mars performed by Tycho Brahe. The first two laws were enunciated in 1607 and the third in 1617. We have pointed out that the usual statement of the third law is slightly incorrect, because of the neglect of the mass of the planet with respect to that of the Sun, although for most applications, the correction is small. Consider then two systems of two particles, e.g., the pairs Sun–Jupiter and Jupiter–Io. The third law gives

$$n_J^2 a_J^3 = G(M_\odot + M_J) = GM_\odot(1 + M_J/M_\odot),$$

$$n_{Io}^2 a_{Io}^3 = G(M_J + M_{Io}) = GM_J(1 + M_{Io}/M_J)$$

With good approximation,

$$n_{Io}^2 a_{Io}^3 / n_J^2 a_J^3 = M_J(1 + M_{Io}/M_J)/M_\odot(1 + M_J/M_\odot) \approx M_J/M_\odot \qquad (12.25)$$

Equation 12.25 provides a direct method to determine the mass of the planet in terms of the mass of the Sun, applicable however only if the planet has a small moon. Therefore, the masses of Mercury and Venus (which have no natural moons) could not be determined with this method. The precision of the masses of these two planets improved when artificial spacecraft were put in orbit around them. Today, all planets, Pluto and Charon, Ceres, several asteroids and comets, have masses determined with great precision by man-made satellites.

Another case where this method cannot be applied with the same confidence is that of the Earth–Moon system, because the mass of the Moon is not negligible with respect to that of the Earth. However, even in this case the correction is small, given the sensible equidistance of the Earth and of the Moon from the Sun. Therefore, the relative orbit continues to be approximately an ellipse, although the deviations are well measurable, as discussed in Chapter 14.

In the astronomical practice, all masses are expressed as fractions of the solar mass taken as unitary mass, the unitary distance is the AU and the unit of time is the mean solar day. The sum of the masses is written as

$$M = M_\odot + m_2 = M_\odot\left(1 + \frac{m_2}{M_\odot}\right) \equiv 1 + \frac{1}{\mu}$$

Several almanacs give the quantity $\mu$ (obviously a very large number), instead of the mass of the planet or of another body of the Solar System.

Using the astronomical system of units, a planet of negligible mass (and therefore unperturbed by external masses), in circular orbit around the Sun at the unitary distance, would have a revolution period $P$ exactly equal to the sidereal year, $P_\oplus = 365.2563835\ldots$; the constant

$$k = 2\pi/P_\oplus = 0.0172021\ldots\text{radians/j} \qquad (12.26)$$

is named Gauss' constant. It formally coincides with the inverse of $G$, expressed though in astronomical, not in physical, units; as previously remarked, the very high precision with which we know $k$ does not transfer to the precision of $G$. The constant $k$ also coincides with the mean diurnal motion:

$$k = 3548''.188\ldots\text{j}^{-1}, \quad 1/k = 58^j.1324\ldots$$

It is also the basis of the Gaussian year introduced in Chapter 10. Gauss' constant was so important in the past for the system of astronomical units to be called a defining constant. Today, the defining constant is the velocity of light. To appreciate the change in emphasis about the constants, until the IAU resolution of 1976 the velocity of light in vacuum was considered only a primary and not a defining constant.

## 12.5  ESCAPE VELOCITY

We define now two useful limiting velocities, namely, the escape velocity and the velocity in the circular orbit. Consider the total energy $E$ of a particle $P_2$ of very small mass $m_2$ at the surface of a nonrotating spherical body $P_1$ of radius $R$ and mass $m_1$:

$$E = \frac{1}{2} m_2 V^2 - G \frac{m_1 m_2}{r}$$

The limiting velocity $V_e$, given by the expression,

$$V_e = \sqrt{\frac{2Gm_1}{R}} \tag{12.27}$$

is the *escape velocity* from body $P_1$. If, by some means we impart to $P_2$ a velocity $V$ greater than $V_e$ in whatever direction, $P_2$ will reach infinity with final velocity greater than zero.

Another useful limiting velocity is that on the circular orbit at distance $r > R$ from the center of $P_1$; from the equilibrium between centrifugal and gravitational forces, we have

$$\frac{m_2 V_c^2}{r^2} = G \frac{m_1 m_2}{r^2}, \quad V_c(r) = \sqrt{\frac{Gm_1}{r}}, \quad V_c(R) = \frac{1}{\sqrt{2}} V_e$$

Table 12.1 provides escape and circular velocities for the eight planets plus Pluto, neglecting their diurnal rotation (see Chapters 13 and 14 for data on inclinations, eccentricities, sidereal periods and rotation characteristics). The third column gives the surface gravity in comparison with that at the Earth's surface (9.78 m/s²). The first two velocities (fourth and fifth columns) pertain to the equator of each body and the other two velocities (sixth and seventh columns) to the circular orbit at the average distance of the body from the Sun.

### TABLE 12.1
### Typical Escape and Circular Velocities in the Solar System

| Body | Distance (AU) | Mass (g) | Radius (km) | $g/g_\oplus$ | $V_e$ (km/s) | $V_c$ (km/s) | $V_e(\odot)$ (km/s) | $V_c(\odot)$ (km/s) |
|------|---------------|----------|-------------|--------------|--------------|--------------|---------------------|---------------------|
| Sun | | $1.99 \times 10^{33}$ | $6.96 \times 10^5$ | 27.9 | 618 | 437 | | |
| Mercury | 0.387 | $3.3 \times 10^{26}$ | 2,439 | 0.3 | 4.3 | 2.5 | 96 | 68 |
| Venus | 0.723 | $4.9 \times 10^{27}$ | 6,051 | 0.9 | 10.4 | 7.3 | 49 | 35 |
| Earth | 1.000 | $6.0 \times 10^{27}$ | 6,378 | 1.0 | 11.2 | 7.9 | 42 | 30 |
| Moon | 1.000 | $7.3 \times 10^{25}$ | 1,738 | 0.2 | 2.4 | 1.6 | 42 | 30 |
| Mars | 1.524 | $6.4 \times 10^{26}$ | 3,393 | 0.4 | 5.0 | 3.6 | 34 | 24 |
| Jupiter | 5.203 | $1.9 \times 10^{30}$ | 71,492 | 2.3 | 59.6 | 42.5 | 18 | 13 |
| Saturn | 9.539 | $5.7 \times 10^{29}$ | 60,268 | 0.9 | 35.5 | 25.0 | 14 | 10 |
| Uranus | 19.191 | $8.7 \times 10^{28}$ | 25,559 | 0.8 | 21.1 | 15.5 | 10 | 7 |
| Neptune | 30.061 | $1.0 \times 10^{29}$ | 24,764 | 1.1 | 23.6 | 16.0 | 7 | 5 |
| Pluto | 39.529 | $\sim 1.3 \times 10^{25}$ | $\sim 1,150$ | $\sim 0.04$ | $\sim 1.1$ | $\sim 0.8$ | 7 | 5 |

These considerations on escape velocities from the planetary surfaces are useful not only for dynamical questions, but also for the understanding of their atmospheres. Let $T$ (in Kelvin) be the temperature of such an atmosphere, supposed in thermal equilibrium; Maxwell's law gives the distribution function of molecules of mass $m_2$ as function of their velocities:

$$dN(V) = \frac{4N}{\sqrt{\pi}} \left( \frac{m_2}{2kT} \right)^{3/2} V^2 e^{-m_2 V^2 / 2kT} dV \tag{12.28}$$

so that the mean square velocity of those molecules will be

$$\frac{1}{2} m_2 V^2 = \frac{3}{2} kT, \quad V = \sqrt{\frac{3}{m_2} kT} \tag{12.29}$$

where $k = 1.38 \times 10^{-16}$ erg/K is Boltzmann's constant. For instance, the mass of the hydrogen atom H is $m_2 \approx 1.6 \times 10^{-24}$ g, so that

$$V_H(T) \approx 1.6 \times 10^{-1} \sqrt{T} \text{ km/s}$$

At the surface of the Earth, assuming $T \approx 290$ K, the thermal velocity of H is $V_H \approx 2.7$ km/s $\ll V_e$. It is 1.4 times less for the molecule $H_2$, which is the most common form of hydrogen in our atmosphere. All other molecules being heavier than $H_2$, we conclude that the Earth is well capable of retaining a massive quasi-stationary atmosphere. However, Maxwell's distribution has a very long tail at high velocity, so that a fraction of the Earth's gases and, in particular, of atomic H will continuously escape to the outer space. The observational evidence of such loss is the so-called *geocorona*, which is visible from orbiting satellites in the H Ly-α spectral line at $\lambda = 1216$ Å, as already discussed in Chapter 11.

Mercury and the Moon do not have such capability of retaining a permanent atmosphere. However, in both cases a tenuous atmosphere has been detected. The continuous loss of gases caused by thermal escape (and by other mechanisms to be discussed in Chapter 17) must be replenished by processes capable of extracting gases from their surfaces. In the absence of volcanic activity, phenomena such as UV solar photons, solar particles impinging on the soil or by meteoroid bombardment have been identified.

In the case of the Sun, the surface gravity is about 28 times that at the surface of the Earth and the photospheric temperature is approximately 5870 K. In higher layers, namely, in the chromosphere and corona, the temperature of the gases rises to tens, hundreds and even millions of degrees, so that the thermal escape becomes conspicuous. However, observations prove that the loss of particles from the Sun (the so-called *solar wind*) is orders of magnitude larger than that accounted for by thermal losses: other more efficient mechanisms, whether magnetic or electric, must act to accelerate the ionized (electrically charged) particles escaping from the Sun.

## 12.6 SOME CONSIDERATIONS ON ARTIFICIAL SATELLITES

Let us launch from the surface of a spherical nonrotating Earth of radius $a_\oplus$ a satellite of mass $m_2$ with initial velocity $V > V_e$. Its energy will be

$$E = \frac{1}{2} m_2 V^2 - G \frac{m_2 M_\oplus}{a_\oplus} \quad (m_2 \ll M_\oplus)$$

At an altitude $H$, the distance from the center becomes $r = a_\oplus + H$, and the energy

$$E = \frac{1}{2}m_2 V_r^2 - G\frac{m_2 M_\oplus}{r}$$

or else, equating the two values for the conservation of the energy,

$$E = \frac{1}{2}m_2 V^2 - G\frac{m_2 M_\oplus}{a_\oplus} = \frac{1}{2}m_2 V_r^2 - G\frac{m_2 M_\oplus}{r},$$

$$V_r^2 = V^2 - 2GM_\oplus\left(\frac{1}{a_\oplus} - \frac{1}{r}\right)$$

At infinity,

$$V_\infty^2 = V^2 - V_e^2 = (V - V_e)(V + V_e) \approx 2V_e\,\Delta V$$

In conclusion, if we launch with $\Delta V = +1$ km/s, the satellite will reach infinity with a velocity of approximately 4.7 km/s (ignoring the very small losses of energy due to the atmospheric drag). There are several practical consequences of this *gain at infinity*; for instance, one has to be careful not to reach the final destination with too high a velocity. We underline the convenience of using in space applications the parameter $\Delta V$ instead of the energy.

The circular velocity at the surface of the Earth is around 8 km/s, which will also be the velocity of low-altitude satellites (e.g., the International Space Station ISS at 400 km). Their period is then of approximately $90^m$. Suppose we place such a satellite in a polar orbit; it will go out of phase with the Sun by about $30^m$ at each orbit; for several orbits, it will see an almost constant illumination (day or night) of its nadir. Therefore, the low polar orbit is used for surveillance. At $H = 36,000$ km the orbital period becomes $24^h$, so that a satellite placed on the equatorial plane at this altitude in a circular orbit (e.g., the Meteosat) will be practically stationary with respect to the ground observer. Actually, several satellites have simply a geosynchronous orbit (that was the case of the International Ultraviolet Explorer IUE), slightly different from the rigorously defined geostationary one. At any rate, the two-body condition is a mathematical abstraction. Several gravitational and non-gravitational perturbing forces (such as the Earth–Moon and solar tides, the non-sphericity of the Earth's potential, and the solar and terrestrial radiation pressure) will act to perturb the orbit; appropriate corrections must be performed to keep the wanted position of the satellite, for instance, by occasional firings of small thrusters.

## NOTES

- An interesting example of non-closed orbit is provided by the Sun under the attraction of the other stars of the Milky Way. According to Oort, the overall force can be expressed as the sum of a point-like mass plus an extended disk of ellipsoidal shape:

$$F = \frac{q}{r^2} + sr \;\; (s \ll q)$$

  The galactocentric movement of the Sun cannot be in a plane, because the force is not a central one. In about 225 million years, the Sun will span a three-dimensional torus of inner radius 8 kpc, outer radius 9 kpc, thickness 0.3 kpc, with a velocity perpendicular to the galactic plane of approximately 7 km/s (see Chapter 9).

## EXERCISE

1. Let us add a small quantity $\delta U$ to the potential energy $U(r) = -U_0/r$. The trajectory will not be closed, so that at each revolution, a small quantity $\delta v$ will be added. Determine $\delta v$ in the two cases:
   1. $\delta U = \beta/r^2$
   2. $\delta U = \gamma/r^3$

   (adapted from Landau and Lifschitz, Bibliography section).

   When the mobile moves from $r_{min}$ to $r_{max}$ and again to $r_{min}$, the radius vector turns by an angle $\Delta\varphi$ given by

$$\Delta\varphi = 2 \int_{r_{min}}^{r_{max}} \frac{\dfrac{K}{r^2}}{\sqrt{2m(E-U) - \dfrac{K^2}{r^2}}} \, dr$$

$$= -2 \frac{\partial}{\partial K} \int_{r_{min}}^{r_{max}} \sqrt{2m(E-U) - \frac{K^2}{r^2}} \, dr \tag{12.30}$$

Putting $U = U_0/r + \delta U$ in Equation 12.30, and developing according to the powers of $\delta U$, the term of order zero gives $2\pi$, and the first term gives the wanted excess to $2\pi$:

$$\Delta\varphi = -\frac{\partial}{\partial K} \int_{r_{min}}^{r_{max}} \frac{2m\delta U}{\sqrt{2m\left(E + \dfrac{U_0}{r}\right) - \dfrac{K^2}{r^2}}} \, dr = \frac{\partial}{\partial K}\left(\frac{2m}{K} \int_0^\pi r^2 \, \delta U \, d\varphi\right)$$

In case (a), the integration gives,

$$\Delta\varphi = -\frac{2\pi}{K^2}\beta = -\frac{2\pi}{U_0 p}\beta$$

In case (b), the integration gives

$$\Delta\varphi = -\frac{6\pi U_0 m}{K^2}\gamma = -\frac{6\pi}{U_0 p^2}\gamma$$

where $p$ is the parameter of the unperturbed ellipse.

# 13 Orbital Elements and Ephemerides

In this chapter, we discuss the problem of referring the theoretical orbit to the observations. First, we will examine how the apparent positions can be determined at every instant from the knowledge of the orbital elements; then, the inverse and more difficult problem of how the orbital elements can be derived from the observed positions will be briefly examined. The treatment will be essentially limited to the case of the Solar System, with few considerations for binary stars.

## 13.1 KEPLER'S EQUATION

Let Q be the mobile particle on the ellipse of focus F and center O, and $P_0$ the perihelion (or the periastron for binary stars, or pericenter in general), where from the true anomalies $v$ are counted (see Figure 13.1).

Draw through O an auxiliary circle of radius $a$ and a new anomaly $E$, named *eccentric anomaly* (from the Latin *ex-centrum*), increasing as $v$. The ellipse can be considered as the projection of the auxiliary circle on the plane passing through the line of the apses and inclined by an angle $\psi$ such as $e = \sin \psi$. Some authors directly write $\sin \psi$ instead of $e$. In the Cartesian system $(x, y)$ centered in F, with $x$ parallel to the major axis and $y$ parallel to the minor axis, the coordinates of Q are

$$x = r\cos v = a(\cos E - e), \qquad y = r\sin v = a\sqrt{1 - e^2}\,\sin E$$

$$r = a(e\cos E), \qquad \cos v = \frac{\cos E - e}{1 - e\cos E}, \tag{13.1}$$

$$\sin v = \sqrt{1 - e^2}\,\frac{\sin E}{1 - e\sin E}$$

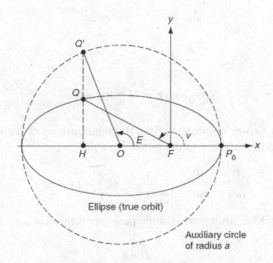

Ellipse (true orbit)

Auxiliary circle of radius $a$

**FIGURE 13.1** Ellipse and auxiliary circle, $v$ = true anomaly, $E$ = eccentric anomaly.

Introducing the semi-anomaly $v/2$, Equation 13.1 can be written as

$$r\left(\cos^2\frac{v}{2}-\sin^2\frac{v}{2}\right)=a(\cos E-e),\quad r\left(\cos^2\frac{v}{2}+\sin^2\frac{v}{2}\right)=a(1-e\cos E)$$

Then,

$$\tan\frac{v}{2}=\sqrt{\frac{1+e}{1-e}}\,\tan\frac{E}{2} \tag{13.2}$$

where $v/2$ and $E/2$ always have the same sign. Recalling the mean anomaly,

$$M=2\pi(t-t_0)/P=n(t-t_0)$$

and the relationship between the angular velocity and the area constant, we obtain

$$C=r^2\dot{v}=na^2\sqrt{1-e^2},\quad C(t-t_0)=\int_{t_0}^{t}C\,d\tau=\int_{0}^{2\pi}r^2\,dv=Ma^2\sqrt{1-e^2} \tag{13.3}$$

Let $Q'$ be the projection of Q on the auxiliary circle; it is also

$$Q'H=a\sin E,\quad QH=a\sqrt{1-e^2}\,\sin E,\quad \frac{Q'H}{QH}=\frac{a}{b}=\frac{1}{\sqrt{1-e^2}}$$

By virtue of the projectivity properties between the auxiliary circle and the ellipse, we obtain

$$\text{area }FP_0Q'=\frac{1}{\sqrt{1-e^2}}\text{ area }FP_0Q=\text{area }OP_0Q'-\text{area }OFQ',$$

$$\text{area }OP_0Q'=\frac{1}{2}aE^2,\quad \text{area }FP_0Q'=\frac{1}{2}\int_{t_0}^{t}C\,d\tau=\frac{1}{2}a^2\sqrt{1-e^2}M,$$

$$\text{area }OFQ'=\frac{1}{2}a^2e\sin E$$

so that

$$E-e\sin E=M=n(t-t_0) \tag{13.4}$$

Equation 13.4, known as *Kepler's equation*, fixes the relationship between eccentric anomaly and time. The same result can be obtained in the following way:

$$\frac{dE}{dM}=\frac{dE}{ndt}=\frac{1}{1-e\cos E}=\frac{a}{r}=1-e\cos v$$

By differencing Equation 13.2, after some simple passages we derive

$$dv=\frac{\sqrt{1-e^2}}{1-e\cos E}dE \tag{13.5}$$

$$\frac{dv}{dE} = \sqrt{1-e^2}\,\frac{a}{r} = \frac{\sqrt{1-e^2}}{1-e\cos E} = \frac{\sin v}{\sin E}, \quad \frac{dE}{dt} = n\frac{a}{r} = \frac{n}{1-e\cos E}$$

Integrating the last differential equation, we obtain again Equation 13.4. In either way, if the four orbital elements $(a, n, e, t_0)$ are known, $E$ is obtained at any instant $t$ by solving Equation 13.4; from the knowledge of $E$, $v$ is known by virtue of Equation 13.2; finally, $r$, $x$ and $y$ can be calculated via Equation 13.1.

The solution of the transcendental Equation 13.4 is by no means a trivial one, unless $e$ is small as in the case of Earth's orbit. More than hundred methods have been proposed in the literature. In the simplest case, Newton's method can be used: take as initial value $E_1 = M + e \sin M$ and calculate

$$M_1 = M + e\sin E_1, \quad \Delta M_1 = M - M_1$$

Remembering that

$$dE = (dE\,/\,dM)dM, \quad dM = (1-e\cos E)dE,$$

we have approximately

$$\Delta E_1 = \Delta M_1\,/\,(1-e\cos E_1), \quad E_2 = E_1 + \Delta E_1$$

Calculate then

$$M_2 = M + e\sin E_2, \quad \Delta M_2 = M - M_2, \quad \Delta E_2 = \Delta M_2\,/\,(1-e\cos E_2),$$

$$E_3 = E_2 + \Delta E_2$$

and so on, until $E$ becomes stable. It is also useful to remember that if $e$ is sufficiently small:

$$\sqrt{1-e^2} \approx 1 - \frac{e^2}{2} - \frac{e^4}{8}$$

To solve Equation 13.2, several developments in series have been derived following d' Alembert, as already discussed in Chapter 2. However, the convergence is not always guaranteed, as shown by Laplace in the case $e > 0.66274$. We quote here the following expressions, written as Fourier series:

$$E = M + \left(e - \frac{1}{8}e^3\right)\sin M + \left(\frac{1}{2}e^2 - \frac{1}{6}e^4\right)\sin 2M + \frac{3}{8}e^3\sin 3M + \cdots$$

$$v = M + \left(2e - \frac{1}{4}e^3\right)\sin M + \left(\frac{5}{4}e^2 - \frac{11}{24}e^4\right)\sin 2M + \frac{13}{12}e^3\sin 3M + \cdots$$

$$\frac{a}{r} = 1 + \left(e - \frac{1}{8}e^3\right)\cos M + \left(e^2 - \frac{1}{3}e^4\right)\cos 2M + \frac{9}{8}e^3\cos 3M + \cdots$$

$$\frac{r}{a} = 1 + \frac{e^2}{2} - \left(e - \frac{3}{8}\right)e^3\cos M + \left(\frac{1}{2}e^2 - \frac{1}{3}e^4\right)\cos 2M - \frac{3}{8}e^3\cos 3M + \cdots$$

The last equation provides the average value of $r$ with respect to time:

$$\langle r \rangle = a\left(1 + \frac{e^2}{2}\right)$$

while the average value with respect to $v$ is $b$ and that with respect to $E$ is $a$. For very small values of the eccentricity,

$$E \approx M + e\sin M, \qquad v \approx M + 2e\sin M$$

The last equation was discussed in Chapter 4 for the geocentric orbit of the Sun as the equation of the center.

Regarding the Cartesian coordinates $(x, y)$, sometimes referred to as *reduced coordinates*, the following expressions are easily found:

$$x = \frac{r}{a}\cos v = \cos E - e$$

$$= -\frac{3}{2}e + \left(1 - \frac{3}{8}e^2\right)\cos M + \left(\frac{1}{2}e - \frac{1}{3}e^2\right)\cos 2M + \frac{3}{8}e^2\cos 3M + \cdots$$

$$y = \frac{r}{a}\sin v = \sqrt{1 - e^2}\,\sin E$$

$$= \left(1 - \frac{5}{8}e^2\right)\sin M + \left(\frac{1}{2}e - \frac{5}{12}e^2\right)\sin 2M + \frac{3}{8}e^2\sin 3M + \cdots$$

## 13.2   EPHEMERIDES FROM THE ORBITAL ELEMENTS

Consider the (quasi) inertial heliocentric reference system $(X_\odot, Y_\odot, Z_\odot)$, with axis $X_\odot$ toward the vernal point $\gamma$ and axis $Y_\odot$ in the ecliptic plane. Let $(X, Y, Z)$ be the instantaneous coordinates of a planet Q (or comet or asteroid). The intersection of the planet's orbital plane with the ecliptic is a line passing through the Sun and called the *line of the nodes*. Let N be the ascending node (that node through which the $Z$ coordinate of the planet passes from negative to positive values along the orbital movement) and $\Omega$ the angle on the ecliptic plane between $\gamma$ and N; this angle is called *longitude of the ascending node*. The angle $i$ between the orbital plane and the ecliptic is said to be the *inclination* of the orbit; it varies between $0°$ and $90°$ for a direct movement, and between $90°$ and $180°$ for a retrograde one. Therefore, the pair $(\Omega\ i)$ gives the position of the orbital plane with respect to the ecliptic. Consider now, in the orbital plane, the angle $\omega$ between the ascending node N and the perihelion $P_\odot$; such angle is named *argument of the perihelion*; it fixes the orientation of the orbit in its plane. There will be cases when $\Omega$ cannot be defined, because $i = 0°$, so that a different choice of arguments must be made (e.g., sometimes, it is useful to consider the angle $\Pi = \Omega + \omega$, even though, in general, this so-called *longitude of the perihelion* is the sum of two angles in two different planes).

We will limit the discussion to the generic orbit for which the seven orbital elements $(a, e, P, n, t_0, \Omega\ i, \omega)$ are known. Actually, by virtue of Kepler's third law, six elements are sufficient, either $P$ or $n$, and only five for a parabolic orbit. In order to fix the position of the planet in the heliocentric system at a given time, several steps are necessary. First, derive the pair $(r, v)$ or the reduced coordinates $(x, y)$ in its orbit. Next, make a rotation of $\omega$ in the plane of the orbit. Then, project the orbit on the ecliptic according to $i$. With a further rotation of $\Omega$, project the position of Q on the heliocentric axes, finally to derive $(X, Y, Z)$. Namely,

$$\left\{\begin{array}{l} X = r\left[\cos(\omega + v)\cos\Omega - \sin(\omega + v)\sin\Omega\sin i\right] \\ Y = r\left[\cos(\omega + v)\sin\Omega + \sin(\omega + v)\cos\Omega\sin i\right] \\ \qquad Z = r\sin(\omega + v)\sin i \end{array}\right. \tag{13.6}$$

Let $(X_\oplus, Y_\oplus, Z_\oplus)$ be the heliocentric coordinates of the Earth at the same instant (with $Z_\oplus \approx 0$); the geocentric coordinates $(x_\oplus, y_\oplus, z_\oplus)$ of the planet Q will be

$$\left\{\begin{array}{l} x_\oplus = X - X_\oplus \\ y_\oplus = Y - Y_\oplus \\ z_\oplus = Z - Z_\oplus \end{array}\right. \tag{13.7}$$

From $(x_\oplus, y_\oplus, z_\oplus)$, the ecliptic coordinates $(\lambda, \beta)$, the equatorial coordinates $(\alpha, \delta)$ and the distance of the planet from the Earth can be derived. This is not the final step of the procedure to calculate the apparent coordinates; we still have to apply the corrections due to aberration, light deflection, precession, nutation, position of the observer (namely, from geocentric to topocentric coordinates) and atmospheric refraction.

Let us briefly discuss the precessional effect. Ignoring the planetary perturbations, the plane of the orbit in the inertial space will be fixed and so will the inclination $i$ and the argument of the perihelion $\omega$; instead, the direction of $\gamma$ is function of time and so will be $\Omega$. Therefore, in first approximation: $di = d\omega = 0$, $d\Omega = -52''.3/$year.

For many newly discovered comets, the orbit is essentially a parabolic one, so that we can put $e \approx 1$, $h \approx 0$ and write

$$r = \frac{p}{1 + \cos v} = \frac{p}{2\cos^2 \dfrac{v}{2}} = q\left[1 + \tan^2 \frac{v}{2}\right] \tag{13.8}$$

The constant $q = p/2$ is the distance to the Sun at the perihelion. It is usual to write

$$s = \tan\frac{v}{2}, \qquad x = r\cos v = q(1 - s^2), \qquad y = r\sin v = 2qs$$

It is easy to find that

$$C = \sqrt{2GM_\odot q} = 0.0243\sqrt{q}, \quad V = \frac{2GM_\odot}{r} = \frac{C}{q\sqrt{1 + s^2}},$$

$$t - t_0 = \frac{1}{3}\sqrt{\frac{2}{GM.}}q^{3/2}(3s + s^2)$$

Therefore, the five orbital elements are $q$, $t_0$, $\Omega$, $i$ and $\omega$. The following relationships are of interest in this case:

$$\dot{v}\sec^2\frac{v}{2} = \sqrt{2GM_\odot}q^{-3/2}, \quad \tan\frac{v}{2} + \frac{1}{3}\tan^3\frac{v}{2} = \sqrt{\frac{GM.}{2q^3}}(t - t_0)$$

For the hyperbolic case, we simply state that the expressions of $r\cos v$, $r\sin v$ contain the hyperbolic functions ($\cosh A$, $\sinh A$) of an appropriate anomaly $A$. As already recalled in Chapter 12, two comets of interstellar origin, coming from infinity and going back with positive energy, and therefore on a hyperbolic orbit, have been detected in 2017 and 2019, respectively.

## 13.3  PLANETARY CONFIGURATIONS AND TITIUS–BODE LAW

The upper part of Figure 13.2 shows the different configurations (projected on the ecliptic plane) with respect to the Sun for an internal planet such as Mercury or Venus, and for an external one such as Mars or Jupiter, with the usual terminology of elongations, conjunctions, quadratures and oppositions. The lower part of the figure shows the apparent paths among the fixed stars, with the predominance of direct movements, occasional points of stationarity, reversal of the motion and even loops. Therefore, for the terrestrial observer the simplicity of the Keplerian orbits is masked by the complexity of the apparent configurations. Undoubtedly, this complexity did not favor the acceptance of the heliocentric Copernican vision of the Solar System.

The maximum elongations of Mercury and of Venus are 23° and 46°, respectively (some authors call elongation the geocentric angle Sun–planet, while strictly speaking the name refers to the distance along the ecliptic, namely, the difference in ecliptic longitudes). Their orbital inclinations are not zero. Therefore, the two planets can be seen only occasionally in front of or behind the solar disk. For Mercury, transits in front of the solar disk can happen every 7 or 13 years in May or November. Recent dates are May 9, 2016 and November 11, 2019 (see the "Notes" section). The next one will occur on November 13, 2032. A typical transit lasts several hours. Transits of Venus are a much rarer phenomenon, but when they happen, there will be 2 passages 8 years apart, separated by long gaps of 121.5 and 105.5 years, respectively. At the present epoch, transits can occur in either June or December, namely, when Venus passes through the nodes of its orbit (descending in June, ascending in December). One such transit was in June 2004, and the next was in June 2012, the last of the 21st century. Such complex pattern is because the orbital periods of Earth and Venus are close to 8:13 and 243:395 commensurabilities (see also Chapter 14).

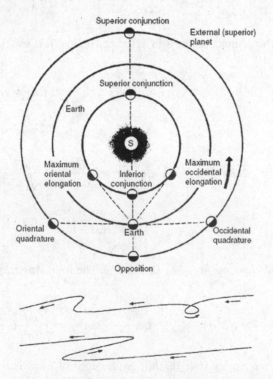

**FIGURE 13.2**  Planetary configurations and apparent paths. Inferior conjunctions are also called transits.

Furthermore, according to the particular configuration, the disks of the two inferior planets can appear partly illuminated, with phases similar to the ones observed on the Moon; the dividing line between the illuminated region and that in shadow is called the *terminator*. Phases of the outer planets can be noticed essentially only for Mars; for Jupiter and Saturn, a modest limb darkening can be seen with careful observations. See the "Notes" section for a list of web sites useful for planetary configurations, occultations and transits.

Other interesting properties of the orbits of the planets are the direct sense of motion, small eccentricities and modest inclination to the ecliptic, the largest values being those of Mercury and Pluto (see Table 13.1). Pluto is kept in this table for continuity, although a resolution of IAU in 2006 classifies it in the new class of dwarf planets. Regarding diurnal rotation, a great variety of situations is encountered, from the very slow retrograde rotation of Venus to the rotation around an axis lying almost in the orbital plane for Uranus. Only the rotations of the Earth and Mars are similar.

None of these properties derives from fundamental physical reasons. They must be associated with the mechanism of formation and subsequent dynamical evolution of the Solar System. Indeed, comets and other minor bodies can have large inclinations and eccentricities and even revolve in retrograde sense (e.g., Halley's). The same characteristics are observed for many moons of the giant planets. For instance, Triton, the large moon of Neptune, is in retrograde orbit, suggesting a capture by Neptune after formation in a different place.

Regarding periods, a distinction must be made between those observed by the terrestrial observer and the true ones with respect to a fixed direction. The first is the so-called *synodic* period, and the second is the *sidereal* one. We can define the first by the interval of time between two similar configurations of Sun, Earth and planet, and the second one by the mean motion $n$ divided by $2\pi$.

If $n_\oplus$ is the mean sidereal motion of the Earth, we have

$$P_{syn} = \pm \frac{2\pi}{n - n_\oplus} \tag{13.9}$$

where the plus sign applies to an internal planet and the negative sign to an external one (see again Table 13.1).

The distances of the planets from the Sun seem to show a remarkable regularity, already noticed by two German astronomers, Titius and Bode, in the 18th century. This regularity was expressed by J. F. Wurm in 1787 (Uranus was discovered in 1781) with the mathematical formula:

$$a_n = 0.4 + 0.3 \times 0.2^n \tag{13.10}$$

## TABLE 13.1
## Orbital Elements of the Eight Classic Planets Plus Pluto

| Planet | a (AU) | e | i (deg) | n ("/j) | $P_{sid}$ (j) | $P_{syn}$ (j) |
|--------|--------|-----|---------|---------|----------|----------|
| Mercury | 0.387 | 0.2056 | 7.004 | 14732.424 | 87.969 | 115.88 |
| Venus | 0.723 | 0.0068 | 3.394 | 5767.740 | 224.698 | 583.92 |
| Earth | 1.000 | 0.0167 | 0.000 | 3548.095 | 365.266 | |
| Mars | 1.524 | 0.0934 | 1.850 | 1886.619 | 686.943 | 779.94 |
| Jupiter | 5.203 | 0.0483 | 1.308 | 299.156 | 4,332.177 | 398.88 |
| Saturn | 9.539 | 0.0560 | 2.488 | 119.993 | 10,800.594 | 378.09 |
| Uranus | 19.191 | 0.0461 | 0.774 | 41.869 | 30,953.629 | 369.66 |
| Neptune | 30.061 | 0.0097 | 1.774 | 21.301 | 60,841.437 | 367.49 |
| Pluto | 39.529 | 0.2482 | 17.148 | 14.194 | 91,304.976 | 366.73 |

where $n = -\infty$ for Mercury, $= 0$ for Venus, $= 1$ for Earth, $= 2$ for Mars, $= 4$ for Jupiter, $= 5$ for Saturn and $= 6$ for Uranus. The lack of a planet with $n = 3$ motivated great efforts to locate the missing one, until finally Father Giuseppe Piazzi in Palermo discovered Ceres on January 1, 1801. Ceres is the largest body inside the main belt of asteroids between Mars and Jupiter. Its discovery prompted Carl Gauss to develop the theory of orbital motions and the method of the least squares for properly treating the observational errors. The discovery of Ceres was followed soon after by that of Pallas and Vesta.

Alternative formulations of Equation 13.10 were proposed after the discoveries of Neptune and Pluto, e.g., $a_n = 1.53^n$ by Armellini ($n = -2$ Mercury, $= -1$ for Venus, $= 0$ for Earth, $= 1$ for Mars, $= 2$ and $= 3$ for two main groups of asteroids, $= 4$ Jupiter, $= 5$ Saturn, $= 6$ Uranus, $= 7$ lacking, $= 8$ Neptune, $= 9$ Pluto).

Of course, there is no reason for the Solar System to stop at Pluto. Several large trans-Neptunian bodies are continuously discovered in the outer regions of the Solar System called Kuiper Belt and scattered disk, respectively (Morbidelli, 2007, Bibliography section). The largest of these bodies has diameters similar or even larger than that of Pluto. The total mass of the trans-Neptunian objects exceeds by at least ten times the total mass of the asteroids in the Main Belt. Interestingly enough, the Titius–Bode relationship seems to maintain its validity even beyond 50 AU.

For more information on Titius–Bode law, see the "Notes" section.

## 13.4 ORBITAL ELEMENTS FROM THE OBSERVATIONS

The inverse problem, of deriving the six orbital elements of a Solar System object from the observations, is more complex than the direct one. Therefore, only some basic notions derived from the methods developed by Laplace and Gauss will be exposed here.

As already said, the knowledge of six initial constants is required. In usual astronomical observations, that means three pairs of angular coordinates $(\alpha_i\ \delta_i)$ at three instants $t_1$, $t_2$, $t_3$ (for sufficiently nearby objects, radar data may instead provide position, distance and velocity at any particular instant). At the beginning of the observations, the distance $\rho$ of the new comet or asteroid is unknown, so it is assumed that the three angular pairs are geocentric ones. A further assumption is that the velocity and acceleration of the body Q remain constant during that interval of time. With reference to Figure 13.3, let $(\xi, \eta, \zeta)$ be the geocentric Cartesian ecliptic coordinates, $(\lambda, \beta)$ be the ecliptic longitude and latitude of Q and $(X, Y, Z)$, $(\lambda_\odot, \beta_\odot)$ be the corresponding heliocentric values.

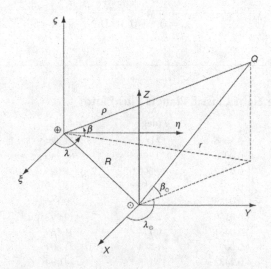

FIGURE 13.3  Heliocentric and geocentric coordinates of a new comet or asteroid, Q.

At each instant $t_i$, we have

$$\xi_i = \rho_i \cos \lambda_i \cos \beta_i, \quad \eta_i = \rho_i \sin \lambda_i \cos \beta_i, \quad \zeta_i = \rho_i \sin \beta_i \qquad (13.11)$$

Each variable can be expressed as

$$\xi_i = \xi(t) + \dot\xi_t (t_i - t) + \frac{1}{2} \ddot\xi_t (t_i - t)^2 + \cdots$$

where $t$ is a generic instant inside the interval $(t_1, t_3)$. It is convenient to start from the intermediate time $t = (t_1 + t_2 + t_3)/3$, because the third-order terms vanish. The development can thus be terminated at the second-order terms:

$$\xi_i = \xi(t) + \dot\xi_t (t_i - t) + \frac{1}{2} \ddot\xi_t (t_i - t)^2 \quad (i = 1,2,3)$$

and similar for the other two coordinates. Therefore, from three observations we can derive the three (supposedly) geocentric components of position, velocity and acceleration of Q at the intermediate time $t$, a part the still unknown distance $\rho$. The heliocentric position of the Earth at the same instant is well known, so that the heliocentric ecliptic coordinates of Q are

$$X = X_\oplus + \xi, \quad Y = Y_\oplus + \eta, \quad Z = Z_\oplus + \zeta \approx \zeta$$

From the last relationship, we obtain $Z/r = \sin \beta_\odot$.

The equations of motion of Q are

$$\ddot X + GM_\odot \frac{X}{r^3} = 0, \quad \ddot Y + GM_\odot \frac{Y}{r^3} = 0, \quad \ddot Z + GM_\odot \frac{Z}{r^3} = 0,$$

where $r$ is the distance of Q from the Sun. Similarly, for the Earth,

$$\ddot X_\oplus + GM_\odot \frac{X_\oplus}{a_\oplus^3} = 0, \quad \ddot Y_\oplus + GM_\odot \frac{Y_\oplus}{a_\oplus^3} = 0$$

Let us introduce three auxiliary variables $(\sigma, \mu, \upsilon)$:

$$\sigma = \rho \sin \beta, \quad \mu = \cos \lambda \cot \beta, \quad \upsilon = \sin \lambda \cot \beta \qquad (13.12)$$

connected to $(\xi, \eta, \zeta)$ by

$$\xi = \mu\sigma = \mu\rho \sin \beta, \quad \eta = \upsilon\sigma, \quad \zeta = \sigma \qquad (13.13)$$

The two variables $(\mu, \upsilon)$ are directly known from the observations, and so are their first and second derivatives at the intermediate time. The equations of motion become

$$\begin{cases} \ddot\mu\sigma + \mu\ddot\sigma + 2\dot\mu\dot\sigma + \ddot X_\oplus + GM_\odot \dfrac{X_\oplus + \mu\sigma}{r^3} = 0 \\[2mm] \ddot\upsilon\sigma + \upsilon\ddot\sigma + 2\dot\upsilon\dot\sigma + \ddot Y_\oplus + GM_\odot \dfrac{Y_\oplus + \upsilon\sigma}{r^3} = 0 \\[2mm] \ddot\sigma_\oplus + GM_\odot \dfrac{\sigma}{r^3} = 0 \end{cases} \qquad (13.14)$$

Multiply the first of Equation 13.14 by $v$ and the second by $\dot{\mu}$, and subtract the 2; taking into account the third, we obtain

$$\rho \sin\beta \left[ \ddot{\mu}\dot{v} - \dot{\mu}\ddot{v} \right] + GM_\odot \left[ X_\oplus \dot{v} - Y_\oplus \dot{\mu} \right] \left( \frac{1}{r^3} - \frac{1}{a_\oplus^3} \right) = 0 \qquad (13.15)$$

The angle $\theta$ between Sun, Earth and Q is known from the observations. It is connected to $r$, $\rho$ by

$$r^2 = \rho^2 - 2\rho a_\oplus \cos\theta + a_\oplus^2 \qquad (13.16)$$

The values of the auxiliary constants $(\mu, v)$ and of their first and second derivatives are known by Equation 13.13; recalling Equation 13.12, we can (at least in principle) derive $(r, \rho)$ from Equations 13.15 and 13.16. The solution is certainly not easy; proceeding in the elementary manner, we would end up with an equation of the eighth degree!

We cannot discuss here the several methods developed for the determination of the possible solutions. Quite often, the orbit remains poorly known, and additional observations are needed to resolve the ambiguities. At any rate, from $\rho$ we have immediately $\sigma$ and then $\dot{\sigma}$ from

$$\begin{cases} \ddot{\mu}\sigma + 2\dot{\mu}\dot{\sigma} + GM_\odot X_\oplus \left( \frac{1}{r^3} - \frac{1}{a_\oplus^3} \right) = 0 \\[2mm] \ddot{v}\sigma + 2\dot{v}\dot{\sigma} + GM_\odot Y_\oplus \left( \frac{1}{r^3} - \frac{1}{a_\oplus^3} \right) = 0 \end{cases} \qquad (13.17)$$

The second derivative of $\sigma$ is known from Equation 13.13.

This approximate procedure provides a first estimate of the position, velocity, acceleration and distance of Q at the intermediate time $t$, in both the heliocentric and geocentric systems. By comparing the heliocentric velocity at distance $r$ with the parabolic value, we can recognize the type of conic and the constant $h$. If the orbit is an ellipse, we have immediately $P$ (or $n$) and $a$. To derive $e$, we take advantage of Kepler's second law:

$$\dot{\varphi}^2 = \dot{\lambda}_\odot^2 + \dot{\beta}_\odot^2, \qquad C = r^2\dot{\varphi}$$

and then derive $E$, $v$, $t_0$. In subsequent iterations of the procedure, the light time corresponding to the geocentric distance and the topocentric position of the observer must be introduced.

As already noted, it is usual to assume at the start of the calculation that the orbit of a new comet is a parabola.

The "Notes" section indicates some software packages, useful to calculate the orbital elements.

## 13.5  APPLICATION TO VISUAL BINARY STARS

In the case of binary stars, the ratio of the two masses can be assumed, at least initially, close to 1; there are cases however where it can be as small as 0.01. For even smaller ratio, one enters the real of brown dwarfs or even exoplanets. Notice that the orbit is projected on the plane tangent to the line of sight and that many binaries have been observed only for small arcs. Therefore, the tridimensional reconstruction of the orbit is not always possible. An additional complication is that many systems have a multiplicity greater than 2 (for instance, $\alpha$ Cen is triple). There are also cases when only one of the two bodies is readily seen; classic in this respect is the case of Sirius, the brightest star in the sky. In 1844, W. Bessel observed that the proper motion of the star is not along a straight line, but instead is undulated; there ought to be a much fainter but massive star, in order to maintain the center of mass along the required straight path (see Figure 13.4).

This fainter star, Sirius B, was discovered three decades later by the telescope maker Alvan Clark. Finally, in 1914, the American astronomer Walter Adams showed spectroscopically that the color and thus the temperature of Sirius B were very similar to those of Sirius A. According to the Stefan–Boltzmann law (see Chapter 16), the very low luminosity of Sirius B had therefore to be attributed to its much smaller radius. Indeed, the radius of Sirius B is even smaller than the radius of the Earth. Sirius B is therefore the prototype of the *white dwarfs*, namely, of stars with radii much smaller than that of normal stars having the same temperatures and masses. According to the theories of the stellar structure, white dwarfs represent the end point of the evolution of stars with solar masses.

We limit the following considerations to well-resolved systems of two self-gravitating stars, observed for the entire orbital period, or at least for a good fraction of it.

Visual observations are typically made comparing the fainter star B (secondary) to the brighter star A (primary), which usually is the more massive one. A polar system ($\rho, \theta$) centered on A is used to measure the relative position of B, with $\rho$ in arcsec and $\theta$ in degrees from north through east. The observations usually take a long span of time, so that precession and proper motion must be removed from the data before analyzing the pairs ($\rho, \theta$) for the orbital analysis. These corrections are of the form (see Chapters 5 and 8, $t$ in years):

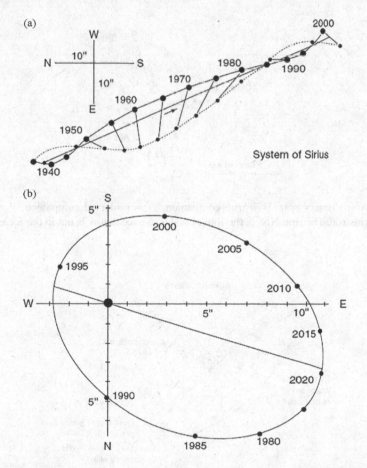

**FIGURE 13.4**    (a) The undulated proper motion of Sirius A. (b) The relative orbit of B with respect to A. The distance between the two stars varies between 8.2 and 31.5 AU, and the orbital period is 50 years.

$$\theta(t) = \theta(t_0) + (\Delta_1 + \Delta_2)(t - t_0), \quad \Delta_1 = -0°.0056 \sin\alpha s\delta/\text{year}$$

$$\Delta_2 = -0°.00417\ \mu_\alpha^s \sin\delta/\text{year}$$

A further correction is due to the motion of the barycenter of the pair, which moves with respect to the observer during an orbital revolution of the secondary. This sort of planetary aberration modifies the apparent time intervals.

The methods of dealing with the observations are not discussed here, as we assume that the apparent orbit has been determined with great accuracy (Kepler's second law plays a very important role in this determination). For its didactic value, we first expound a graphical method, essentially due to the Dutch astronomer Hendrikus Zwiers in the 19th century, to derive the orbital elements. We have already underlined that the relative apparent orbit is an ellipse, which is the projection of the true one on the plane tangent to the celestial sphere (see Figure 13.5).

In other words, the observed elements are connected to the true ones by a projective transformation, which conserves the center C of the ellipse, Kepler's second law, the period $P$ and the instant of passage through the periastron, but not the focus of the ellipse, nor the orthogonality of the axes nor the area constant.

In Figure 13.6, let C be the center of both apparent and true ellipses, and A be the primary (namely, the focus of the true orbit); the direction CA is therefore that of the true major axis, and point $A_1$ on that direction is the position of the true periastron. The ratio $CA/CA_1 = ae/a$ immediately

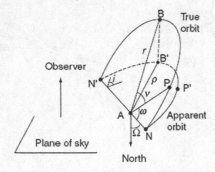

**FIGURE 13.5**   A = Primary star, B = true companion, B′ = projected companion, P = periastron and P′ = projected periastron. The line NN′ is the line of nodes. Notice that A is not in the focus of the apparent ellipse.

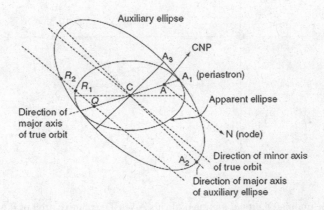

**FIGURE 13.6**   Apparent and auxiliary ellipses. CNP = celestial north pole.

gives the true eccentricity $e$. Let us now take the conjugate direction of $CA_1$ in the projectivity: draw any chord parallel to $CA_1$, and then, identify its middle point and join it with C. This is the direction of the true minor axis.

Then, draw a line parallel to the true minor axis through a generic point Q belonging to the line CA; this parallel intersects the apparent ellipse in $R_1$. Determine a segment $QR_2$ of such length that $QR_2/QR_1 = 1\sqrt{1 - e^2}$.

By repeating this construction for several points Q along CA, a series of points $R_2$ is obtained, which define an auxiliary ellipse, tangent to the true one in the periastron and apoastron. This auxiliary ellipse is also the projection on the sky of the auxiliary circle we have used to define the eccentric anomaly $E$. Therefore, the semi-major axis of the auxiliary ellipse is also the true semi-major axis:

$$CA_2 = a$$

The normal to $CA_2$ defines on the auxiliary ellipse the semi-minor axis and the point $A_3$: the ratio $CA_3/CA_1$ gives therefore $\cos i$ and finally the modulus of $i$ (but not its sign). The direction $CA_2$ is also the direction around which we have performed the projection, namely, the line of nodes of Figure 13.5. Now, let CNP be the direction of the celestial north pole through A: the parallel to $CA_2$ through A points to the node N, so that the angle in A between the directions of CNP and N gives $\Omega$. However, we do not know which node is the ascending one, so that $0° < \Omega < 180°$. The ambiguity can be resolved in the favorable case when the radial velocity $V_r$ of B is known: at the descending node, B is approaching the observer and $V_r(B) < 0$.

In order to derive the angle $\omega$ between the ascending node and the periastron, consider both Figures 13.5 and 13.6. $P'$ being the projection of P from the pole of the apparent orbit, the spherical triangle P'NP is rectangle in $P'$ and the angle $P'NP = i$. On the other hand, angle NP' in Figure 13.5 is also angle $A_1AN$ in Figure 13.6, so that

$$\tan NP' = \tan NP \cos i = \tan \omega \cos i \tag{13.18}$$

We have thus derived $e$ and the geometric elements $(i, \Omega, \omega)$, which are called Campbell's elements.

Regarding the time arguments: in Figure 13.6, it is

$$\text{angle } (CNP - A - A_1) = \theta(t_0), \quad \text{angle}(NA_1) = \theta(t_0) - \Omega,$$

$$\text{angle}(NAP) = \omega$$

For any position of the true secondary B and projected secondary B',

$$\text{angle } (CNP - B' - A_1) = \theta(t), \quad \text{angle}(NB') = \theta(t) - \Omega,$$

$$\text{angle}(NB'P) = \omega + v$$

Finally,

$$\tan NB' = \tan NP \cos i, \quad \tan(\theta - \Omega) = \tan(\omega + v)\cos i \tag{13.19}$$

$$\rho \sin(\theta - \Omega) = r \sin(v + \omega)\cos i, \quad \rho = \frac{a(1 - e^2)}{1 + e \cos v}\frac{\cos(v + \omega)}{\cos(\theta - \Omega)} \tag{13.20}$$

From the knowledge of $\theta(t)$, we derive $v$ and $E$. By solving Kepler's equation, the mean motion $M = 2\pi(t-t_0)/P$ is calculated. From two values of $M$, we immediately know $t_0$:

$$t_0 = \frac{M_1 t_1 - M_2 t_2}{M_1 - M_2}$$

and thus $P$.

Several analytical approaches have been developed to replace the above geometrical procedure. We quote some elements of the method named after Thorvald Thiele and Robert Innes. As seen earlier, the primary star A is not in the focus of the projected ellipse, while the fundamental direction in the observations is the axis through A toward CNP. In other words, the apparent ellipse, referred to a system of axes centered in A, with axis $x$ directed to the CNP and axis $y$ toward the East, can be expressed as

$$\alpha x^2 + 2\beta xy + \gamma y^2 + 2\delta x + 2\varepsilon y = 1, \quad x = \rho\cos\theta, \quad y = \rho\sin\theta$$

where the five constants $(\alpha, \beta, \gamma, \delta, \varepsilon)$ are known from the observations. On the plane of the true orbit, $(X, Y)$ are the reduced coordinates

$$X = \frac{r}{a}\cos v = \cos E - e, \quad Y = \frac{r}{a}\sin v = \sqrt{1-e^2}\sin E \tag{13.21}$$

whose origin is A, while the direction of $X$ is that of the periastron. Therefore, the apparent coordinates $(x, y)$ are obtained by a rotation followed by a projection, so that they are linear combinations of the former:

$$x = \rho\cos\theta = AX + FY, \quad y = \rho\sin\theta = BX + GY$$

The four quantities $A$, $F$, $B$ and $G$ are the Thiele–Innes constants, given by

$$A = a(\cos\omega\cos\Omega - \sin\omega\sin\Omega\cos i)$$

$$B = a(\cos\omega\sin\Omega + \sin\omega\cos\Omega\cos i)$$

$$F = a(-\sin\omega\cos\Omega - \cos\omega\sin\Omega\cos i)$$

$$G = a(-\sin\omega\sin\Omega + \cos\omega\cos\Omega\cos i)$$

The coordinates of the periastron and of the point having $v = 90°$ on the true orbit are, respectively,

$$\begin{cases} X_0 = 1-e \\ Y_0 = 0 \end{cases} \quad \begin{cases} X_1 = 0 \\ Y_1 = 1-e^2 \end{cases},$$

while on the apparent one,

$$\begin{cases} x_0 = A(1-e) = a(1-e)(\cos\omega\cos\Omega - \sin\omega\sin\Omega\cos i) \\ y_0 = B(1-e) = a(1-e)(\cos\omega\sin\Omega + \sin\omega\cos\Omega\cos i) \end{cases}$$

$$\begin{cases} x_1 = F(1-e^2) = a(1-e^2)(-\sin\omega\cos\Omega - \cos\omega\sin\Omega\cos i) \\ y_1 = G(1-e^2) = a(1-e^2)(-\sin\omega\sin\Omega + \cos\omega\cos\Omega\cos i) \end{cases}$$

After several manipulations,

$$\tan(\Omega+\omega) = \frac{B-F}{A+G} \quad (0° < \Omega < 180°)$$

$$\tan(\Omega-\omega) = \frac{B+F}{A-G} \quad (0° < \omega < 360°)$$

$$\tan^2\frac{i}{2} = \frac{A-G}{A+G}\frac{\cos(\Omega+\omega)}{\cos(\Omega-\omega)} = \frac{B+F}{B-F}\frac{\sin(\Omega+\omega)}{\sin(\Omega-\omega)}$$

Let C be the area integral on the true orbit and $c$ that on the projected one:

$$c = C\cos i = x\dot{y} - \dot{x}y = (X\dot{Y} - \dot{X}Y)(AG - BF), \quad a^2 = \frac{AG - BF}{\cos i}$$

It is customary to provide, in addition to Campbell's and Thiele–Innes' elements, also the following:

$$C' = a\sin\omega\,\sin i, \quad H = a\cos\omega\sin i, \quad z = C'X + HY = r\sin(v+\omega)\sin i$$

where $z$ is along the line of sight. Its derivative is the radial velocity (in arcsec/year) of the secondary with respect to the primary:

$$\dot{z} = C'\dot{X} + H\dot{Y} = n\left[C'\frac{dX}{dM} + H\frac{dY}{dM}\right]$$

If the parallax $\pi$ of the binary is known, then it is possible to derive the radial velocity in km/s:

$$V_r = 4.74\left(\frac{2\pi}{P}\right)\frac{1}{\pi}\frac{a\sin i}{\sqrt{1-e^2}}[e\cos\omega + \cos(v+\omega)] \tag{13.22}$$

This expression can also be read in the other way, in order to derive the dynamical parallax of the binary (see Chapter 8).

The knowledge of the two individual masses, and not only of their sum, must come from additional information, e.g., from the motion of the barycenter (such as for Sirius), or from the radial velocities of both stars.

The study of binary stars has made great progress by virtue of modern optical techniques, such as adaptive optics and multiple telescope array interferometry (see Chapter 11 and the "Notes" section). The angular resolution that can be achieved is approximately $\lambda/D$ radians, where $\lambda$ is the wavelength and D the diameter of the mirror or the total length of the interferometric array; the resolution varies from a few hundredths to a few thousandths seconds of arc. However, the great majority of binary (or multiple) stars remain unresolved. Their multiplicity can be recognized by photometric or spectral variations.

## NOTES

- The Web Sites section provides the address of several sites dedicated to ephemerides of planets, comets and asteroids for each location on Earth, eclipse, occultations, transits of Mercury and Venus over the solar disk, and orbital determination. The transit of Mercury in November 11, 2019 has been followed for the first time by a spacecraft, namely, by the extreme ultraviolet telescopes on board NASA's Solar Dynamics Observatory (https://sdo.gsfc.nasa.gov/).
- Although the Titius–Bode law tends to be considered not more than numerology, an interesting book about it was written by Nieto (1972, Bibliography section). A variant of the Armellini's formula, namely, the geometric progression $a_n = r_0 K^n$, has been discussed by Graner and Dubrulle (1994a,b). The validity of the law for planets of other stars has been discussed by several authors. For instance, Huang and Bakos (2014) found a null result for the systems discovered by NASA's Kepler telescope.
- For a classic book on binary stars, see Aitken (1935) in the Bibliography section.

# 14 Elements of Perturbation Theories

This chapter provides elementary notions on several topics pertaining to the general theme of orbital perturbations of bodies of the Solar System. After the statement of the general problem of perturbations, the relativistic precession of the perihelion of Mercury will be treated. Then, the following cases will be examined: a planet plus a small moon; the inequalities of the lunar movement; the Jupiter–Saturn interaction; the restricted circular three-body problem; the case of two bodies, one of which with finite dimensions; and the lunar librations. The present treatment will not consider other very interesting problems, such as the case of multiple stars and stellar systems, treated in specialized literature.

From a historical perspective, Newton's *Principia* paved the way to many investigations. For instance, Clairaut examined the three-body interaction, d'Alembert started the study of the dynamics of extended bodies, and Euler considered, in addition to the polar motion of the Earth discussed in Chapter 6, many other questions of celestial mechanics, such as the interaction Jupiter–Saturn and the secular variation of the orbital elements. Other great mathematicians, Lagrange, Laplace, Gauss, Poisson, Hamilton, Jacoby and Poincaré, to name a few, made fundamental contributions during the whole 18th and 19th centuries, arriving to Albert Einstein who changed the very foundations of Mechanics with his theory of general relativity in 1915.

Today, precise ephemerides of the planets and their moons are calculated with very refined methods, which take into account a large number of perturbing factors and general relativity corrections. Much used are the ephemerides of the Sun, Moon and many other bodies of the Solar System (including spacecraft) constructed at the Jet Propulsion Laboratory (JPL, see Standish, 1982, 1990). The mathematical model used at JPL includes point mass interactions among the Sun, planets and their moons; Newtonian perturbations of selected asteroids; the Moon and Sun's actions on the figure of the Earth; the Earth and Sun's actions on the figure of the Moon; Earth tides on the Moon; physical librations of the Moon; and general relativity. See also Chapter 8 of the *Explanatory Supplement* (Urban and Seidelmann, 2012). Other useful references are quoted in the Web Sites section.

## 14.1 PERTURBATIONS OF THE PLANETARY MOVEMENTS

When $N$ point-like bodies are under their mutual gravitational attraction, the difficulty of describing their movements greatly increases, even when excluding external forces. Consider three bodies having masses $m_0$, $m_1$ and $m_2$, respectively. Let $\mathbf{r}_i$ be the vectors from the barycenter B and $\mathbf{d}_{ij}$ the mutual distances ($i, j = 0, 1, 2, i \neq j$). We have three vector differential equations:

$$
\begin{cases}
m_0 \ddot{\mathbf{r}}_0 = G\left( m_0 m_1 \dfrac{\mathbf{d}_{01}}{d_{01}^3} + m_0 m_2 \dfrac{\mathbf{d}_{02}}{d_{02}^3} \right) \\[2ex]
m_1 \ddot{\mathbf{r}}_1 = G\left( m_1 m_0 \dfrac{\mathbf{d}_{10}}{d_{10}^3} + m_1 m_2 \dfrac{\mathbf{d}_{12}}{d_{12}^3} \right) \\[2ex]
m_2 \ddot{\mathbf{r}}_2 = G\left( m_2 m_0 \dfrac{\mathbf{d}_{20}}{d_{20}^3} + m_2 m_1 \dfrac{\mathbf{d}_{21}}{d_{21}^3} \right)
\end{cases}
\qquad (14.1)
$$

By projecting these equations on the three coordinate axes, we obtain $3 \times 3$ scalar differential equations of the second order, whose solution requires the knowledge of 18 initial constants (in general, for $N$ bodies we would derive $3N$ scalar equations). The existence of three integrals of motion, namely, the rectilinear motion of the barycenter, the conservation of the energy and the conservation of the total angular momentum, can be easily demonstrated. The latter integral insures the existence of a so-called *invariable plane* (Laplace plane) of the system passing through the barycenter. In the case of the Solar System, the invariable plane is inclined by $1°.5$ to the ecliptic, intermediate between the orbital planes of Jupiter and Saturn. Those three integrals are, however, only equivalent to $6 + 1 + 3 = 10$ initial constants, while 18 are needed. Therefore, the general case of three bodies cannot be solved in an analytical way. However, analytical solution has been found in special cases, like the restricted circular three-body problem treated in a later section.

Let us now identify the Sun with the first body of mass $m_0$ ($= M_\odot$), and draw through it a non-rotating heliocentric reference system ($X, Y, Z$), parallel but slightly accelerated with respect to the barycentric inertial one. Let us call $\Delta_1, \Delta_2$ the distances of two planets from the Sun and $d_{12}$ their mutual distance. The equations for the relative movement of the first planet are

$$\ddot{X}_1 = -G(M_\odot + m_1)\frac{X_1}{\Delta_1^3} + Gm_2\left[\frac{X_2 - X_1}{d_{12}^3} - \frac{X_2}{\Delta_2^3}\right]$$
(14.2)

and similar for $Y_1, Z_1$ and for the second planet. It can be demonstrated that

$$\ddot{X}_1 = -\frac{\partial V_1}{\partial X_1} = -\frac{\partial}{\partial X_1}\left\{G(M_\odot + m_1)\frac{1}{\Delta_1} + R_{12}\right\}$$
(14.3)

The first term represents the gravitational interaction between the Sun and the first planet. The second term $R_{12}$ is the *perturbing function* of the second planet on the first: its first term is the gravitational interaction between the two planets, while the second term is inversely proportional to the cube of the distance of the perturbing planet from the Sun:

$$R_{12} = Gm_2\left[\frac{1}{d_{12}} - \frac{X_1X_2 + Y_1Y_2 + Z_1Z_2}{\Delta_2^3}\right]$$
(14.4)

Several different cases can then occur, according to the relative distances of the three bodies, as we will see in the following paragraphs.

Suppose that at a certain instant $t_0$ the perturbing function vanishes. The first planet would then follow a two-body trajectory whose initial conditions were also determined by the perturbation due to the other planet. Such hypothetical trajectory is called (with slight imprecision) *osculating* orbit. True and osculating orbits are actually only tangent to each other, having at that instant identical positions and velocity vectors. After a short time $\Delta t$, the positions on the two orbits will be identical, but not so the force nor the velocity. Therefore, at $t_0 + \Delta t$ we can calculate a new osculating orbit, in other words a new set of osculating Keplerian orbital elements, slightly different from the previous set. The variations of the orbital elements can be expressed by the following Lagrange's planetary equations:

$$\dot{a} = \frac{2}{na}\frac{\partial R}{\partial M}, \quad \dot{M} = n - \frac{2}{na}\frac{\partial R}{\partial a} - \frac{1}{na^2}\frac{1-e^2}{e}\frac{\partial R}{\partial e}$$

$$\dot{e} = -\frac{1}{na^2}\frac{\sqrt{1-e^2}}{e}\frac{\partial R}{\partial \omega} + \frac{1}{na^2}\frac{1-e^2}{e}\frac{\partial R}{\partial M}$$

$$\dot{i} = -\frac{1}{na^2}\frac{1}{\sqrt{1-e^2}\sin i}\frac{\partial R}{\partial \Omega} + \frac{1}{na^2}\frac{\cos i}{\sqrt{1-e^2}\sin i}\frac{\partial R}{\partial i} \tag{14.5}$$

$$\dot{\Omega} = \frac{1}{na^2}\frac{1}{\sqrt{1-e^2}\sin i}\frac{\partial R}{\partial i}$$

$$\dot{\omega} = \frac{1}{na^2}\frac{\sqrt{1-e^2}}{e}\frac{\partial R}{\partial e} - \frac{1}{na^2}\frac{\cos i}{\sqrt{1-e^2}\sin i}\frac{\partial R}{\partial i}$$

where $M$ is the mean anomaly. Notice that the variable $n$ on the perturbed orbit is only the designation of $(GM_\odot)^{1/2}a^{3/2}$, because $a$ is continuously varying. The six nonlinear equations above must be solved numerically, but a solution is not always possible; there are, indeed, cases where the denominator becomes a very small number or is even zero (e.g., a circular orbit or one exactly on the ecliptic). Therefore, while it is very intuitive to regard the true orbit as a Keplerian one, with continuously changing osculating elements, from the mathematical point of view there are more convenient methods and variables. We can only quote here the variables associated with the names of the French mathematicians Delaunay and Poincaré. For more clarity, the osculating elements change from time to time, and so they do not represent the average orbit described by the body.

The case of a second body having negligible mass (e.g., a small comet) and a third body being a large planet, e.g., Jupiter, provides a simple and instructive situation. The planet will follow an essentially unperturbed Keplerian orbit, while the comet acceleration will be given by Equation 14.3. In such equation, $(X_1, Y_1, Z_1)$, $(X_2, Y_2, Z_2)$, $\Delta_1$ and $\Delta_2$ are the heliocentric coordinates and distances from the Sun of comet and Jupiter, respectively; $d_{12}$ is the comet–Jupiter distance.

The perturbing force can be conveniently broken down into different directions, for instance, the radial one, the tangential one or the one perpendicular to the orbital plane. Let us discuss only the radial perturbation, and write

$$U = GM\left[\frac{1}{r} + g\frac{1}{r^2}\right] \tag{14.6}$$

with $g$ being very small. The perturbed orbit will have the expression (see Chapter 12):

$$\frac{1}{r} = GM\frac{1}{C^2(1-q)^2} + A\cos(1-q)(\varphi - \varphi_0), \qquad (1-q)^2 = 1 - \frac{gM}{C^2}$$

namely, an ellipse rotating in its plane by $2\pi q$ at each revolution.

The difficulty of taking radial perturbations into account led several times to the suspicion that Newton's law is not fully representative of the reality. Clairaut in 1750 was probably the first to question the validity of the $1/r^2$ law, owing to his inability to reproduce the motion of the line of apses of the Moon. In the following century, it was the turn of Le Verrier, because of the unexplained advance of Mercury's perihelion. Another example is the discrepancy of the ephemerides of Uranus: 60 years after its discovery by Herschel in 1781, the error, already larger than 2′, was tentatively attributed to a failure of Newton's law.

More recently, doubts were raised in order to account for the minute acceleration of the spacecraft Pioneer 10 and 11 beyond 20 AU (Anderson et al., 2002). Finally, the anomalous acceleration has been traced to non-symmetric thermal effects. Even the behavior of the rotation curves of some galaxies has been attributed to a modified Newtonian dynamics (MOND, Milgrom, 1983). Scarpa et al. (2003) proposed to use globular clusters to test gravity in the weak acceleration regime. See also Sanders and McGaugh in the Bibliography section.

The discrepancy between theory and observations that led to the most fruitful results was the excess of the advance of Mercury's perihelion. The observed advance is approximately 573″/century, of which 43″/century cannot be explained by planetary perturbations. Le Verrier suspected the existence of another planet internal to the orbit of Mercury (often called Vulcan; volcanoids is a term used to indicate still unknown small bodies possibly orbiting inside Mercury's orbit). Dicke and Goldenberg (1974), in a remarkable paper, discussed the effect of the oblateness of the Sun, namely, of a possible quadrupole moment, on the excess, with reference to earlier ideas of S. Newcomb. Finally, the effect was convincingly explained by general relativity (e.g., Nobili and Will, 1986) and actually the excess represented one of the decisive proofs for the correctness of the theory. According to general relativity, the force can be written as

$$F(r) = -GM \frac{1}{r^2} \left[ 1 + \frac{3C^2}{c^2 r^2} \right] \tag{14.7}$$

where $C$ is the area constant, and $c$ is the velocity of light in vacuum, namely, a physical entity completely extraneous to classical celestial mechanics. Instead of the instantaneous action-at-distance of the Newtonian mechanics, general relativity foresees a transmission of forces with the velocity of light (and the radiation of gravitational waves by suitably accelerated bodies or by a body collapsing in a strongly asymmetric configuration). From Equation 14.7, after several elaborated calculations, one can derive the following variation of $\omega$:

$$\dot{\omega} = 6\pi \frac{GM_\odot}{Pa\left(1 - e^2\right)c^2} \text{ (radians/s)}$$

or else

$$\Delta\omega = 6\pi \frac{n^2 a^2}{\left(1 - e^2\right)c^2} \text{ (radians/revolution)}$$

The value of the excess is 8″.6/century for Venus, 3″.8/century for the Earth, and 1″.4/century for Mars.

The "Notes" section gives additional information on relativistic effects on the orbit of Mercury and artificial satellites.

Range measurements to the NASA Messenger spacecraft in orbit about Mercury have provided new data permitting to estimate the precession of Mercury's perihelion with unprecedented precision (Park et al., 2017). Their estimate gives a rate of (575″.3100 ± 0″.0015)/century and a value of the quadrupole moment of the Sun $J_2 = (2.25 \pm 0.09) \times 10^{-7}$. In conjunction with the already quoted Shapiro delay measured by the Cassini spacecraft (see Chapter 7), the post-Newtonian parameter $\beta$ turns out to be essentially zero.

A perturbation tangent to the orbit, and so parallel to **V**, would alter the velocity modulus but not its direction; this is the case, for instance, of a body traveling in a resisting medium (e.g., an artificial satellite in the Earth's atmosphere). The consequence would be a gradual diminution of the semi-major axis and a corresponding (and at first sight counterintuitive) increase of the mean motion. This situation of movement in a resisting medium was suspected in the past for comets, when several of them appeared to increase their mean motion; but on the contrary, other comets displayed a decrease. Finally, an internal source for these unexplained accelerations was found, i.e., sudden directional losses of mass, and of gaseous and dusty jets (a sort of a rocket effect). More properly, we can speak of non-gravitational forces. Owing to the rotation of the body, to the inclination of the rotation axis to the orbital plane and to the thermal inertia, these jets can occur in any direction. In general, a tangential perturbation will alter the elements ($a$, $e$, $n$, $\omega$).

A perturbation perpendicular to **V** cannot alter the modulus, but only the direction of the velocity vector. Therefore, $e$ and $\omega$ will be affected, but not $a$ nor $n$. If this perturbation has a component normal to the orbital plane, then $(i, \Omega)$ will also change.

## 14.2   PLANET PLUS SMALL MOON

Consider, for instance, Jupiter of mass $M_J$ and one of its small and distant (in order to remain in the point-like approximation) moons of mass $m \ll M_J$. The distance from the planet to its moon $d$ is much smaller than the distance to the Sun, so that both Jupiter and its moon can be considered at equal distances from the Sun, $\Delta \approx \Delta_J$. Therefore, we can translate the origin of the reference system to the center of Jupiter and consider the perturbations to the orbit of the moon caused by the Sun, assumed in a fixed Keplerian orbit around Jupiter.

Let $(x, y, z)$ be the planetocentric coordinates of the moon; the perturbing function then becomes

$$R = GM_\odot \left[ \frac{1}{\Delta} - \frac{x_\odot x + y_\odot y + z_\odot z}{\Delta_J^3} \right] \tag{14.8}$$

We also have

$$\Delta^2 = (x - x_\odot)^2 + (y - y_\odot)^2 + (z - z_\odot)^2 = \Delta_J^2 - 2(xx_\odot + yy_\odot + zz_\odot) + d^2$$

Call $\theta$ the angle of Sun–Jupiter–moon:

$$\cos\theta = \frac{xx_\odot + yy_\odot + zz_\odot}{\Delta_J d}$$

From the binomial theorem,

$$\frac{1}{\Delta} = \frac{1}{\Delta_J} \left[ 1 - 2\frac{d}{\Delta_J}\cos\theta + \frac{d^2}{\Delta_J^2} \right]^{-1/2}$$

$$\approx \frac{1}{\Delta_J} \left[ 1 + \frac{d}{\Delta_J}\cos\theta + \frac{d^2}{\Delta_J^2}\left( -\frac{1}{2} + \frac{3}{2}\cos^2\theta \right) + \cdots \right] \tag{14.9}$$

Notice, however, that in the equations of motion, the partial derivatives of $R$ contain only the terms depending on the coordinates of the perturbed body, so that we can conclude

$$R = GM_\odot \frac{1}{\Delta_J}\left[ \frac{d^2}{\Delta_J^2}\left( -\frac{1}{2} + \frac{3}{2}\cos\theta \right) + \frac{d^3}{\Delta_J^3}\left( -\frac{3}{2}\cos\theta + \frac{5}{2}\cos^2\theta \right) + \cdots \right] \tag{14.10}$$

The complete expression can be given in terms of Legendre polynomials $P_n$ of argument $\cos\theta$ (see Equation 6.9). The leading term is

$$R_1 = GM_\odot \frac{d^2}{\Delta_J^3}\left( -\frac{1}{2} + \frac{3}{2}\cos\theta \right) \tag{14.11}$$

which is sufficiently small to leave almost Keplerian the planetocentric orbit of the moon.

## 14.3  CASE EARTH–MOON

The theory of the movement of the Moon is one of the most challenging problems of celestial mechanics. To the names already quoted in the first paragraph, we can add Plana, Hansen, Hill, Brown and von Zeipel, who made great contributions to this subject. It is truly admirable that Newton was able to derive a fairly good theory of the lunar orbit by means of purely geometric reasoning.

In the following, we consider the case of the Moon perturbed by a hypothetical Sun describing a fixed Keplerian orbit around the Earth. Such approximation is of such interest to be called the *principal problem* of the lunar orbit.

With reference to Figure 14.1, consider the spherical triangle Sun–ascending node *N*–Moon. After several tedious but not difficult passages, one can derive the approximate expression:

$$R_1 = n_\odot^2 a^2 \left( \frac{a_\odot}{\Delta_\odot} \right)^3 \left( \frac{d}{a} \right)^2 \left\{ \begin{array}{l} -\dfrac{1}{2} + \dfrac{3}{8}\left[ \cos 2(\omega + v) + \cos 2(\omega_\odot + v_\odot - \Omega)\sin^2 i \right] \\[2mm] +\dfrac{3}{4}\left[ 1 + \cos 2(\omega - \omega_\odot + v - v_\odot + \Omega) \right]\cos^4 \dfrac{i}{2} \\[2mm] +\dfrac{3}{4}\left[ 1 + \cos 2(\omega + \omega_\odot + v + v_\odot - \Omega) \right]\sin^4 \dfrac{i}{2} \end{array} \right\} \quad (14.12)$$

We can still simplify Equation 14.12 by assuming a circular orbit for the Sun, namely, $a_\odot/\Delta_\odot = 1$ and $v_\odot = M_\odot$ (mean anomaly, do not confuse it with the mass of the Sun). Furthermore, because $i_\leftmoon$ is fairly small ($i_\leftmoon \approx 5°.2$), we put $\sin i = i_\leftmoon$ (in radians) and ignore all terms in $i_\leftmoon^4$; moreover, we can ignore all terms in $e_\leftmoon^2$ and higher because $e_\leftmoon$ is also small ($e_\leftmoon \approx 0.055$). With such approximations, the non-periodic part of $R_1$ becomes

$$R = n_\odot^2 a_\leftmoon^2 \left[ \frac{1}{4} + \frac{3}{8}\left( e_\leftmoon^2 - i_\leftmoon^2 \right) \right] \quad (14.13)$$

The complete expression contains the mean motion, the argument of the perigee, the mean anomaly of the Sun (all quantities known with great accuracy) and the six osculating elements of the Moon. The planetary equations of Lagrange would thus allow the calculation of the variation of those elements with time. For instance, the secular motions of the perigee and of the nodes would have periods of 9ʸ.21 and 17ʸ.86, respectively. The observations provide, instead, the values 8ʸ.85 and 18ʸ.61. Such discrepancies give an indication of the amount of approximations made in the previous discussion. Already after few days, the true orbit is appreciably different from an arc of ellipse. Therefore, the method of osculating elements is not appropriate for the precise calculations of lunar

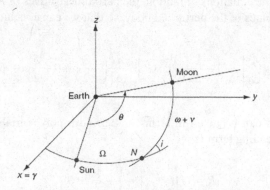

**FIGURE 14.1**  Geocentric elliptic orbit of the Moon in the geocentric ecliptic reference frame. The plane *xy* is the ecliptic. N = ascending node, $\Omega$ = angle $\gamma$N, $\omega + v$ = angle (N - Moon), $\omega_\odot + v_\odot$ = angle ($\gamma$ - Sun).

ephemerides. However, it provides a convenient and intuitive visualization, namely, of an elliptical orbit with $e_\mathbb{D} = 0.0549$ undergoing a double rotation, one in its plane around its focus (retrogradation of the perigee) with a period of $8^y.85$ and one in the orbital plane around the normal to the ecliptic (retrogradation of the line of nodes) with a period of $18^y.61$. Since the eccentricity $e_\mathbb{D}$ is so much larger than that of the Sun, the corresponding equation of the center is also extremely large: $EC_\mathbb{D} \approx 2e_\mathbb{D} \approx 6°17'$. A whole spectrum of perturbing terms with different periods and amplitudes (terms called *inequalities* and mostly of solar origin) deforms this basic orbit. For instance, the passage of the Sun through the lines of the apses occurs every 206 days. Consequently, $e_\mathbb{D}$ varies between 0.045 and 0.065 within the same period. In addition, the inclination $i$ does not remain constant; it varies by approximately $\pm 9'$ around the mean value of $5°8'43''$, with the period of $173^j$ corresponding to the passage of the Sun through the nodes (it is slightly shorter than 6 months because of the retrogradation of the nodes). When the Sun passes through the nodes, the inclination displays its largest value.

We can visualize the total effect in longitude by superimposing to the equation of the center $EC_\mathbb{D}$ a number of inequalities of different periods and amplitudes:

- The largest inequality, namely *evection*, is so great, $\pm 1°.274$, that it was already detected by Hipparchus. Its expression is $1°.274 \sin [(M - 2M_\odot) + 2(\varpi - \varpi_\odot) + 2\Omega)]$. The evection displaces the Moon in longitude by 2.5 times its diameter every 31.812 days.
- Tycho Brahe detected the second largest inequality, namely, the *variation*, of amplitude $\pm 39'.5$ and period of $14^j.78$ (half the lunar month, namely, half the synodic period). This inequality depends on the angle Sun–Earth–Moon, having as its argument the angle $2[(M - M_\odot) + (\varpi - \varpi_\odot) + \Omega]$. Therefore, it is zero both at the so-called *syzigies* (new Moon and full Moon) and at the quadratures; it is maximum at the octants.
- The *parallactic* inequality, with an amplitude of $\pm 125''$, is a function of the varying distance to the Sun. It has twice the period of the variation; therefore, its argument is $[(M - M_\odot) + (\varpi - \varpi_\odot) + \Omega]$. The name derives from the fact that by measuring the varying angular dimension of the Moon, the mean distance to the Sun can be derived, which constitutes another method of obtaining the solar parallax. The parallax of the Moon is therefore composed of a constant mean value: $\pi_\mathbb{D} = 3422''.70$ ($a_\mathbb{D} = 384,000$ km, the corresponding average angular diameter is $\alpha_\mathbb{D} = 932''$, i.e., 1738 km for a circular disk), plus a number of periodic terms. At the new moon, the parallax is approximately $1'$ (110 km) smaller than the average and correspondingly larger at the full moon. We shall consider again the apparent diameter in Chapter 15, when discussing eclipses and occultations. Notice that the monthly variation of the geocentric diameter depends on the motions of the two bodies around the common barycenter, in other words on the ratio of the two masses. This ratio also appears in the equations of motion of the Earth–Moon system around the Sun. Therefore, the agreement between theory and observations constitutes another proof of the equality between gravitational and inertial masses.
- The *annual equation* has an amplitude of $\pm 12'$ and a period of 1 year (argument $M_\odot$). This term is due to the periodic variation of the distance of the Earth–Moon system to the Sun caused by the eccentricity of the Earth orbit. Tycho Brahe discovered such inequality by noticing a periodic variation of the instants of the eclipses. From January to July, eclipses occur with a noticeable delay with respect to the expected time, the maximum deviation of $20^m$ occurring in April. The phenomenon reverses from July to January, with eclipses advancing by the same amount. In order to describe in more depth the occurrence of eclipses during the recorded millennia, one must consider two additional points. First, the secular variation of the terrestrial eccentricity $e_\oplus$ causes a secular (and not only periodic) increase of the distance to the Sun and therefore a secular decrease of the mean motion of the Moon. The second one is the slowing down of the Earth's rotation, as already noted in Chapter 10. In 1693, Halley noticed that the mean motion of the Moon at that time

had to be smaller than in past centuries; otherwise, the instants of the eclipses measured by Ptolemy and later by Arab astronomers could not be reproduced. Lalande estimated the amount of this acceleration to be $10''$/century$^2$ (the modern value is more like $12''.4$/century$^2$). In 1787, Laplace made a theoretical discussion of the lunar acceleration as being entirely due to the secular variation of $e_\oplus$, coming very close to Lalande's value. However, Laplace's conclusions were questioned by Delaunay and Adams in the following century, because no more than $6''$/century$^2$ can be due to that reason. Energy dissipation by the tides, which lengthen the day, must play an equally important role. Furthermore, the other planets of the Solar System exert both a direct perturbation of the lunar orbit and an indirect one through the changes of the Earth orbit. So to speak, the Earth amplifies the planetary perturbations on the lunar orbit.

Like the line of the apses, the line of the nodes presents several inequalities. The largest has the same period ($173^j$, namely, the passage of the Sun through the nodes) of the variation of inclination. The double inequality of the nodes and of the inclination is reflected by a great variation in latitude, with amplitude $\pm5'12''$ ($\pm550$ km) and period $32^j.38$, very close to that of evection but not to be confused with it.

The number of measurable periodic terms in longitude, latitude and distance has grown from the several hundreds used by mathematicians such as Plana and Delaunay in the 18th and 19th centuries, to several thousands in contemporary computer programs. Today, observations have greatly improved the precision of the third dimension. While in the past, the distance to the Moon was known with far lesser accuracy than the angular coordinates, today radar and laser lunar ranging give the radial coordinate with a precision better than $\pm1$ cm (e.g., the already quoted Apollo program, see the Web Sites section). This precision is of the order of the value of the Schwarzschild radius for the Earth, so that general relativity considerations must appear in present theories.

## 14.4   THE LUNAR MONTH AND THE LIBRATIONS

The very complex situation discussed in the previous section influences both the definition of the lunar month and the visibility of the lunar surface from a terrestrial site. We shall discuss in Chapter 15 the eclipses. Regarding the lunar month, we have the following definitions:

- *Synodic month:* the apparent geocentric period, namely, the interval of time between two identical phases of the Moon, in particular between two new moons (Sun and Moon in conjunction). This is also the lunar months in civil calendars. The mean value is $P_{syn} = 29^j12^h44^m = 29^j.53059$. A more refined value is given by Chapront-Touzè and Chapront (1983, 1988):

$$P_{syn} = 29^j.5305888531 + 0.000000216211T$$

where $T = (JD - 2451545.0)/36{,}525$. Any particular cycle can vary around this average value by up to $7^h$.
- *Sidereal month:* by the knowledge of the solar mean motion, we derive the lunar mean sidereal motion:

$$n_{syn} = n_\text{☾} - n_\odot = 1296000/29.5305881 = 43886.6979''/j$$

$$n_\text{☾} = 43886.6979 + 3548.1928 = 47434.8907''/j$$

The sidereal period $P_{\mathrm{\circleddash}}$ is therefore the mean interval of time needed for the lunar longitude to increase by 360° with respect to the fixed equinox, $P_{\mathrm{\circleddash}} = 1296000/n_{\mathrm{\circleddash}} = 27^{\mathrm{j}}07^{\mathrm{h}}43^{\mathrm{m}}11^{\mathrm{s}} = 2\,7^{\mathrm{j}}.3216609$. Consequently, the Moon moves $13°10'25''$ eastwards each day with respect to a fixed star. This is the fundamental period in lunar occultations of celestial sources.

- *Tropical month*: the interval of time needed for the mean lunar longitude to increase by 360° with respect to the mean equinox. Recalling the value of the general precession, namely, $0.1376''/\mathrm{j}$, we derive $P_{\mathrm{trop}} = 27^{\mathrm{j}}07^{\mathrm{h}}43^{\mathrm{m}}05^{\mathrm{s}} = 27^{\mathrm{j}}.3215816$, which is also the interval of time for the right ascension of the Moon to increase by $24^{\mathrm{h}}$.

- *Anomalistic month:* the interval of time between two passages through the mobile perigee. Its value is $P_{\mathrm{anom}} = 27^{\mathrm{j}}13^{\mathrm{h}}18^{\mathrm{m}}33^{\mathrm{s}} = 27^{\mathrm{j}}.5545502$. Therefore, the mean anomaly of the Moon increases by 360° in a time slightly greater than the period of sidereal revolution.

- *Draconic (or Draconitic) month:* the time between two passages of the Moon through the ascending node. It regulates both the declination and the eclipses. Owing to the secular retrogradation of the nodes, its value is the shortest: $P_{\mathrm{drac}} = 27^{\mathrm{j}}05^{\mathrm{h}}05^{\mathrm{m}}35^{\mathrm{s}} = 27^{\mathrm{j}}.2122178$. Due to the irregularities of the retrogradation (the largest having the same period of $173^{\mathrm{j}}$ affecting the inclination), the duration of draconic month has appreciable variations.

We close this section with a discussion regarding the visibility of the Moon's surface by a terrestrial observer. The rotation of the Moon is approximately synchronous with the orbital period, so that from the Earth, we always see one face, but not exactly so, as telescopic observations have proved since the time of Galileo Galilei. Cassini made very accurate measurements of the lunar surface. His name is associated to three basic laws, enunciated in 1693, regarding the lunar figure. For the dynamical justification of the laws, see, for instance, Danby (1998, in the Bibliography section).

The basic observational fact is that every lunar month, the surface appears to nod, in both north–south and east–west directions; that is, the lunar figure exhibits a phenomenon named *libration*.

Consider three vectors through the center O of the Moon: the rotation axis OR (inclined by $1°32'.1$ to the ecliptic), the normal to the orbit OP (inclined on average by $5°8'.7$ to the ecliptic and undergoing a retrograde precession of $\pm9'.0$ in $18^{\mathrm{y}}.61$) and the normal to the ecliptic OE, as in Figure 14.2.

According to the first of Cassini's laws, the rotation axis is fixed in the lunar body and the period of rotation is equal to the sidereal period (the Moon is in synchronous rotation with the Earth). According to the second law, the angle EOR (the inclination of the spin axis to the ecliptic) is constant, about $1°32'$. According to the third law, while the node of the orbit regresses, the inclination EOP remains constant at about $5°9'$, and the three vectors always lie in the same plane; in other words, OR precesses around OE in $18^{\mathrm{y}}.61$. Therefore, every 27.2 days (2 days shorter than the lunation) the Moon's north and south poles are alternatively tilted toward the Earth by $6°40'.8$. This is the libration in *latitude* (see Figure 14.3a). Owing to the strong variation of the lunar orbital velocity,

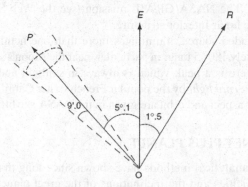

**FIGURE 14.2** The three vectors connected with the lunar figure. OP is perpendicular to the orbit, OE is perpendicular to the ecliptic and OR is the direction of rotation (angles not in scale).

(a)                                                      (b)

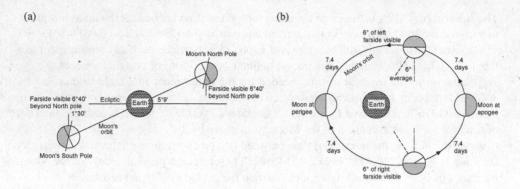

**FIGURE 14.3**    The libration in latitude (a) and in longitude (b).

while the rotational velocity is essentially constant, the part of the lunar surface visible from the Earth undergoes slight changes during the lunation (see Figure 14.3b).

The greatest variations occur halfway between the perigee and apogee. The mean value of this libration in *longitude* is 6°9′, but orbital perturbations can raise this value to approximately 8°. The mean period is 27.6 days, somewhat greater than that of the libration in latitude. The combination of the two librations is analogous to the combination of two periodic motions at right angles with slightly different periods; the maximum displacement of the center of the lunar disk can amount to about 10°.5.

Furthermore, due to his diurnal rotation the terrestrial observer moves his vantage point from moonrise to moonset. At both these times, almost 1° above the limb can be seen, but the exact amount depends on the lunar altitude (this is the so-called *diurnal* libration).

Finally, because of the unsymmetrical distribution of internal masses, the whole body of the Moon undergoes very small oscillations (*physical* librations), never exceeding 0°.04.

The overall result is that approximately 59% of the lunar surface can be seen from the Earth, although in an irregular manner, while 41% remains invisible to the terrestrial observer.

The first images of the lunar far side were delivered by the Soviet spacecraft *Luna* 3 in 1959. The first complete map came from the American *Lunar Orbiters*. Later on, several space missions from NASA, ESA and JAXA in Japan, ISRO in India and CNSA in China have provided a wealth of detailed images of both faces. Together with the surface details, we now know, with great precision, both the altimetry and the internal distribution of masses. The barycenter is displaced by some 2 km toward the Earth with respect to the geometrical center. Just below the surface, on the Earth side there are strong concentrations of masses, the so-called lunar *mascons*, causing strong gravitational anomalies on its surface. In some locations, in particular in large impact basins such as Mare Imbrium, a plumb line on the lunar surface would be inclined by approximately 20 arcminutes to the geographic vertical. The NASA GRAIL mission (see the Web Sites section) has provided detailed information on the lunar interior structure.

The Sun, being an extended source, illuminates more than one hemisphere at each time. Light extends for arcs approximately 180°.5 long; in particular, near the Moon's south pole, inside the gigantic Aitken impact basin, there is a peak which is always in sunlight and has been given the fairly pompous name of *peak of eternal light* by the popular French writer Camille Flammarion. Useful animations regarding the lunar aspect and orbit are given in the NASA site https://svs.gsfc.nasa.gov/4442.

## 14.5   THE CASE PLANET PLUS PLANET

As a preliminary remark, analytical methods have shown since long that the semi-major axes, the mean motions, the eccentricities and the inclinations of the great planets do not have appreciable secular inequalities. Therefore, the present configuration of the Solar System must be stable for long times; this result is associated with the names of Delaunay and Tisserand. However, the stability cannot be demonstrated for arbitrarily long periods, as was already shown by Poincaré. See, for instance, Nobili et al. (1989). In particular, the earliest phases of the Solar System, approximately 4.5 billion years ago, witnessed a very rapid evolution of the overall structure, as described by the so-called *Nice* model (Tsiganis et al., 2005).

In the present configuration of the large planets, consider Jupiter and Saturn of masses $m_J$ and $m_S$ respectively; for simplicity, let us assume that the two planets have the same orbital plane. Their mutual distance $d$ can be larger than their distances to the Sun $\Delta_J$, $\Delta_S$. If $\varphi$ is the angle between the two planets as seen from the Sun, the perturbing function of Saturn on Jupiter can be written as

$$R = Gm_S \left[ \frac{1}{d} - \frac{\Delta_J \cos\varphi}{\Delta_S^2} \right]$$

$$= Gm_S \Delta_S \left[ \frac{1}{\sqrt{1 + \left(\dfrac{\Delta_J}{\Delta_S}\right)^2 - 2\dfrac{\Delta_J}{\Delta_S}\cos\varphi}} - \frac{\Delta_J}{\Delta_S^3}\cos\varphi \right] \tag{14.14}$$

where $\Delta_J < \Delta_S$.

This treatment can be extended to the perturbation of an external planet on an inner one, for instance, of Jupiter upon Mars or upon a Main Belt asteroid. Appropriate methods have been developed for the symmetric case, e.g., perturbation of Jupiter on Saturn.

$R$ is a periodic function of the true anomalies of the two planets. Indeed, after $p_J$ laps of Jupiter and $p_S$ laps of Saturn, the position will be identical to the starting one. Equation 14.14 is formally identical to that of the Moon, but there are important differences. First, in no case the ratio $\Delta_J/\Delta_S$ can be considered arbitrarily small. Second, the secular motions of the nodes and perihelia of the two planets are fairly slow, with periods exceeding 18,000 years. Therefore, we are justified to search an approximate solution valid for few centuries and to assume that the perturbing planet moves on a fixed Keplerian ellipse. Thus, each orbital element of the perturbed body can be expressed by a series of secular, long- and short-period inequalities.

Some periodic terms can become of very high amplitude (*resonances*) when the mean motions $n_1$, $n_2$ of the two planets are commensurable. A large amplitude resonance occurs when two integer numbers $i, j$ (negative or positive) can be found for which

$$in_1 + jn_2 = n_r \approx 0 \tag{14.15}$$

In the case of Saturn–Jupiter,

$$n_J = 299''.1283\,j^{-1}, \quad n_S = 120''.4547\,j^{-1}, \quad 5n_J - 2n_S = n_r \approx 4''.0169\,j^{-1}$$

which corresponds to a period of 883 years. The largest effect is in longitude and amounts to approximately 20′ for Jupiter and 50′ for Saturn. The amplitude is so large that Kepler was able to detect it; he interpreted the effect as a secular acceleration of Jupiter and deceleration of Saturn. Laplace found the correct explanation in 1783.

An immediate application of the theory of perturbations was the discovery of Neptune in 1846. The situation was, so to speak, the reverse of the one previously discussed: from the observed perturbations of a given body (Uranus), the mass, position and orbital elements of a still unseen perturbing body were inferred. Adams in England and Le Verrier in France investigated the large inequalities of Uranus almost contemporarily. Le Verrier's calculations were completed in August 1846, and by September 23,

---

**TABLE 14.1**

**Orbital Elements of Neptune**

| Element | Adams | Le Verrier | True |
|---|---|---|---|
| $a$ | 37.2 | 36.1 | 30.1 |
| $e$ | 0.12 | 0.11 | 0.009 |
| $\omega$ | 299 | 284 | 46 |
| $M$ | 0.00015 | 0.00011 | 0.00005 |

---

Galle in Berlin discovered Neptune at only 52′ from the predicted position. The discovery was regarded as a great triumph for celestial mechanics. However, the true orbital elements of Neptune are appreciably different from those calculated by both Le Verrier and Adams, as seen in Table 14.1. Notice in particular the small mass of Neptune. The discovery was therefore largely accidental and due to the presence of favorable circumstances: Neptune and Uranus are in heliocentric conjunction (when the mutual perturbations are the largest) every 171 years; the 1822 occurrence was only 24 years before the discovery. Therefore, any reasonable mean motion and small inclination would have produced a position correct to approximately 1°. The question of the perturbations of Uranus is therefore not entirely explained by Neptune. The next (formerly) planet, Pluto, discovered by Clyde Tombaugh in 1938, has a mass far too small to induce such large perturbations (see also the "Notes" section). From time to time, the existence of a tenth planet (actually, ninth, since the reclassification of Pluto as dwarf planet; confusion might arise by calling Planet X this unseen ninth body) with sufficiently high mass has been revived, but none has been found beyond doubt. Alternatively, we are led to suspect strong and systematic errors in the timescales used in the 17th and 18th centuries.

## 14.6  THE RESTRICTED CIRCULAR THREE-BODY PROBLEM

This very special case was solved in analytical form by Lagrange (1772, see Bibliography section). Consider two bodies of mass $m_1$ and $m_2$, respectively, ($m_1 > m_2$), on a relative orbit of circular form, and therefore at a constant relative distance. It is customary to put the gravitational constant $G = 1$, to use as unit of distance the fixed distance between the two bodies and let the total mass $m_1 + m_2 = 1$. Thus, the first (heavier) body has indicative mass $\mu_1 = 1 - \mu_2$ and the second one $\mu_2 < 1/2$. It also follows that the common mean motion $n$ is unity.

Let $(x, y)$ be a barycentric reference system in the orbital plane, with the $x$-axis always coincident with the line joining the two bodies (in other words, a synodic reference system). The invariable coordinates of the two masses in it will be $(x_1, y_1) = (-\mu_2, 0)$ and $(x_2, y_2) = (\mu_1, 0)$. Therefore, this system rotates with respect to the inertial barycentric system with an angular velocity equal to the mean motion of the two bodies $n$ (which is equal to 1 in the present unit system, but that we keep explicit for clarity).

Let us add to the system a third body of negligible mass $m'$ (so that it is gravitationally attracted by the other two, but does not exert any attraction on them), orbiting in the same plane of the relative orbit and therefore with zero velocity component perpendicular to that plane. Its distances from the two particles will be, respectively,

$$r_1^2 = \left(x + \mu_2\right)^2 + y^2, \qquad r_2^2 = \left(x - \mu_1\right)^2 + y^2$$

It can be demonstrated that the accelerations of the test particle are given by

$$\ddot{x} - 2n\dot{y} = \frac{\partial U}{\partial x}, \qquad \ddot{y} + 2n\dot{x} = \frac{\partial U}{\partial y} \tag{14.16}$$

where $U$ is a pseudopotential:

$$U(x,y) = \frac{n^2}{2}(x^2 + y^2) + \frac{\mu_1}{r_1} + \frac{\mu_2}{r_2} \qquad (14.17)$$

As expected, the introduction of a non-inertial system has produced the presence of the Coriolis terms ($-2n\dot{y}$ and $+2n\dot{x}$ in Equation 14.16) and of the centrifugal acceleration (the terms in $n^2x$ and $n^2y$ in the equations of movement, which can be derived from Equation 14.17).

After some manipulation and integration of the previous equations, the modulus of the total velocity $V$ of the test particle in the rotating frame can be written as

$$V = \sqrt{\dot{x}^2 + \dot{y}^2} = 2U - C \qquad (14.18)$$

Equation 14.18 is the so-called *Jacobi integral*; the constant $C = 2U - V$ is the Jacobi *constant* (notice that the Jacobi integral is not an energy integral). It follows that the movement of the test particle in the rotating frame can occur only in the regions of the $(x, y)$ plane for which $2U > C$; inside the line for which $2U = C$, the particle has a stable position.

Without entering into the very interesting but very complex discussion of the topology of the curves $2U = C$ as a function of the value of $C$ (usually referred to as Roche's curves), we state the following results:

- five points of stability (zero velocity in the rotating frame) can be identified; these points are called Lagrangian points, and are designated $L_1$, $L_2$, $L_3$, $L_4$ and $L_5$;
- $L_1$, $L_2$ and $L_3$ are collinear on the line joining the two bodies (they all have $y = 0$). $L_2$ is outside the two bodies and opposite to that of the smaller mass, $L_1$ is between the two bodies and nearer to that of the smaller mass, and $L_3$ is outside the two bodies and opposite to that of the larger mass (be advised that some authors interchange the designation of $L_1$ and $L_2$). The precise derivation of their $x$-coordinates is not a trivial exercise; approximately, their distances from the two masses are

$$r_2(L_2) \approx \alpha + \frac{1}{3}\alpha^2, \quad r_2(L_1) \approx \alpha - \frac{1}{3}\alpha^2$$

$$r_1(L_3) \approx 1 - \frac{7}{12}\left(\frac{\mu^2}{\mu_1} - \frac{\mu_2^2}{\mu_1^2}\right).$$

where the parameter $\alpha$ is defined by

$$\alpha = \left(\frac{\mu_2}{3\mu_1}\right)^{1/3} \qquad (14.19)$$

Therefore, $L_1$ and $L_2$ are almost symmetric with respect to the smaller body;
- $L_4$ and $L_5$ are at the vertexes of the equilateral triangle for the two bodies, $L_4$ on the preceding (leading) point and $L_5$ on the following (trailing) one. Their coordinates are

$$x = \frac{1}{2} - \mu_2, \quad y = \pm\frac{\sqrt{3}}{2}$$

Several noticeable examples of Lagrangian points can be given:

- in the Sun–Jupiter system, the regions around $L_4$ and $L_5$ are occupied by many small bodies, collectively called Jupiter's *Trojans*; more than 7000 of them are known, but perhaps millions of smaller bodies occupy the two regions. Jupiter is not the only planet to have satellites in the two Lagrangian points; Mars, Uranus and Neptune too are known to have "trojans" in $L_4$ and $L_5$. Even the Earth has one. Only Saturn at moment is deprived of "trojans";
- in the Sun–Earth system (see Figure 14.4), $L_1$ is occupied by spacecraft such as the Solar Heliospheric Observatory (SOHO), while $L_2$ was occupied by NASA WMAP and ESA Planck (all satellites dedicated to the cosmological background radiation). At present, ESA GAIA is there and so will be in the future NASA JWST, the successor of the HST;
- on closer examination, those mathematical "points" of stability become regions of small extent, inside which the test particle moves with respect to the ideal point (the particle *librates* inside this region). The stability inside the libration regions can be weak and readily perturbed, so that orbital corrections at appropriate intervals are needed to keep the satellite parked around the $L_2$ point of the Sun–Earth system;
- a relay satellite allowing radio communication with the Chinese Chang'è 4 lander parked on the rear side of the Moon librates around the Earth–Moon system $L_2$.

The same considerations can be applied to the planets of other stars (exoplanets). For binary stars, there is the important difference that the two stellar masses can be equal and their ratio is rarely lower than 1/100. Examination of the Roche curves allows identification of the regions through which the two stars can exchange mass.

We will briefly examine three consequences of the restricted three-body problem, namely, Tisserand's criterion for the stability of the orbital elements, the concept of the Hill sphere and the technique of spacecraft gravitational assist:

1. *Tisserand criterion:* the Jacobi integral can be used to follow the changes of the orbital elements of a lighter particle after an encounter with a heavier body, for instance, a comet after a near pass with Jupiter. Call $(a, e, i)$ the initial elements and $(a', e', i')$ those after the encounter; it can be shown that

$$\frac{1}{2a} + \sqrt{a\left(1 - e^2\right)}\cos i = \frac{1}{2a'} + \sqrt{a'\left(1 - e'2\right)}\cos i'$$

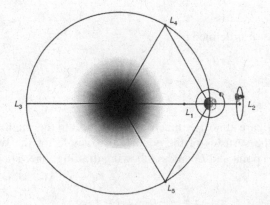

**FIGURE 14.4**   The Lagrange points of the Sun–Earth system. A libration region has been sketched around $L_2$.

where the unit of length is the Sun–Jupiter distance, the unit of mass is the Sun's mass, and the timescale is such that Jupiter's angular velocity is unity. The Tisserand criterion is not a rigorous one nor provides a definite proof that the comet observed after the encounter is the same observed before; nevertheless, if the orbital elements are widely different, it can be concluded that the two sets do not belong to the same comet.

2. *Hill sphere:* the Hill sphere can be loosely defined as the volume around a planet where its gravitational influence overcomes that of the Sun. If $\Delta$ is the distance of the planet from the Sun, then the radius of its sphere $R_h$ can be expressed as

$$R_H = \alpha\Delta \tag{14.20}$$

where the parameter $\alpha$ is given in Equation 14.19. Actually, several definitions of $R_H$ are found in the literature, with different expressions of the parameter. However, all have a linear dependence on the distance to the Sun, which leads to an increasing distance between the planets (in qualitative agreement with the Titius–Bode law), and only a weak dependence on the mass of the planet. For the Earth, $R_H \approx 0.01$ AU, for Jupiter it is approximately 30 times larger. Should a comet pass inside Jupiter's Hill sphere, different outcomes are possible: (1) the comet falls inside Jupiter, as it happened to Shoemaker-Levy 9 in July 1994; (2) the comet is captured in an orbit inside the inner Solar System; (3) it is gravitationally scattered outwards. Indeed, we know of a Jupiter family of comets, like comet 67P/Churyumov-Gerasimenko, target of the ESA Rosetta mission. The same qualitative considerations can be applied to Saturn and other large planets. During the initial formation phases of the Solar System, both mechanisms of accretion, inward and outward scattering of the original mass elements (known as planetesimals) by the large planets, have occurred, thus contributing to shape its present structure, which includes an outside cloud of comets (Oort's cloud) beyond 50,000 AU, populated by billions of small bodies.

3. *Gravitational assist:* the previous considerations about the Tisserand criterion and the Hill sphere can be applied to the so-called gravitational slingshot effect or gravitational assist. This is a navigational technique, by which the initial orbit of a spacecraft is purposely strongly perturbed by a passage inside the Hill sphere of a planet. Following this close encounter, the spacecraft is directed to reach another planet, where the rocket energy would not be sufficient to go directly. Tisserand's criterion gives an indication of the orbital elements after the passage. Many missions to planets and comets (e.g., Galileo, Cassini, Rosetta) took advantage of this technique. A noticeable example is the trajectory of Ulysses, a spacecraft which orbited between Jupiter and the Sun along an orbit essentially perpendicular to the ecliptic plane, namely, on an orbit where previously only comets of the Jupiter family were known (see Figure 14.5). As already noted, the energy of the third body is not conserved in the general three-body problem. The same violation occurs in the gravitational assist: the increase of the semi-major axis of the third body must be compensated by a decrease of the semi-major axis of the second one. However, in practice, the mass of the spacecraft is so small that no effect can be detected on the planet orbit.

## 14.7   A NON-SPHERICAL BODY PLUS A SMALL NEARBY SATELLITE

In the case of an artificial satellite in low Earth's orbit, the real shape of the Earth cannot be ignored. The following considerations will apply to other similar configurations, e.g., the two small moons Phobos and Deimos close to Mars or an inner moon of Jupiter. For simplicity, let us approximate the Earth as an elliptical body of equatorial radius $a_\oplus$ and total mass $M_\oplus$ having rotational symmetry. Its gravitational potential on an external point-like body at distance $r$ and latitude $\varphi$ with respect to the Earth equator can be expressed as (see also Chapter 6)

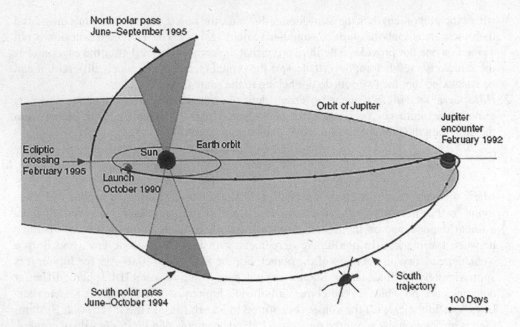

**FIGURE 14.5**   The initial orbit of Ulysses.

$$U(r) = G\frac{M_\oplus}{r} \times \left[ 1 + J_2 \frac{1}{r^2}\left(\frac{1}{2} - \frac{3}{2}\sin^2\varphi\right) + J_4 \frac{1}{r^4}\left(\frac{3}{8} - \frac{15}{4}\sin^2\varphi + \frac{35}{8}\sin^4\varphi\right) + \cdots \right]$$

$$= G\frac{M_\oplus}{r}\left[ 1 - \sum_{n=1}^{\infty} \frac{1}{r^{2n}} J_{2n} P_{2n}\left(\sin\varphi\right) \right] \qquad (14.21)$$

where the $P_{2n}$ are Legendre's polynomials of degree $2n$.

In our qualitative discussion, we can truncate the development after the second term, writing the perturbing function as

$$R = G\frac{M}{2r^3} J_2 \left[ 1 - 3\sin^2\varphi \right] \qquad (14.22)$$

where

$$J_2 = -(C - A)/M_\oplus = -\frac{3}{2} J a_\oplus^2, \quad J = \frac{3}{2}\frac{C - A}{M_\oplus a_\oplus^2}.$$

The perturbing force will contain radial and non-radial terms. In other words, the force is a non-central one, with the accompanying consequences of non-closed orbit of the satellite, of orbital plane precession and advance (or retrogradation) of the pericenter on the orbital plane. According to Lagrange's planetary equations, we must expect the presence of secular, long- and short-term periods in the orbital elements. The previously cited theorems of Delaunay and Tisserand insure that the semi-major axis, the inclination and the eccentricity do not have secular terms.

The largest long-term periods have amplitudes of the order $J_2$ and periods half of that of the pericenter; higher-order terms in $J_2$ exist, with periods 1/4, 1/6, etc. of the fundamental one.

The short-term periods will deform the basic orbit, which turns and undergoes long-term periodic perturbations. In particular, the line of the nodes has a retrograde motion with a period given by

$$P_{\text{retr}} = \frac{4}{3}\pi \frac{1}{J_2} \frac{\left(1-e_0^2\right)^2 a_0^2}{n_0 \cos i_0} \tag{14.23}$$

where $a_0$, $e_0$, $i_0$ are the initial orbital elements of the satellite. The motion is more rapid when the inclination tends to zero, and it does not exist for polar orbits; it is slower for large eccentricities and for large semi-major axes.

Let us consider, e.g., the motion of the perigee according to Lagrange's Equation 14.5. Even in the present simplified treatment, to the first order in $J_2$ we would find a fairly elaborate expression for $\omega(t)$:

$$\omega(t) = \omega_0 + \frac{3n_0 J_2}{4a_0^2}\left(-1 + 5\cos^2 i_0\right)t + f\left(\frac{J_2}{a_0^2}, \omega_0, n_0(t-t_0), e_0\right)$$

where $f$ is an appropriate function of its arguments. In the coefficient of the secular term (the second one), there is a particular inclination $I_1$ for which $(-1 + 5\cos^2 I_1) = 0$, i.e., $I_1 = 63°26'$, dividing two cases: for $i_0 > I_1$, the movement of the pericenter is retrograde, for $i_0 < I_1$, it is direct. Should we push the treatment to the second order in $J_2^2$, we would find long-term periods with amplitudes of the order of $J_2$: this is a major problem in celestial mechanics, and the long-term periods of the first order are found only in the second approximation.

Regarding the mean motion, it can be shown in the same manner that there is a second inclination $I_2$ for which $(-1 + 3\cos^2 I_2) = 0$, i.e., $I_2 = 35°15'$, again dividing into two cases: if $i_0 > I_2$, the mean motion of the satellite is slower than the Keplerian value $2\pi/n_0$, and if $i_0 < I_1$, the mean motion is faster.

In the case of an equatorial satellite (or small moon) in circular orbit, a simple formula expresses the influence of the non-sphericity of the planet of mass $m$ on Kepler's third law; if $f$ is the flattening as in Chapter 2, then

$$\frac{P^2}{a^3} = \frac{4\pi^2}{Gm}(1-\kappa), \qquad \kappa = \left(\frac{R}{a}\right)^2\left(f - \frac{\sigma}{2}\right)$$

where $R/a$ is the ratio between the radius of the planet and the semi-major axis of the orbit of the small moon, and $\sigma$ is the ratio between the centrifugal force and the gravity at the equator of the planet ($\sigma = \omega^2 R^3/Gm$). Consider, for instance, the case of Mars and Phobos:

$$f = 1/192, \qquad \sigma = 1/218, \qquad a(\text{Phobos})/R(\text{Mars}) \approx 2.8, \qquad \kappa \approx 1/2400$$

a small deviation indeed. This result can be derived from Clairaut's theorem, which expresses the gravity at the distance $a$ from a rotating body of ellipsoidal shape (see Chapter 6).

In addition to the low precision of the qualitative discussion dealt with in the present section, the case of a real artificial satellite is physically more complex. Non-gravitational forces, such as atmospheric drag, terrestrial and solar radiation pressure (perturbations proportional to the ratio of the surface to the mass of the satellite), and the Earth's magnetic field, cannot be ignored.

In order to complete the information on the Solar System, Table 14.2 provides some data on the flattening and rotation characteristics of the Sun, the planets and the Moon. Note that the giant gaseous planets have different rotation periods according to how they are measured; that is, the period measurable by the visible surface details is not the same given by the internal rotation, manifested through the magnetic field and associated radio-frequency emission (system III).

By definition, the north pole of any planet is that above the ecliptic plane; therefore, the rotation is considered retrograde for Venus, Uranus and Pluto.

### TABLE 14.2
### Data on Flattening $f$ and Rotation Characteristics of the Sun, Planets and Pluto

| Body | $f$ | $P_{rot}$ | $P_{rot}$ (Internal) | Obliquity (°) |
|---|---|---|---|---|
| Sun | 0.0 | 25.4[ja] | | 7.25 |
| Mercury | 0.000 | 58.65[j] | | 0.034 |
| Venus | 0.000 | (R)243.01[j] | | 177.36 |
| Earth | 0.0034 | 23.9345[h] | | 23.44 |
| Moon | 0.0020 | 27.32[j] | | 6.67 |
| Mars | 0.0069 | 24.623[h] | | 25.19 |
| Jupiter | 0.0649 | 9.841[h] | 9.925 h (system III) | 3.12 |
| Saturn | 0.0980 | 10.233[h] | 10.656 h (system III) | 26.73 |
| Uranus | 0.0229 | (R) 17.9[h] | (R) 17.240 h | 97.86 |
| Neptune | 0.0170 | 19.0[h] | 16.11 h | 29.56 |
| Pluto | 0.000 | (R) 153.2928[h] | | 122.53 |

(R) Means retrograde rotation; the obliquity is the tilt of the equator with respect to the orbital plane (for the Sun with respect to the ecliptic).

[a] This is the adopted period at 16° latitude. The actual rotation rate varies with the latitude $\Lambda$ as $(14.37 - 2.33 \sin 2\Lambda - 1.56 \sin 4\Lambda)$ degree/day.

## 14.8   OTHER INTERESTING CASES

The Solar System offers many other interesting problems of gravitational perturbations and interaction, among planets, their moons and asteroids. Among the vast literature on such rich field of investigation, we quote in the Bibliography section the books by Danby, Celletti and Perozzi, Shirley and Fairbridge and Lissauer and de Pater. Exoplanets offer other important case studies. Few examples are given in the following.

Greek name of *Mercury* is Hermes, so that the adjective Hermean is often used to refer to its characteristics. The terrestrial optical telescopes are usually limited to observe the surface of the planet at maximum elongations, of which six take place, three more favorable for observers in the northern hemisphere and three for the southern ones. The main orbital characteristics, sidereal period of 87.97 days, high inclination of 7° to the ecliptic and high eccentricity $e \approx 0.206$, were known since long. Comparing the Hermean revolution period to the terrestrial one, it was recognized that the planet is locked into a quasi-commensurability with the Earth: 54 Hermean orbital sidereal periods are equal to $\approx 13$ terrestrial years. Instead, it was only in the 1960s that a reliable determination of the rotation period, 58.65 days, could be made, thanks to radar signals sent from Arecibo. From the observational point of view, such period implies that northern telescopes can observe Mercury's surface centered at meridian longitudes between 90° and 270°, and southern telescopes most of the other regions, except longitudes 90° and 270°. From the dynamical side, a very stable spin-orbit resonance 3:2 between rotation and revolution periods was recognized; thus, the rotation velocity (10.8 km/hour at the equator) is 1.5 times higher than the mean orbital velocity. The stability of the 3:2 resonance might be due to a slight internal asymmetry of mass distribution. Other interesting consequences of the high eccentricity and spin-orbit coupling were ascertained. For instance, at alternate perihelion passages, the same hemisphere faces the Sun; the same is true for alternate aphelion passages, with drastic longitude-dependent temperature effects. Another consequence is the duration of the solar day (one full day–night cycle), about 0.646 as long as the year. That happens because 4 days before perihelion, the rotational angular velocity is equal to the orbital angular velocity, causing the apparent westward motion of the Sun to stop. At perihelion, the angular orbital velocity exceeds the angular rotational velocity, and the Sun begins to move in reverse. In other words, one Mercury solar day equals 176 Earth days, slightly longer than 2 years on Mercury. Another unique feature among the large planets is the lack of seasons, because Mercury's axis of rotation is tilted by not more than 2° with respect to the plane of its orbit around the Sun.

As already mentioned, given the difficulties of optical observations from the Earth, radar data were of paramount importance. NASA's space missions Mercury 10 and Messenger have greatly enhanced the understanding of such complex planet. The ESA BepiColombo (see the Web Sites section) is on its way, with an expected arrival in December 2025.

*Venus–Earth:* Venus has a very slow retrograde rotation of 243 days, while revolving in 225 days. At each inferior conjunction, Venus presents the same face to the Earth. See also Chapter 13.

*Other resonances.* There are many examples of resonances in the Solar System, for instance, the ratio 3:2 between the orbital periods of Pluto and Neptune. Other resonances involve three bodies and are called Laplace resonances. For instance, the periods of the Medicean moons Io, Europa, and Ganymede are in the ratio 4:2:1. Noticeable resonances are revealed by studies of exoplanets.

*Synchronous rotation:* As in the case Moon–Earth, most natural satellites are in synchronous rotation, always keeping the same face pointed toward the planet (the so-called spin-locking effect). Most of the observational evidence came from the Voyagers, the twin space mission to the outer Solar System. This co-rotation situation is probably not the initial one: satellites were trapped in the synchronous rotation, in a time much shorter than the age of the Solar System. Thus, Cassini's laws could be generalized to any spin-locked satellite undergoing orbital precession at a uniform rate. However, for the general case, particular values of the inclinations are required, the so-called Cassini's states (see, e.g., Colombo, 1966). In Cassin's state 1, both the spin axis and the orbit normal axis are on the same side of the normal to a so-called Laplace plane (not to be confused with the Solar System invariable plane). In Cassini state 2, the spin axis and the orbit normal axis are on opposite sides of the normal to such plane. If the satellite is very close to its planet, the Laplace plane practically coincides with the equatorial plane of its planet. If the satellite is far away from its planet, the Laplace plane coincides with the orbital plane of the planet. Examples of satellites whose Laplace plane is close to their planet's equatorial plane include Deimos and Phobos around Mars and the inner satellites of the giant planets. Therefore, the Moon, in Cassini state 2, is an example of the second case of the position of the Laplace plane. There are intermediate cases, such as Saturn's Iapetus, having their Laplace plane midway between the equatorial plane and the orbital plane of their planet.

*Tumbling moons and asteroids.* Among all the moons, Hyperion is certainly a very peculiar one. A major collision in the past with another body possibly blew part of Hyperion away. Its orbit is fairly eccentric ($e = 0.1$), and its radial distance from Saturn is large, approximately 25 Saturn's radii, so as to make unlike a tidal trapping into co-rotation. To complicate the matter, the revolution of Hyperion is in 4:3 resonance with Titan. The rotational period varies from one orbit to the next, a situation described as *chaotic*. Even the attitude of the moon is unstable, the spin axis tumbles in space (see, e.g., Black et al., 1995). Tumbling has also been observed on a few asteroids and probably results from collisions (e.g., Paolicchi et al., 2002).

## NOTES

- The relativistic effects on the orbits of Mercury and other bodies, including artificial satellites, are not limited to those previously described. According to the Lense–Thirring effect, a precession of the plane of the geodesic orbit of a test particle around a rotating mass might also be detected. It arises from the coupling of the rotation of the central mass with the orbital angular momentum of the test particle. This precession is the result of the dragging of inertial frames. The theory is reviewed, for instance, by Ciufolini and Wheeler (1995) and Mashhoon (2008, both in the Bibliography section). The Lense–Thirring effect is also present in the Earth's artificial satellites. For its measurement, see, for instance, Ciufolini (2000). The effect was detected using the Lageos satellite (see the Web Sites section).

- In 1979, Pluto was discovered to be a double planet, with a fairly large moon named Charon. Since the discovery in 2002 of Eris, another large body beyond Pluto, quickly followed by the discovery of numerous similar trans-Neptunian bodies, in 2006 the IAU instituted the new class of dwarf planets. The NASA mission New Horizons (see the Web Sites section) discovered several more moons around Pluto, before flying-by the trans-Neptunian object (486958) Arrokoth (nicknamed Ultima Thule) in early 2019.

## EXERCISE

1. An artificial satellite is usually launched with an easterly component in order to obtain a speed advantage from the Earth's rotation from west to east. Equation 14.23 show that for a circular orbit, the node line regresses westward by

$$\psi = \frac{6\pi}{5}\left(\frac{C-A}{C}\right)\left(\frac{R}{a}\right)^2 \cos i = 2\pi J_2 \left(\frac{R}{a}\right)^2 \cos i$$

$$\approx 0.01\left(\frac{R}{a}\right)^2 \cos i \,(\text{radians/revolution})$$

where $\psi$ is measured with respect to a fixed direction in the equatorial plane.

# 15 Eclipses, Occultations and Transits

In this chapter, we examine in some detail the solar and lunar eclipses and the occultations of stars by the Moon. Stellar occultations by asteroids and trans-Neptunian objects and transits of planets over the disk of their host star (exoplanets) will be briefly mentioned. Transits are indeed the most effective way to discover exoplanets, both from well-equipped ground telescopes even of modest size, for instance, the TRAPPIST telescope, and from space, as attested by the NASA satellites Kepler and TESS (see Web Sites section).

The treatment will be limited essentially to the geometric aspect, even if eclipses have important physical implications. For instance, when the Moon enters the terrestrial shadow, the flux of solar photons on its surface is suddenly turned off; the thermal balance is altered and so is the very tenuous transient atmosphere, observable from the Earth especially in the light of the yellow Na doublet (see, e.g., Verani et al., 2001). We have already commented on the usefulness of ancient eclipses to determine the rate of change of the Earth's diurnal rotation.

The solar eclipses deserve the merit of having permitted the discovery of the hot solar chromosphere and corona by visual observations. For instance, Halley in 1715 noticed both the corona and red prominences, but he tended to attribute the luminous gases to the heated lunar surface. In 1836, Francis Baily discovered, during an annular eclipse, the bright rays coming from the valley separating the mountains on the lunar limb (the so-called *Baily's beads*); in 1842, he noticed again the corona and the prominences. Finally, during the eclipse of August 18, 1860, Father A. Secchi, S.J. and W. de la Rue for the first time could photograph the corona from two separate places, and thus it became clear that the gases were indeed around the Sun.

Of course, one cannot forget that the $6^m51^s$ long solar eclipse of 1919 allowed Dyson, Eddington and Davidson to confirm Einstein's theory of general relativity, as already mentioned in Chapter 7. See, for instance, the commentary by Longair (2015, Bibliography section).

The generic term *occultation* indicates the situation of an opaque body passing across the line of sight to a planet or to a star; therefore, the term *transit* (say of Mercury or Venus in front of the solar disk, see also Chapter 13) means actually a partial occultation; a solar eclipse too is a form of occultation. Occultations can provide very important information on the occulted body, such as its position and structure. Lunar occultations were, indeed, the most precise way to determine the positions of radio or X-ray sources in the early days of these new astronomies. Two examples are the FK4 position of 3C 273B by Hazard et al. (1971, see "Notes" section of Chapter 6) and the European satellite EXOSAT, launched with the specific purpose to observe occultations of X-ray sources. Furthermore, occultations allow to determine the structure of both the occulted source (such as duplicity, apparent size, limb darkening) and the occulting body. For instance, the rings of Uranus were first detected observing the occultation of a star and subsequently confirmed by Voyager's images (see "Notes" section).

## 15.1 MOON'S PHASES

The Moon is an opaque body illuminated by the Sun, with a day side and a night side. When the angle between the Moon and the Sun is at its minimum value as seen by the geocentric observer, in other words when the night side faces the Earth, it is the astronomical *new moon*. Following the new moon, we see a gradual increase in the illuminated portion of the disk (*waxing moon*). The crescent moon is seen at sunset on the western horizon. After approximately 7.4 days, the ecliptic longitude of the Moon is 90° greater than that of the Sun; this is the astronomical *first quarter*. The Moon is in the meridian of each place at approximately 6 P.M. local time. The first quarter marks the end of the crescent phase and the start of the gibbous phase. After another 7.4 days, the longitude of the Moon is 180° larger than that of the Sun; the lunar disk is fully illuminated (*full moon*); it rises when the Sun sets, reaches the meridian around local midnight and finally sets when the Sun rises. Notice that full moon is not equal to phase angle (namely, the angle Earth–Moon–Sun) $\alpha = 0°$; otherwise, an eclipse would occur; the phase angles of full moons are usually larger than 1°.5, because the lunar ecliptic latitude is different from zero. The next two periods of 7.4 days lead, respectively, to the last quarter and again to the new moon. This half lunation is also called the *waning moon*. We have already noted in Chapter 14 that the solar perturbations influence the duration of the lunations; for instance, in January 1974 there was one of the longest, $29^{j}19^{h}55^{m}$, while in June 2035, there will be one of the shortest, $29^{j}06^{h}39^{m}$ (see Meeus, Bibliography section).

The disk of the Moon can be taken for simplicity as circular, so that the illuminated surface will appear as the intersection between the circular limb and an elliptical *terminator*, which is the border of the circular illuminated area as projected on the surface. Let us call $k$ the fraction of the illuminated area and $\alpha$ the phase angle. Then,

$$k = \frac{1}{2}(1 + \cos\alpha) \tag{15.1}$$

Notice that $k$ is also the fraction of the illuminated diameter perpendicular to the terminator. According to the formulae of Chapter 3, the phase angle $\alpha$ can be derived from the geocentric equatorial or ecliptic coordinates of the Sun and of the Moon, by the intermediate of the angular distance $\vartheta$ between the two bodies (which we can call the elongation of the Moon from the Sun along the great circle joining them):

$$\cos\vartheta = \sin\delta_{\odot}\sin\delta_{\mathrm{D}} + \cos\delta_{\odot}\cos\delta_{\mathrm{D}}\cos(\alpha_{\odot} - \alpha_{\mathrm{D}})$$

$$= \cos\beta_{\mathrm{D}}\cos(\lambda_{\odot} - \lambda_{\mathrm{D}}) \tag{15.2}$$

$$\tan\alpha = \frac{\sin\vartheta}{\Delta - \cos\vartheta} \tag{15.3}$$

$\Delta$ being the distance Earth–Moon normalized to the instantaneous Earth–Sun distance.

The cusps (popularly, horns) of the terminator always point away from the Sun. Therefore, if C is the mid-point of the illuminated disk, its position angle $p$ (from North through East, and paying attention to the true quadrant) is given by:

$$p = \frac{\cos\delta_{\odot}\sin(\alpha_{\odot} - \alpha_{\mathrm{D}})}{\sin\delta_{\odot}\cos\delta_{\mathrm{D}} - \cos\delta_{\odot}\sin\delta_{\mathrm{D}}\cos(\alpha_{\odot} - \alpha_{\mathrm{D}})} \tag{15.4}$$

The cusps are then at $p \pm 90°$. For any particular observer, the zenith distance of C, $z_{\mathrm{C}}$, will be given by $z_{\mathrm{C}} = p - q$, where $q$ is the parallactic angle (see Equations 3.18 and 3.19). The best chance to see the very first crescent phase after the new moon is when the ecliptic is as perpendicular as possible to the horizon, namely, in March for a Northern observer; for similar reasons, the winter full moon will be higher in the sky for the same observer.

## 15.2   CONDITIONS FOR THE OCCURRENCE OF AN ECLIPSE

Solar eclipses occur when the Moon passes in front of the Sun (Moon and Sun in conjunction), therefore necessarily around new moon phases. In the same manner, lunar eclipses happen when the shadow of the Earth falls on the Moon (Moon and Sun in opposition), namely, around full moon.

However, because of the inclination $i$ of the lunar orbit on the ecliptic, the conditions of new or full moon are not sufficient for the occurrence of an eclipse. In addition, the Sun (or the anti-solar point) and the Moon must be close to the same orbital node. Therefore, the Moon must be as close as possible to the ecliptic (hence, the name *ecliptic,* the locus around which eclipses can occur). In other words, the ecliptic latitude of the Moon, the distance of the Sun from one of the nodes and the angular distance between the two bodies (or between the Moon and the anti-solar point) must all be smaller than given maximum values. The exact amount of such value varies from eclipse to eclipse, essentially due to the varying conditions of the lunar orbit.

From Equation 15.2, at new moon (or full moon) the angular distance $\vartheta$ between the Moon and the Sun (or the anti-solar point) will reach a minimum value $\sigma$ given approximately by

$$\sigma \approx |\beta_{\mathbb{D}}| \tag{15.5}$$

In order to verify the possibility of an eclipse, this minimum angular separation must be compared to the angular diameters of the Sun, of the Moon and of the Earth as seen from the Moon. Actually, the Sun is an extended source, so that more precise considerations are needed regarding the shadows projected by the Moon and by the Earth: the shadows are composed of two cones, one of total obscuration and one of partial obscuration (in Latin, *umbra* and *penumbra*, respectively). In the following section, we demonstrate that a solar eclipse cannot take place if the latitude of the Moon is larger than $1°35'$, or else if the distance of the Sun from the node is greater than approximately $18°$, while a lunar eclipse cannot occur if the distance of the Sun from the node exceeds approximately $10°$. These angles (and the angular diameters of the two bodies) are small enough that plane trigonometry is sufficient for the present approximate treatment.

## 15.3 SOLAR ECLIPSES

To gain more insight into the circumstances of a solar eclipse, let us derive some preliminary indications from a very simple *ad hoc* fictitious model, in which the observer is on the equator, the Sun and the Moon both are on the celestial equator describing circular geocentric orbits and the eclipse happens at noon. The observer would then see the disk of the Moon moving eastwards and approaching the western limb of the Sun, with an angular velocity whose value can be easily computed in the following ways:

- the Sun moves eastward with respect to the fixed stars with an angular velocity $n_{\odot} = 0°.9856/j^{-1}$; the Moon moves in the same direction with sidereal velocity $n_{\mathbb{D}} = 13°.1763/j^{-1}$, so that the Moon overtakes the Sun with a relative angular velocity of $0'.5079/min$. In order to cross the entire solar disk of $32'$, the eastern limb of the Moon will then take approximately $62^m$; in total, the eclipse will last about $2^h$, of which only a few minutes are of totality;
- the shadow projected by the eastern border of the Moon will run eastward above the Earth's surface with a linear velocity of approximately $0.93$ km/s, but the equatorial observer is moving in the same direction with a velocity of $0.46$ km/s, so that the apparent velocity is $0.47$ km/s, again toward the east;
- if the Sun was a point source at infinity, the shadow of the Moon (here, a disk with a diameter of $3480$ km) would be a disk with a cross section decidedly smaller than the Earth itself, so that a particular solar eclipse would be seen only by a fraction of terrestrial observers, namely, by those located inside this circle. In reality, the Sun is a source at finite distance with an angular extension $\theta_{\odot}$ of approximately $32'$, so that the shadow can be divided into two conical regions of different obscuration. The vertex of the umbra is located at approximately $400,000$ km from the lunar center, a value very close to the Earth–Moon distance; owing to the strong ellipticity of the lunar orbit at certain times the vertex will be just inside the surface of the Earth, and at other times, it will be just outside it. In its turn,

the Earth can appear in these cones in a variety of relative positions (see Figure 15.1, all angles are exaggerated). Therefore, a total solar eclipse will be a rare event; more common will be the occurrence of a partial eclipse.

To carry out precise calculations of the circumstances and times of a particular eclipse, several additional factors must be taken into account:

- the Sun moves on the ecliptic, and the Moon on a plane inclined to it;
- their orbits are elliptical, and consequently, their apparent diameters vary with the date. The apparent solar semi-diameter $\theta_\odot/2$ varies between $15'44''$ and $16'18''$, and the apparent lunar semi-diameter $\theta_\)/2$ between $14'41''$ and $16'44''$;
- the position of the observer on the Earth's surface. The linear velocity of the shadow on the tangent plane through the observer increases with the latitude $\phi$ from 2009 km/h at the equator, to 2514 km/h at $\phi = 45°$, and to 2844 km/h at $\phi = 60°$, so that the eclipses are longer at the equator. Furthermore, at noon the velocity will be smaller because the Sun is higher in the sky. From these numbers, it appears that the velocities of the shadow on the ground are comparable to those of high-speed airplanes, so that the lunar shadow can be followed for a considerable time by flying telescopes.

The projection of the shadow over the Earth's surface is therefore a locus of complex shape (see, for instance, the *Astronomical Almanac,* Section A), according to several possible configurations. If the instantaneous direction joining the centers of the Moon and of the Sun intercepts the Earth, the eclipse is said a *central* one. However, even a central eclipse does not necessarily imply totality, because the disk of the Moon can be slightly smaller than that of the Sun (Moon at apogee, Earth outside the vertex of the umbral cone); a bright rim is then observed (*annular* eclipse). Nor is a central eclipse mandatory for a total eclipse; there are rare circumstances when observers around the poles of the Earth can see a total non-central eclipse. The width of the strip of total eclipse usually varies between 40 and 100 km, but it can shrink to zero or rise to 700 km. We have already seen that the duration of totality is maximum at the equator around noon, when the Earth is at aphelion and the Moon at perigee. The maximum possible duration is $7^m31^s$, but this rare event will not happen before 2186. In July 2009, a solar eclipse occurred with a duration of more than $6^m$. As already mentioned, the total solar eclipse on May 29, 1919 had a duration of totality of $6^m51^s$. This eclipse was a member of a so-called *semester series*. An eclipse in a semester series of solar eclipses repeats approximately every $177^d$ and $4^h$ (a semester) at alternating nodes of the Moon's orbit.

Other points are worth mentioning:

- the lunar barycenter does not coincide with the center of the disk; there are minute differences between the times calculated from dynamical considerations and the actual times;

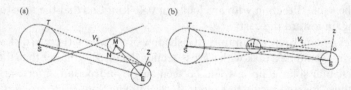

**FIGURE 15.1**   Different circumstances of a solar eclipse as seen by the observer O having his zenith in Z. (a) The Moon occults the limb of the Sun in $T$ (on the horizon of O). (b) The Moon is fully inside the solar disk. $V_1$ is the vertex of the penumbral cone and $V_2$ the vertex of the umbral one. The angle ESO is the solar parallax $\pi_\odot$, the angle EMO is the lunar parallax $\pi_\)$, the angle SOT is the angular semi-diameter of the Sun $\theta_\odot/2$ and the angle MON is the angular semi-diameter of the Moon $\theta_\)/2$.

- the calculations of an eclipse are usually performed several years in advance; dynamical calculations use TD as the time argument, but observations refer to UT1. The difference $\Delta T = TD - UT1$ is therefore only tentative (see Chapter 10);
- a source of larger time differences is the atmospheric refraction (see Chapter 11), which is usually not included in the published circumstances of an eclipse.

## 15.4 LUNAR ECLIPSES

Many of the previous considerations can be repeated in the present section. The Earth produces a cone of total shadow extending for approximately 0.01 AU in the anti-solar direction (therefore, to the Lagrangian point L2) and a region of penumbra, as shown in Figure 15.2 (all angles are exaggerated).

Analogous to Figure 15.1, the angle OME is the horizontal parallax $\pi_{\mathrm{D}}$ of the Moon, the angle SET is the apparent angular radius $\theta_{\odot}/2$ of the Sun, the angle ETO is the solar horizontal parallax $\pi_{\odot}$ and the segment ES is the geocentric distance to the Sun. All these quantities are known with great precision. Calling $f_1$ the angle LEM on the left panel (namely, the angular radius of the penumbra as seen from the geocenter) and $f_2$ the angle $V_2$EM (the corresponding radius for the umbra), we obtain

$$f_1 = \pi_{\mathrm{D}} + \pi_{\odot} + \theta_{\odot}/2, \qquad f_2 = \pi_{\mathrm{D}} + \pi_{\odot} - \theta_{\odot}/2 \qquad (15.6)$$

Hipparchus knew these two relations.

To be more precise, the disk of the umbra is slightly more flattened than that of the Earth. To compensate for this effect, the equatorial parallax $\pi_{\mathrm{D}}$ is substituted with that at 45° latitude, which is 0.998 $\pi_{\mathrm{D}}$. Furthermore, the extinction caused by the terrestrial atmosphere enlarges the width of the angles $f_1$ and $f_2$ by approximately 2%.

The angular width of the Earth's shadow at the distance of the Moon varies from 75' to 90', so that a lunar eclipse can be seen by all observers having the Moon above the horizon. Moreover, the duration of totality can exceed $2^h$. These two aspects are greatly different with respect to the solar case. The particular geometric circumstances produce total or partial penumbral, and total or partial umbral eclipses. For the Moon to be totally occulted by the umbra, the Sun cannot be more than 4°.6 from the node. During a penumbral eclipse, the dimming of the lunar disk is usually so slight to go unnoticed by visual observations.

The terrestrial atmosphere causes another remarkable phenomenon, namely, the faint illumination of the lunar disk seen during a total eclipse (the color of this faint light varies appreciably from eclipse to eclipse, but it is usually a deep red). The reason is the diffusion of solar radiation by the lower atmospheric layers, where there is a variable content of clouds, haze, pollutants, ozone, meteoric dust, etc. Scattering of the solar light by clouds of the Earth's atmosphere is obviously present also outside the eclipses; it is seen as a faint gray illumination beyond the terminator. Because of its color, it is usually called ashen light (in Latin, *cinerea lux*). A first correct explanation was given by Leonardo da Vinci at the end of the 15th century.

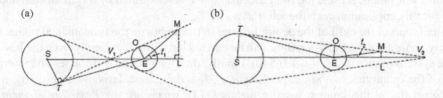

**FIGURE 15.2** Penumbra (a) and umbra (b) cast by the Earth. S = center of the Sun, E = center of the Earth, M = point on the penumbral or umbral cone at the distance Earth–Moon (not necessarily the center of the Moon) and LM = semi-diameter of the cone at the distance of the Moon. Points $V_1$ and $V_2$ are both on the ecliptic.

## 15.5   BESSELIAN ELEMENTS AND MAGNITUDE OF THE ECLIPSE

The precise calculation of the circumstances of the solar and lunar eclipses can be made with a method devised by Bessel. It can be adapted to calculate the circumstances of a stellar occultation by the Moon or by a planet.

Firstly, let us discuss a solar eclipse. Consider the line between the center of the Sun and the center of the Moon, namely, the axis of the shadow and its intersection with a perpendicular plane passing through the Earth's center. This is Bessel's *fundamental plane*; on it, the umbra and the penumbra are projected as two concentric circles, which are at a given distance and position angle from the geocenter. A so-called *fundamental* Cartesian geocentric system $(x, y, z)$ is then instituted, with the $z$-axis parallel to the shadow axis and positive toward the Moon, the $x$-axis parallel to the equator and positive toward east, and the $y$-axis positive toward North. At every instant, the geocentric equatorial coordinates of the Moon $\mathbf{R}_{\mathrm{☾}}$ and of the Sun $\mathbf{R}_{\odot}$ are known with great precision. For construction, the shadow axis (and thus also the $z$-axis) is parallel to the vector $\mathbf{G} = \mathbf{R}_{\odot} - \mathbf{R}_{☾}$. It intercepts the celestial sphere in a point Z having equatorial coordinates $(\alpha_Z, \delta_Z)$. Denoting as usual with the letter $\pi$ the horizontal parallaxes and normalizing to the Sun–Earth distance, we also have

$$r_{☾} = \frac{R_{☾}}{R_{\odot}} = \frac{\sin \pi_{\odot}}{R_{\odot} \sin \pi_{☾}}, \quad \mathbf{g} = \frac{\mathbf{G}}{R_{\odot}} = \mathbf{r}_{\odot} - \mathbf{r}_{☾}, \quad g = \frac{G}{R_{\odot}} \tag{15.7}$$

$$\mathbf{r}_{\odot} = \begin{pmatrix} \cos\alpha_{\odot} \cos\delta_{\odot} \\ \sin\alpha_{\odot} \cos\delta_{\odot} \\ \sin\delta_{\odot} \end{pmatrix}, \quad \mathbf{r}_{☾} = \frac{\sin\pi_{\odot}}{R_{\odot}\sin\pi_{☾}} \begin{pmatrix} \cos\alpha_{☾} \cos\delta_{☾} \\ \sin\alpha_{☾} \cos\delta_{☾} \\ \sin\delta_{☾} \end{pmatrix},$$

$$\mathbf{g} = g * \begin{pmatrix} \cos\alpha_Z \cos\delta_Z \\ \sin\alpha_Z \cos\delta_Z \\ \sin\delta_Z \end{pmatrix} \tag{15.8}$$

The vector distance of the Moon in the fundamental system, in units of the terrestrial radius, is then given by

$$\mathbf{r}_{☾\mathrm{Fund}} = \frac{1}{\sin\pi_{☾}} \begin{pmatrix} \cos\delta_{☾} \sin(\alpha_{☾} - \alpha_Z) \\ \sin\delta_{☾} \cos\delta_Z - \cos\delta_{☾} \cos(\alpha_{☾} - \alpha_Z)\sin\delta_Z \\ \sin\delta_{☾} \sin\delta_Z + \cos\delta_{☾} \cos(\alpha_{☾} - \alpha_Z)\cos\delta_Z \end{pmatrix} \tag{15.9}$$

The $(x, y)$ components of this vector are the intersection of the axis of the shadow with the fundamental plane. For practical uses, the right ascension of Z is substituted by its Greenwich hour angle, indicated for this application with the letter $\mu$: $\mu_Z = ST_{\mathrm{Greenw}} - \alpha_Z$.

The calculation of the radii of the penumbra and of the umbra on the fundamental plane, $l_1$ and $l_2$, respectively, in units of the Earth's radius, can be carried out without difficulty from the basic geometry of the system (notice that $l_1 = l_2 + 0.546$). Finally, the almanacs provide tables and interpolating formulae of the quantities $x$, $y$, $\sin\alpha_Z$, $\cos\delta_Z$, $\mu$, $l_1$, $l_2$ and the time derivatives of $\alpha_Z$, $\delta_Z$, indicated with $\alpha'_Z$, $\delta'_Z$ (time in this context usually means UT1), which are the *Besselian elements of the eclipse*. At this stage, we insert into the procedure the position of the observer in the fundamental reference system, which can be derived by means of a suitable rotation of its ellipsoidal coordinates $(\rho, \Lambda, \phi')$ given in Chapter 2 and the velocity with respect to the shadow. This relative velocity can be computed by knowing the Earth's diurnal rotation and the lunar motion. The fairly intricate procedure (see the *Explanatory Supplement*) is carried out through a set of auxiliary Bessel elements. Regular and auxiliary Bessel elements are the basis for deriving the general circumstances of a given solar eclipse in a given place on the Earth's surface.

For lunar eclipses, the procedure can be considerably simplified, because the observer is on the body casting the shadow, and the circumstances are identical for all terrestrial observers having the Moon above the horizon. Therefore, the $z$-axis is not used, and the origin of the fundamental coordinates is the center of the umbra. The Besselian elements are not even published by the *Astronomical Ephemerides*. See an example in the "Exercises".

For a lunar occultation of a star, the fundamental plane is geocentric and perpendicular to the line joining the center of the Moon and the star, the star being essentially a point source at infinity. The lunar shadow is a cylinder having cross section equal to the diameter of the Moon. The method can be adapted to the occultation of a star by a planet or asteroid, by scaling down the size of the shadow.

Let us now discuss the so-called *magnitude* of an eclipse (the term *magnitude* here has a different meaning than that employed for stellar magnitudes, see Chapter 16). In the case of a solar eclipse, the magnitude is the fraction of the solar diameter covered by the Moon at the instant of greatest phase, as seen by a given observer (the term *obscuration* means more rigorously the fraction of covered area, but the two terms are loosely used as synonyms). Notice that for total eclipses, the magnitude can be larger than 1.

The same can be said for lunar eclipses. Let us introduce the quantity:

$$g_i = \frac{f_i - (\sigma - \theta_{\mathbb{D}}/2)}{\theta_{\mathbb{D}}} \qquad (i = 1, 2)$$

where $f_{1,2}$ are the quantities defined in Section 15.6, $g_1 = 0$ at the beginning of the partial phase, $0 < g_1 < 1$ during partiality, $g_2 = 1$ at the beginning of totality, and $g_2 > 1$ during totality. Therefore, $g_1$ corresponds to the eclipsed fraction of the lunar diameter during partiality and $g_2$ to the penetration inside the umbra during totality. The maximum of the eclipse occurs at the instant of minimum distance $\sigma$, in the middle of the eclipse; $g$ is then called magnitude of the lunar eclipse, and its value is published in the almanacs.

## 15.6 NUMBER AND REPETITIONS OF ECLIPSES

To determine the yearly number and repetition cycles of eclipses, we have to recall the values of the several lunar months discussed in Chapter 14. Eclipses are possible around the beginning of every draconitic month, which is the time interval of 27.21 days between two passages through the node.

Suppose a lunar eclipse (phase of full moon) occurred. The Sun moves away from the node with a velocity of $30°.67$ per synodic month of $27^j.21$, so that after six synodic months ($177^j.18$), the Sun has moved in longitude by about $174°.64$. On the other hand, the node regresses at a speed of $-0°.53j^{-1}$, corresponding to $-9°.38$ after six synodic months, while the opposite node is located at longitude $180° - 9°.38 = 170°.62$. Therefore, the full moon is at $174°.64 - 170°.62 = 4°.02$ from the node, and another lunar eclipse can occur. If all circumstances are favorable, a third lunar eclipse can happen before the end of that year. Notice though that there is no warranty that the first eclipse happens; there are indeed years with no umbral eclipses.

Now, consider solar eclipses (phase of new moon). The motion of $30°.67$ per synodic month of the Sun away from the node is less than twice the minimum ecliptic distance, so that at least one solar eclipse is inevitable at each node. Therefore, the minimum number of eclipses in 1 year is two; in that case, both will be solar eclipses. At the other extreme, if all orbital circumstances are favorable, two solar eclipses can take place at each node. Actually, if the first is in early January, a fifth can occur, because there can be 13 lunations in that year (this rare circumstance will not happen before the year 2160). Therefore, 2, 3, 4 or even 5 solar and 3 lunar eclipses can occur in 13.5 lunations. This interval of time is larger than the Julian year, so that in conclusion, the maximum yearly number is 7, either 5 solar and 2 lunar, or 4 solar and 3 lunar.

Regarding the repetitions, the Chaldean and Babylonian astronomers had detected a cycle of 18 years 11 days, after which the eclipses would repeat themselves in the same sequence and at the same node under almost identical conditions. This basic cycle is called Saros (a name possibly derived from the Chaldean word indicating the number 3600). To understand its value of $18^y11^j$, consider that the ratio between the durations of the draconitic and the synodic months is numerically very close to the ratio of the integer numbers 242 and 223: if a given date is full moon, after 223 synodic months, namely, after $6583^j.32$, the Moon again will be full. Those 6585.32 days correspond to $18^y11^j.3$ if there were 4 bissextile years, or to $18^y10^j.3$ if the number of bissextile years was 5. This time interval is practically equal to 242 draconitic months; although there is a difference of $0^j.46$, the Sun will not have moved out of the node by more than 29′. Furthermore, the cycle also matches 239 anomalistic months, which guarantees almost the same position of the Moon in its orbit.

Table 15.1 shows an example of three total solar eclipses of similar duration occurred at intervals of one Saros. The spacing among them is $18^y11^j$ because the eclipses occur at the same ascending node. Each eclipse moves about 120° west of its predecessor, because the fraction $0^j.32$ corresponds to a diurnal rotation of 120°, so that the zones of totality move westward by the same quantity during the Saros.

After three Saros periods, namely, after 54 years plus 34 days, the eclipse returns approximately to the same longitude. Such cycle is called in Greek "exeligmos", which liberally can be translated as "repetitions". Furthermore, the totality moves south (it would move north for an eclipse at the descending node). Figure 15.3 shows all eclipses from 1980 to 2010, namely, over 1.5 Saros cycles. By examining these patterns, we can see that eclipses occur every year at an earlier date due to the retrogradation of the nodes. The same sequence repeats after one Saros.

It is possible to find other periodicities and commensurabilities between the lunar months and the year. For instance, 235 lunations closely correspond to 19 tropical years, a period called the Metonic cycle. It enters into the determination of the Christian Easter date. The stability of the Metonic cycle is weaker than that of the Saros, but it can persist for many centuries.

Hipparchus is credited with having discovered cycles of 20.2, 345 and 441.3 years. Many more can be found by moving to high integer numbers. These regularities are not simply numerology, because they can be used in dynamical theories of the Sun–Earth–Moon system (see, e.g., Steves, 1998 in References section; Roth and Valsecchi both in Bibliography section. Useful sites are quoted in Web Sites section). An additional consideration, put forward by Fiala et al. (1994), is that after a Saros, the librations of the Moon are identical, producing a similar pattern of the already mentioned Baily beads; a possible application is the measurement of the solar diameter.

**TABLE 15.1**

**Three Solar Eclipses at Intervals of One Saros**

| Date | Type | Duration | Central Point of Path | |
|------|------|----------|-----------|----------|
| | | | Longitude | Latitude |
| June 11, 1983 | Total | $5^m11^s$ | −126°.91 | −4°.70 |
| June 21, 2001 | Total | $4^m56^s$ | −8°.73 | −10°.97 |
| July 02, 2019 | Total | $4^m32^s$ | +104°.94 | −17°.48 |

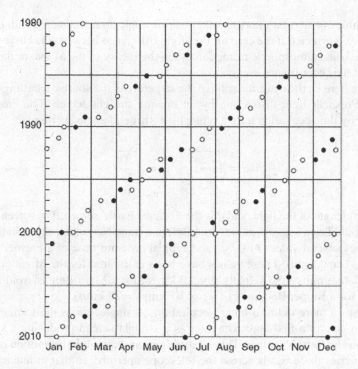

**FIGURE 15.3** The eclipses from 1980 to 2010 arranged along 1.5 Saros period. Filled circles: lunar eclipses; open circles: solar eclipses.

## 15.7 STELLAR OCCULTATIONS

When a star is occulted by the Moon, one can determine its position and structure with great accuracy, at least in the direction along the path of disappearance and reappearance (or immersion and emersion). Actually, the method can be used to determine the position of the Moon, or the longitude of the observer, to improve the lunar orbit theory or to ameliorate the knowledge of the rotation of the Earth. Consider indeed the spherical triangle between the celestial North pole P, the center of the Moon M and the contact point X between the star of coordinates $(a, \delta)$ and the lunar limb (namely, the immersion or emersion point). The following conditions apply:

$$\text{arc PN} = 90 - \delta_{\leftmoon}, \quad \text{arc XN} = 90 - \delta, \quad \text{arc MX} = \alpha_{\leftmoon} - \alpha$$

Let the position angle of M with respect to X be $p$. From Equation 15.4, we have

$$\frac{1}{2}\theta_{\leftmoon}\sin p = (\alpha - \alpha_{\leftmoon})\cos\delta, \quad \frac{1}{2}\theta_{\leftmoon}\cos p = \delta - \delta_{\leftmoon} \tag{15.10}$$

$$\frac{1}{2}\theta_{\leftmoon} = (\alpha - \alpha_{\leftmoon})\cos\delta\sin p + (\delta - \delta_{\leftmoon})\cos p$$

$$\tan p = \frac{(\alpha - \alpha_{\leftmoon})\cos\delta}{\delta - \delta_{\leftmoon}} \tag{15.11}$$

If the observed values of angular diameter and position angle do not coincide with those calculated by the ephemerides (corrected if the case for the slight difference between the barycenter and center of the apparent disk and atmospheric refraction), then the theory of the Moon, or the position of the observer, or the stellar coordinates are in error.

We concentrate here on the determination of the structure of the source, recalling the well-known phenomenon of Fresnel's light diffraction by an opaque straight screen. The Fresnel diffraction fringes produced by the occultation have a typical length $x$ and angle $\psi$ given by

$$x = \sqrt{\frac{\lambda d}{2}}, \qquad \psi = \sqrt{\frac{\lambda}{2d}}$$

where $\lambda$ is the wavelength of the light, and $d$ is the distance Earth–screen. The screen is not necessarily the Moon; it could be an asteroid in the Main Belt or a trans-Neptunian object in the Kuiper Belt.

Table 15.2 gives several values of $(x, \psi)$ and typical crossing time of a Fresnel fringe for three wavelengths (0.5, 1 and 5 nm). These values have been calculated for the Moon ($d \approx 3.8 \times 10^5$ km, with a typical relative transverse velocity $v \approx 0.9$ km/s), for a Main Belt asteroid ($d \approx 3 \times 10^8$ km, $v \approx 15$ km/s) and for a Kuiper Belt body ($d \approx 6 \times 10^9$ km, $v \approx 25$ km/s).

Let us examine in more detail a lunar occultation. A single star is the source at infinity; the limb of the Moon acts in a first approximation as a straight screen producing a system of bright and dark bands on the Earth's surface. During disappearance, the relative motion of the Moon and of the observer carries these bands across the telescope aperture, so that in few milliseconds, the observer will notice regular fluctuations of intensity before the geometrical condition of dark is reached. The effect begins to occur when the star is about 0″.01 from contact. Approximately 20 ms elapses between the first and the second diffraction minimum; the contrast between the first bright and the first dark band reaches 50%. The phenomenon repeats in reverse order at emersion. Should the source be a double star, or an extended disk, the diffraction pattern will be decidedly different, because the fringe modulation decreases to zero. Therefore, from the observations one can determine the separation of the binary or the diameter of the star, or even its limb darkening (stars are not necessarily uniformly illuminated disks; our Sun is decidedly darker at the limb than at the center and so are all normal stars). Moreover, the observed fringe patterns are sensitive to the actual shape

**TABLE 15.2**

**Fresnel Fringe Length, Angular Size and Crossing Times**

|  | 0.5 nm | 1.0 nm | 5.0 nm |
|---|---|---|---|
| | | **Moon** | |
| $x$ (m) | 10 | 14 | 32 |
| $\psi$ (arcsec) | 0″.005 | 0″.007 | 0″.016 |
| $\tau$ (s) | 0.012 | 0.017 | 0.038 |
| | | **Main Belt Asteroids** | |
| $x$ (m) | 273 | 383 | 857 |
| $\psi$ (arcsec) | 0″.00019 | 0″.00027 | 0″.00059 |
| $\tau$ (s) | 0.018 | 0.256 | 0.058 |
| | | **KBO** | |
| $x$ (m) | 1220 | 1710 | 3833 |
| $\psi$ (arcsec) | 0″.000043 | 0″.000060 | 0″.00013 |
| $\tau$ (s) | 0.1776 | 0.216 | 0.39 |

of the lunar limb; mountains or even boulders only few meters high above the lunar surface can cause noticeable distortions of the light curve. In other words, the proper reduction of lunar occultations requires an accurate model of the lunar limb at the time and circumstances of the observations.

Figure 15.4 provides two examples of lunar occultations at disappearance. Data were obtained at the 1.2m Galileo and 1.8m Copernicus Asiago telescopes with the photon counting detectors Aqueye[+] and Iqueye (Zampieri et al., 2019). The left panel shows two fits of the light curve pertaining to the K3-K4 red giant $\mu$ Psc (Brown et al., 2018). The upper line is a fit with a point-like source; the lower line is the best fit with a uniform disk of $3.14 \pm 0.06$ mas, decidedly better than the first solution. Using the GAIA stellar parallax of $9.85 \pm 0.32$ pc, the radius of the disk turns out to be $32 \pm 1.2$ solar radii. However, this value can be larger than the photospheric diameter, because the data were obtained with a filter centered on the hydrogen Balmer line H$\alpha$, and thus the light curve can be affected by chromospheric emission. The right panel shows the light curve of the double star SAO 92922 of spectral type K0. The secondary star B, approximately 1/5 the brightness of the primary A, was detected at a separation of $13.7 \pm 1.0$ mas and position angle of $7°.3 \pm 2°.8$, as in the insert.

An additional interesting phenomenon occurs when a circular planet with an atmosphere occults a star, namely, a so-called central flash of the star caused by the stellar light refraction and focusing at the planet's limb.

It is worth noting that the occultation technique does not require large telescopes to achieve high-spatial resolution, because the diffraction takes place not in the telescope but at the screen. Of course, a larger telescope provides a larger number of photons and therefore a better signal to noise ratio. Furthermore, the achievable resolution does not critically depend on the local atmospheric seeing. Occultations provide not only astrometric information; they can be a powerful instrument for astrophysical considerations. For instance, Gies (2004) obtained low-resolution spectra with a time sampling rate of 7 ms of the H$\alpha$ emission line in the spectrum of the Be star Pleione. See also Richichi (2004 and 2019, in Bibliography section).

The obvious disadvantages are that the occultation circumstances are not under the control of the observer; for instance, the Moon covers no more than 10% of the sky.

## 15.8   TRANSITS OF EXOPLANETS

After the radio detection of a planet around a pulsar in 1992, the first optical exoplanet orbiting the solar-type star 51 Pegasi was discovered in 1995, thanks to the periodic radial velocity variations of its host star (Mayor and Queloz, 1995; for this discovery, the two scientists have been awarded the

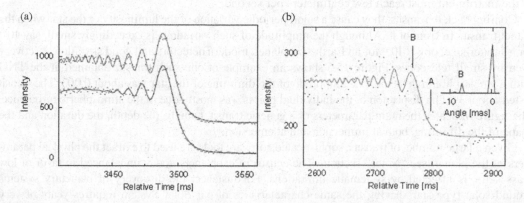

**FIGURE 15.4**   (a) The lunar occultation of $\mu$ Psc. The upper line is a fit with a point-like source; the lower line is the best fit with a uniform disk of $3.14 \pm 0.06$ mas. (b) Light curve of the double star SAO 92922. The insert shows the reconstruction of the duplicity. (Adapted from Zampieri et al., 2019, © AAS. Reproduced with permission).

Nobel Prize in 2019). Since then, the number of exoplanets grew continuously; today, their census surpasses 4000. Most of those exoplanets are gaseous giants similar to Jupiter or even larger, but a sizeable fraction has Earth-size diameter. Transits cause a minute dimming of the stellar light lasting say few hours and repeating at each orbital passage. Such dimming can be discovered by ground telescopes of modest size, equipped with specific detectors, such as TRAPPIST (see Jehin et al., 2011). The most prolific discoverers are however dedicated space telescopes, such as the NASA Kepler satellite and its successor TESS (see Web Sites section). The astrometric satellite GAIA is already contributing information about hundreds of exoplanets; its importance, perhaps in conjunction with Hipparcos data obtained more than 25 years ago, will increase with time, due to the steady improvement of the accuracy of measurements of the proper motions of the host stars and of the tiny wobble induced by the planet. For instance, the determination of the mass of the planet will become better and better.

The nearest exoplanet known today is in the Southern Milky Way, inside the triple system of Alpha Centauri four light-years away (see also Chapter 8). Proxima has one or probably more planets; the first to be discovered is named Proxima-b; it shows some similarities to our Earth, although it orbits much closer to its faint cool star (spectral type M, see Chapter 17) with an orbital period of 11 days. In addition to Proxima-b, two more cases are worth mentioning:

- a super Earth orbiting Barnard's star (Ribas et al., 2018), only 6 light-years away and with the highest known proper motion (about 10 arcsec/year), as already mentioned in Chapter 9;
- Trappist-1, a system of seven almost coplanar planets orbiting an M-star 39 light-years away (Gillon et al., 2017).

Although the present section is mostly devoted to transits, we give some more detail about the radial velocity method.

Variations of the radial velocity of the star are due to its orbit around the barycenter of the system star-planet(s). Such variations are very small, not more than few meters per second, and only sophisticated equipment at large telescopes permits their measurement. Great advancements came from the technical and scientific achievements of the HARPS (High Accuracy Radial velocity Planet Searcher) instrument built for the 3.6m ESO telescope in Chile (Mayor et al., 2003). The already quoted 3.5m Telescopio Nazionale Galileo (TNG) is one of the best telescopes for such measurements in the Northern hemisphere. Figure 15.5 shows the TNG radial velocity variation due to the first discovered planet around 51 Pegasi. The case of 51 Pegasi is one of the easiest, because of the high amplitude of the velocity variation. To reach a substantial number of exoplanets, the sensitivity of the instrument must reach few centimeters per second.

Coming back to transits, they cause a small periodic variation of the luminosity of the star when the planet transits in front of it. Although the amplitude of such variations is exceedingly small, say $10^{-3}$ for a Jupiter-size, or $8 \times 10^{-5}$ for an Earth-size planet, modern detectors and good sky allow discovery even by small telescopes. Figure 15.6 shows an example of one such transit, obtained at the TNG with a single-photon detector. The amplitude of the dimming of the star was about 0.003. The transit lasted about $3^h$. The dispersion of the individual points was mostly due to the atmospheric turbulence. The figure indicates the main parameters of a generic transit, namely, the depth, the duration and the shape of the dimming, both at immersion and at emersion.

Given a large number of transits, sophisticated software tools are used to extract the physical parameters of the planet (see "Notes" section). The census of exoplanets, today unbalanced in favor of low mass stars, is affected by systematic limitations. For instance, the discovery of planetary systems around solar-type stars having the same characteristics of our Solar System requires years of very accurate observations. The day–night sampling limitation of ground telescopes may introduce spurious frequencies (for instance, our Earth employs about $14^h$ for an equatorial transit across the solar surface).

The capabilities of space telescopes are much better, thanks to the lack of atmospheric disturbances, the much darker sky background and the possibility to observe continuously for dozens

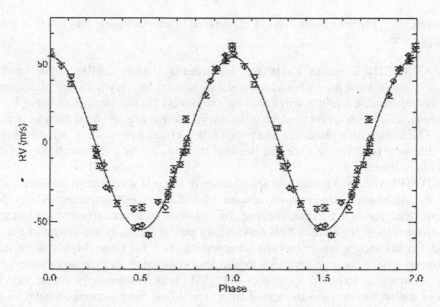

**FIGURE 15.5** The radial velocity curve of 51 Pegasi obtained at the TNG. The orbital period of the planet is 4.2 days. (Courtesy R. Gratton, INAF OAPd; Adapted from Barbieri, 2019, Bibliography section).

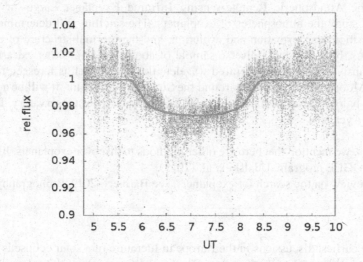

**FIGURE 15.6** Transit of an exoplanet over the disk the star WASP 6. High-time resolution data obtained with the fast photometer Iqueye at the TNG. (Courtesy M. Barbieri, INAF OAPd; Adapted from Barbieri, 2019, Bibliography section).

of hours. Those are the key reasons why CNES COROT (the first mission entirely dedicated to exoplanets) and even more NASA KEPLER were so successful. The same principle of detection by luminosity variation has been adopted by the successor of Kepler, namely, the Transiting Exoplanet Survey Satellite (TESS), which started operating from a very peculiar orbit in the Earth–Moon system in mid-2018.

While radial velocity variations provide essentially the mass, transits can give the radius and the density. Moreover, the atmospheres of transiting exoplanets can be studied spectroscopically. Such line of investigations awaits the completion of space telescopes such as the NASA James Webb.

In addition, the European Space Agency is engaged in exoplanet searches and characterization with three projects:

- CHEOPS (**CH**aracterising **ExOP**lanet **S**atellite) is a small satellite meant to characterize already known exoplanets around bright stars by very accurate photometry in the visual band. CHEOPS will focus on exoplanets having sizes from super-Earth to Neptune, aiming to a precise determination of their characteristics, such as bulk density. The launch took place on December 18, 2019 from Kourou on a Soyuz rocket. The satellite will orbit for an expected lifetime of 3.5 years on a sun-synchronous orbit at 700 km altitude.
- PLATO (**PLA**netary **T**ransits and **O**scillations of stars) is a very large satellite, meant to discover and characterize exoplanets around solar-like stars, with emphasis on the properties of terrestrial planets in the habitable zone. Twenty-seven telescopes arranged in a battery will constitute a most powerful and photometrically precise measurement system in the visual band. For instance, accuracy on radii is expected around 3%. Ground-based radial velocity follow-up observations are envisaged for the determination of the planetary masses (around 10% accuracy). Another scientific goal of PLATO is asteroseismology, for the determination of stellar masses, radii and ages. Launch toward the outer Lagrangian point L2 of the Sun–Earth system is foreseen by 2026; the minimum expected lifetime is 4 years. Why the name PLATO? Because exoplanets are discovered by their shadows, reminding the ideas of the great Greek philosopher.
- ARIEL, the **A**tmospheric **R**emote-sensing **I**nfrared **E**xoplanet **L**arge-survey, aims to detect and study the atmospheres of exoplanets. The satellite will determine the chemical composition, the formation and evolution, and the thermal structure of exoplanetary atmospheres, by surveying a diverse sample of about 1000 transiting extrasolar planets, simultaneously in visible and infrared wavelengths. The launch is foreseen for 2028 into an Earth–Moon eclipse-free orbit around the Sun–Earth L2 point. It will be a large amplitude quasi-halo orbit, as was the case for ESA's Herschel Space Observatory. The expected lifetime is 4 years

For completeness, we mention that here are other methods to discover exoplanets, like gravitational lensing. See the OGLE program, Udalski et al. (2015).

For a recent review on the search for exoplanets, see Barbieri (2019, Bibliography section).

## NOTES

- One of the earliest discussions in the European literature of a solar eclipse is the letter by the Irish monk Dungal to the Emperor Charles the Great, about the solar eclipse of year 810. See Zanna and Sigismondi (2004).
- A computational masterpiece for eclipses was the *Canon der Finsternisse* (*Catalog of Eclipses*) produced by von Oppolzer (1887). The catalog contained detailed information about 8000 solar and 5200 lunar eclipses between 1208 B.C and 2161 A.D.
- For lunar occultations in the visible, see, for instance, Evans D. (1971); Richichi et al. (2002). A summary paper is provided by Richichi (2019, Bibliography section).
- For the Uranian rings, discovered for the first time in 1977 from the Airborne Kuiper Observatory when the planet occulted the star KM 12, see the paper by Elliot et al. (1981).
- On February 20, 1998 Sco X-1, the brightest X-ray source in the sky, was eclipsed by the Moon. The event was captured by the German satellite ROSAT (see Web Sites section).
- For occultations by Kuiper Belt Objects (KBOs), see, for instance, Cooray (2003).

## EXERCISES

1. Calculate the arc on the spherical surface of the Earth corresponding to the linear diameter of the Moon if projected by a point-like source at infinity.
2. From a table of solar and lunar ephemerides, verify that the product between the angular diameters and the angular velocities of each body is approximately constant (Kepler's third law), and discuss why the constancy is better for the Sun than for the Moon. This exercise indicates how to use apparent diameters as distance indicators.
3. Here, we sketch a method for calculating a lunar eclipse by the assumed known instantaneous values of the umbral semi-diameter $\theta_{1/2}$, and of the lunar and solar apparent semi-diameters and distances (or parallaxes), without recurring to the Besselian elements. Call $l$ the instantaneous distance between the center of the Moon M and the center of the umbra O; at the external and internal contacts, we have, respectively,

$$l = \sigma_{ext} = \frac{1}{2}\theta_1 + \frac{1}{2}\theta, \quad l = \sigma_{int} = \frac{1}{2}\theta_1 - \frac{1}{2}\theta_{\leftmoon}$$

while

$$\frac{1}{2}\theta_1 = \pi_{\leftmoon} + \pi_{\odot} - \frac{1}{2}\theta_{\odot}$$

Then, the two values of $\sigma$ can be derived. Now, let $(\alpha_1, \delta_1)$ be the coordinates of the center of the umbra, which can be derived from those of the Sun:

$$\alpha_1 = \alpha_{\odot} + 12^h, \quad \delta_1 = -\delta_{\odot}$$

From the spherical triangle POM (where $P$ is the celestial North pole), with a procedure analogous to that employed to derive Equations 15.10 and 15.11, we also obtain the position angle of the center of the umbra with respect to the center of the Moon:

$$\sin l \sin p = \sin(\alpha_1 - \alpha_{\leftmoon})\cos\delta_1,$$

$$\sin l \cos p = \sin\delta_1 \cos\delta_{\leftmoon} - \cos(\alpha_1 - \alpha_{\leftmoon})\sin\delta_{\leftmoon}\cos\delta_1$$

All angles, except $p$, being small quantities, the following approximations will be sufficiently good:

$$x = l\sin p = (\alpha_1 - \alpha_{\leftmoon})\cos\delta_1, \quad y = l\cos p = \delta_1 - \delta_{\leftmoon} \tag{15.12}$$

The two auxiliary variables $(x, y)$ can be calculated for a set of instants around opposition, and so their time derivatives $(x', y')$ can be obtained (*note*: it is customary in the almanacs to indicate time derivatives with primed symbols). Now, insert in Equation 15.12 the two previously found values of $\sigma$. After some manipulation, we will obtain the position angle and the instant of the external and of the internal contact, respectively. It might well happen that no solution can be found for the position angle; if this happens for $\sigma_{ext}$, then no eclipse can occur; if it happens only for $\sigma_{int}$, then the eclipse will be a partial one.
4. Calculate the magnitude of a central lunar eclipse ($\sigma = 0$) when

$$\pi_{\leftmoon} = 57', \quad \theta_{\leftmoon}/2 = 16', \quad l_1 = 63', \quad l_2 = 41'$$

# 16 Elements of Astronomical Photometry

The relative apparent luminosities of the heavenly bodies can be measured even with the naked eye. Hipparchus and then Ptolemaeus subdivided the visible stars in six classes of decreasing splendor (more properly *magnitudes*), indicated by the Greek letters α, β, γ, δ, ε and ζ followed by the abbreviated Latin name of the constellation, e.g., α Cyg or β Lyr. The ideal criterion is that passing from one class to the next, the eye senses the same difference of visual stimuli. However, the human eye can reliably judge this difference only over small angular distances and for objects of approximately identical color, so that the classification is not coherent over large arcs on the celestial sphere. This problem is not completely solved even by modern detectors; great caution must be exerted in transferring a photometric sequence from one area of the sky to a distant one, especially for ground telescopes affected by atmospheric extinction, as discussed later on in this chapter.

In the following sections, we shall examine point-like, self-luminous objects (the stars), with only brief considerations of extended luminous objects such as nebulae or galaxies, and of illuminated objects, either point-like, such as asteroids, or extended, such as comets and planets. Furthermore, the treatment will be limited mostly to the "extended visible" spectral band from 3000 to 10,000 Å (300 to 1000 nm), with short excursions into the near UV and near infrared (IR). It must be underlined that in other spectral bands, the celestial bodies can change appreciably their photometric characteristics. For instance, Jupiter in the far-IR emits more radiation than it reflects from the Sun, due to an internal source of energy.

Photometric quantities will be defined according to the astronomical tradition and nomenclature, which is not always the same as that used by physicists or radiometrists. Furthermore, no account will be made of a possible refracting medium, so that wavelength $\lambda$ and frequency $\nu$ are related through the velocity of light $c$ by

$$\lambda = \frac{c}{\nu}, \quad d\lambda = -\frac{c}{\nu^2}d\nu = -\frac{\lambda^2}{c}d\nu \tag{16.1}$$

As unit of wavelength, the angstrom will be generally employed (1 Å = $10^{-10}$ m = 0.1 nm), while the frequency will be expressed in hertz (Hz) and multiples.

## 16.1 VISUAL MAGNITUDES

For millennia, and until less than 200 years ago, the human eye, firstly naked and after Galileo Galilei in 1609 at the focus of a telescope, was the only receptor of radiation. If $S_1$ and $S_2$ are the luminous fluxes produced by two stars, the eye produces a difference of visual stimuli $V_1 - V_2$, which is approximately a logarithmic function of the ratio of the two fluxes:

$$V_1 - V_2 = k \log \frac{S_1}{S_2} \tag{16.2}$$

This physiological law was found by Fechner and Weber in the early 19th century. The true behavior of the eye is not described so simply by a logarithmic function, but Equation 16.2 has been assumed correct in all astronomical applications. In particular, the English astronomer N. Pogson (1856, see also Hearnshaw, 1992, both in References section) found that the law reproduced the astronomical system of magnitudes, provided the constant k was made equal to −2.5 and the logarithms were on the decimal base (see the "Notes" section). Thus, by definition, the difference of magnitude between two stars whose intensities are $S_1$ and $S_2$ is given by

$$m_1 - m_2 = -2.5 \log_{10} \frac{S_1}{S_2} \tag{16.3}$$

Suppose then that a particular star, for instance, Polaris, is used as common reference. For any other star, we can write

$$m = m_0 - 2.5 \log_{10} S \tag{16.4}$$

where $m_0$ is the magnitude of the reference star, and $S$ is the ratio between the intensity of the star and the reference one. Such equation must be treated with some caution in order not to forget that $S$ must be a dimensionless quantity. Notice also that the logarithm is in base 10, not $e$ as the natural one ($\log_{10} e = 0.43429...$, $\log_e 10 = \ln 10 = 2.30259...$); throughout this chapter, for simplicity we drop the suffix 10 from the logarithm; natural logarithm will be indicated, when need, with ln.

Therefore, $\Delta m = -1$ corresponds to $S_2/S_1 = 10^{0.4} = 2.51188...$, not to be confused with Pogson's constant 2.5 which has no decimals. For the differential or a small increment, $dm = 1.086 \, dS/S$: in other words, a small difference of magnitude corresponds approximately to the percent variation of the luminous intensity. Engineers use a quantity analogous to the magnitude, namely, the decibel (dB), where the constant is 10 instead of 2.5 (4 dB = 1 mag).

A ratio of 10 in intensity produces a difference of magnitudes $\Delta m = -2.5$, a ratio of 100 a difference $\Delta m = -5.0$ and so on. The minus sign means that the fainter is the star, the larger is its magnitude. On the scale with Polaris at a visual magnitude $m = +2.1$, Sirius, the brightest star in the sky, has approximately $m = -1.5$. Only three other stars have negative visual magnitudes, namely, α Car (Canopus, $m = -0.74$), α Cen (Rigil Kent, $m = -0.27$) and α Boo (Arcturus, $m = -0.05$); Vega (α Lyr) has $m = +0.03$. The unaided eye, in normal conditions, can see stars brighter than approximately the 6th; their number is around 6000, evenly distributed on each hemisphere (for a classic catalog of bright stars, see Hoffleit and Jaschek, Bibliography section). A dark-adapted and trained eye, in very dark sites, can discern the 7th or even the 8th magnitude; at the focus of a large telescope, it can reach the 20th (Bowen, 1947, References section). The faintest stars visible on Hubble Space Telescope (HST) long exposures can be as faint as the 29th. In other words, the observable stars span an impressive range of approximately 30 magnitudes, which means a ratio of intensities $S_1/S_2 \approx 10^{12}$. These 12 orders of magnitude are due to several factors, namely, the different intrinsic luminosities, the widely different distances and the absorption by a diffuse material (dust) in the interstellar space. For objects at cosmological distances, the expansion of the Universe causes an additional dimming, because the light is not only red-shifted, but also spread over a larger spectral band (see Chapter 9).

Regarding Solar System objects, the Sun has $m = -26.7$ and the full moon $m = -12.7$. At their maximum brightness, the visible planets reach the following approximate values: Mercury $m = -0.1$, Venus $m = -4.4$, Mars $m = -2.8$ (in rare circumstance outshining Jupiter, as it happened in 2003 when Mars was at the closest distance in approximately 60,000 years), Jupiter $m = -2.6$ and Saturn $m = +0.4$. Io, moon of Jupiter, is seen at $m = +5.5$ and Ceres at $m = +6.8$.

Planets, comets, nebulae and galaxies are extended sources. Therefore, their apparent splendor is obtained by integrating the light over their finite area: one has to distinguish between the total magnitude and the magnitude per unit area (the surface brightness). We shall come back to this point in a later paragraph.

## 16.2 EXTENSION OF THE DEFINITION OF MAGNITUDE

The concept of magnitude, which originated from visual observations, has been generalized to other types of detectors, such as the photographic emulsion and solid-state devices; moreover, it has been extended to other spectral regions. Let us suppose that a particular detector is sensitive to radiation in the spectral band $(\lambda_1, \lambda_2)$, with $\lambda_1 < \lambda_2$. Let $T(\lambda)$ be the function describing the sensitivity of the apparatus in that band and $f(\lambda)$ the quantity of stellar radiation impinging on the detector per unit time per unit band pass. The total energy $i$ collected per unit time will be

$$i = \int_{\lambda_1}^{\lambda_2} f(\lambda) T(\lambda) \mathrm{d}\lambda \approx i_0 f(\lambda_c) \tag{16.5}$$

where $\lambda_c$ is a wavelength inside the spectral band $(\lambda_1, \lambda_2)$, usually close to the midpoint, and $i_0$ is a normalization constant which also takes care of the physical units of $i$. We can therefore define magnitude at $\lambda_c$ the quantity:

$$m(\lambda_c) = m_0(\lambda_c) - 2.5 \log i_0 f(\lambda_c) \tag{16.6}$$

where the constant $m_0$ changes with $\lambda_c$. Notice that $m(\lambda_c)$ is only apparently a *monochromatic* quantity, and it is actually derived from the integration of the luminous flux, weighted by the instrumental sensitivity, over a spectral band of finite width; the designation *heterochromatic* is often employed to indicate this situation. With an ideal detector having 100% efficiency over the entire electromagnetic spectrum, one could measure the *bolometric* magnitude, namely, the quantity:

$$m_{\mathrm{bol}} = m_0 - 2.5 \log \int_0^\infty f(\lambda) \mathrm{d}\lambda \tag{16.7}$$

The measurement of such quantity could be carried out by an ideal bolometer outside the terrestrial atmosphere. In general, one must be content with a bolometric correction $C_V$ applied to the heterochromatic visual magnitudes, assuming that the shape of $f(\lambda)$ is well known from zero to infinity: $C_V = m_{\mathrm{bol}} - m_{\mathrm{vis}}$.

The practical realization of $T(\lambda)$ implies the consideration of several factors, such as

- the reflectivity (or transmissivity) of the optics of the telescope, $R(\lambda)$;
- the transmissivity of the filters in front of the detector, $K(\lambda)$;
- the efficiency of the detector, $Q(\lambda)$;
- the transmissivity of the terrestrial atmosphere, $A(\lambda)$,

so that $T(\lambda) = R(\lambda) \cdot K(\lambda) \cdot Q(\lambda) \cdot A(\lambda)$. The first three functions are under the direct control of the astronomer; the fourth depends on the characteristics of the site and of the particular observational circumstances; obviously, it is absent in space astronomy. We will now briefly examine these factors.

### 16.2.1 THE REFLECTIVITY OF THE OPTICS AND TRANSMISSIVITY OF FILTERS

The basic optical elements of large telescopes are mirrors, whose surfaces are made reflective by coating them with a thin metallic film. From an historical point of view, silver (Ag) coatings were employed before 1930. Since then, the use of aluminum (Al) coatings deposited in vacuum became widespread in the astronomical field, thanks to the good reflectivity of Al over a wide spectral band. The reflectivity $r(\lambda)$ of Al is indeed very high, approximately 90%, from the UV to the near IR. Therefore, it is used in most ground telescopes and in space telescopes such as the HST, where the

Al is additionally protected with a thin coating of magnesium fluoride $MgF_2$. In most telescopes, there are several mirrors (two at the Cassegrain focus, three at the Nasmyth focus, etc.), so that the overall reflectivity becomes $R(\lambda) = r^n(\lambda)$, where $n$ is the number of mirrors. Moreover, $R(\lambda)$ depends on the age of the coating. Figure 16.1 shows both the aging effect and the wide dip of reflectivity around $\lambda = 8000\,\text{Å}$ of the Al coating.

Silver reflects very well from the green to the IR and lacks the dip at 8000 Å; therefore, it is used in telescopes optimized for IR observations, such as the two Gemini telescopes (see the Web Sites section). However, its blue and violet response is decidedly worse than that of Al; caution must be exerted when comparing older blue magnitudes with modern ones. Other coatings can be employed for specific applications. For instance, the forthcoming James Webb Space Telescope (JWST, Web Sites section), optimized for the IR, has the primary and secondary mirrors coated with a gold (Au) layer. A thin layer of amorphous silicon dioxide ($SiO_2$) is deposited on top of the gold to protect it.

Often, telescopes employ refracting elements, either as primary optical elements or inside their instrumentation. For instance, a corrector of optical aberrations can be made of several lenses, whose transmissivities must be factored in the overall $R(\lambda)$. Note that the percentage of light transmitted by a refractive element depends on its chemical composition and thickness, but also on the reflection at the air-glass surface. For instance, the absorption by 1-cm thick Crown glass is 3% at 3600 Å, rapidly decreasing to less than 1% going to the red, while the reflection from the first surface is around 4%, unless a special antireflective coating is applied.

Regarding filters, there are several possible choices, from colored glasses to gelatin films to interferential devices. All colored glasses have a drawback; they tend to transmit the IR even if their color is blue or yellow. This so-called red-leakage can cause considerable problems with solid-state detectors such as charge-coupled devices (CCDs). Therefore, each filter must be characterized in the laboratory in a spectral band as wide as possible before being used at the telescope.

In the following, we will consider in detail the photometric system UBV introduced by Johnson and Morgan (Johnson and Morgan, 1951 and 1953, References section) and its extensions to the red (Johnson, 1966). This system was widely used in the past. Other systems have been developed, according to the availability of new detectors, to specific scientific interests, to space applications such as HST and GAIA, as discussed in the following. However, definitions and considerations based on the UBV system maintain their basic validity; UBV magnitudes of several stars are reported still today in the *Astronomical Almanac*. Therefore, we base on it the photometric discussion of this chapter.

The original realization of the UBV employed the following glass filters: U: Corning 9863; B: Corning 5030 plus Schott GG13; and V: Corning 3384. The detector was a blue–green-sensitive photomultiplier RCA 1P21, so that red-leakage was not important. The relative sensitivity stated in

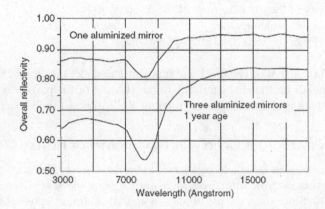

**FIGURE 16.1**   The overall reflectivity of one Al-coated mirror (upper curve) and of a three-mirror telescope (such as the TNG) 1 year since aluminizing. The Al reflectivity is very good also in the UV part of the spectrum, down to approximately 1000 Å.

the original paper is given in Figure 16.2. The telescope had two aluminized mirrors, whose reflectivities were not included in the graph.

Owing to its usefulness, the UBV system was largely adopted by other observers. However, by necessity many astronomers had to use different filters (e.g., the Schott filters, such as Schott UG2 for the U, Schott BG12 + plus Schott GG13 for the B, Schott GG11 or GG14 for the V) and different detectors (e.g., blue–green-sensitive photographic emulsions such as the several O–types produced by Eastman Kodak). Obviously, the telescopes and sites were also different. Therefore, each UBV realization has its own peculiarities, so that care must be exerted to compare the magnitudes if great precision, for instance, at the level of few hundredths of magnitude, is required. Notice also that the colors of the stars play a role in Equation 16.5, so that stars of different colors will define a different $\lambda_c$. In other words, the relations between a given photometric system and another one usually contain color terms that must be accurately calibrated.

Table 16.1 shows a set of indicative characteristics of the generalized UBV system. The second column gives the wavelength interval over which the transmission is different from zero. The third column of the table shows the largely used parameter FWHM (full width at half maximum), namely, the width in angstrom of the filter at 50% transmissivity. The next three columns give three

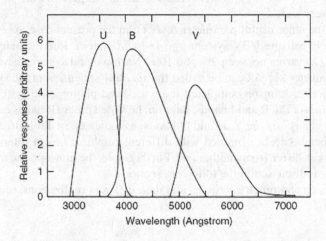

**FIGURE 16.2**   Response of the UBV photometer to an ideal lamp emitting equal energy at all wavelengths.

---

**TABLE 16.1**

**Indicative Characteristics of the UBV System**

| Band | $\Delta\lambda$ | FWHM | $\lambda_{max}$ | $\langle\lambda\rangle$ | $\lambda_{eff}$ |
|------|------|------|------|------|------|
| U | 3100–4000 | 600 | 3670 | 3680 | 3550 at $T = 2.5 \times 10^4$ K |
| | | | | | 3650 at $T = 1.0 \times 10^4$ K |
| | | | | | 3800 at $T = 4.0 \times 10^3$ K |
| B | 3750–5350 | 1,000 | 4295 | 4450 | 4330 at $T = 2.5 \times 10^4$ K |
| | | | | | 4400 at $T = 1.0 \times 10^4$ K |
| | | | | | 4500 at $T = 4.0 \times 10^3$ K |
| V | 4950–6350 | 850 | 5450 | 5460 | 5470 at $T = 2.5 \times 10^4$ K |
| | | | | | 5480 at $T = 1.0 \times 10^4$ K |
| | | | | | 5510 at $T = 4.0 \times 10^3$ K |

definitions of the indicative wavelength of the system; $\lambda_{max}$ is where the efficiency of the system is maximum, while $\langle\lambda\rangle$ is defined as

$$\langle\lambda\rangle = \frac{\int \lambda T(\lambda)\,d\lambda}{\int T(\lambda)\,d\lambda}$$

Some authors call $\langle\lambda\rangle$ *isophotal* wavelength. Both $\lambda_{max}$ and $\langle\lambda\rangle$ can be calibrated in the laboratory. Instead, $\lambda_{eff}$ depends on the color (namely, on the temperature) of the particular star:

$$\lambda_{eff} = \frac{\int \lambda T(\lambda)f(\lambda)\,d\lambda}{\int T(\lambda)f(\lambda)\,d\lambda}$$

as shown in Table 16.1 by changing the stellar surface temperature from $2.5 \times 10^4$ to $4.0 \times 10^3$ K; $\lambda_{eff}$ is therefore more representative of a real observation. The differences between the three wavelengths and the shift to the red of $\lambda_{eff}$ with decreasing stellar temperature are so important because the UBV filters are very wide.

Let us introduce the other useful parameter $\Delta\lambda/\langle\lambda\rangle$; in the present case, $\Delta\lambda/\langle\lambda\rangle$ is greater than 10%, so that we might call the UBV system a *wide-band* system. Intermediate-band systems are those for which $\Delta\lambda/\langle\lambda\rangle$ varies between 2% and 10%. Narrow-band systems have $\Delta\lambda/\langle\lambda\rangle$ less than 2%. The inverse parameter $\langle\lambda\rangle/\Delta\lambda$ can be called the *spectral resolution* of the system.

As already mentioned, Johnson completed the wide-band photometric system in the extended visible with the addition of the R and I bands; later on, he added other IR bands called JKLMN. The current scheme also comprises the Y, H and Q bands, as indicated in Table 16.2 (see Figure 16.7). These photometric bands can be obtained with different combinations of filters and detectors, so that each realization can differ from another one. Furthermore, the atmospheric transmission is very severe and selective, as discussed in the following sections.

For further details on photometric systems and their different realizations, see the papers quoted in "Notes" section.

## 16.2.2 THE EFFICIENCY OF THE DETECTORS

The human eye, actually its retina, is a very good detector of light, even though it is a nonlinear device. Furthermore, it has a limited spectral sensitivity, without the capability to integrate the flux for long times and store the information. The retina, having a total area of approximately $1 \times 10^3\,mm^2$, contains two types of photoreceptors, rods and cones. The rods, distributed all over the retina except in the central area, are more numerous (some 120 millions), about three times more

### TABLE 16.2
### Red and IR Wide Band Photometric Systems

| Band | $\lambda_c$ (nm) | $\Delta\lambda$ (nm) | Band | $\lambda_c$ (μm) | $\Delta\lambda$ (μm) |
|------|------|------|------|------|------|
| R | 700 | 220 | K | 2.2 | 0.48 |
| I | 900 | 240 | L | 3.4 | 0.70 |
| Y | 1,000 | 250 | M | 5.0 | 1.20 |
| J | 1,250 | 380 | N | 10.2 | 5.70 |
| H | 1,630 | 370 | Q | 20.1 | 7.80 |

sensitive than the cones, but color-blind; the four to six million cones provide the eye's color sensitivity. The cones are mostly concentrated in the 3 mm wide central spot known as the *macula*, whose central area is called *fovea centralis*. The fovea has a diameter of about 1.5 mm, and it is densely packed with thin cones and deprived of rods. Low-light level vision (scotopic vision) is insured by the rods, while the cones function in bright-light level vision (photopic vision). See Figure 16.3 for a schematic representation of the eye color sensitivity. As already said, only the cones are color sensitive, and among them, there are three different types of color receptors. With the naked eye, we can clearly see the colors of bright stars (Sirius, Arcturus, Betelgeuse, etc.), but all faint stars look gray. The rods can detect flashes of even few photons; they also determine the acuity of vision (the smallest angle the eye can resolve is typically from 40″ to 60″). However, their response time is decidedly slower than that of the cones. For maximum sensitivity, the averted vision technique can be tried, directing the sight to approximately 18° from the optical axis of the eye (namely from the fovea). In reality, each observer has his own preferred angle.

As already recalled, Bowen discussed how faint is the faintest star the human eye can see at the focus of a telescope. According to his formula, the limiting magnitude is $m_{lim} = 7.5 + 2.5 \log D + 2.5 \log M$, where $D$ is the diameter of the telescope in centimeters and $M$ the magnification of the eyepiece. For $D = 150$ cm and $M = 500$, the formula gives $m_{lim} = +19.7$. However, the seeing conditions and the brightness of the night sky can lead to differences of more than one magnitude with respect to the formula.

The photographic emulsion was introduced in the astronomical field in the second half of the 19th century. It played a major role until about 1990, when solid-state detectors became available. The great advantages of photographic emulsion were its capability to integrate luminous flux (exposures of many hours could be achieved), the large sensitive area (plates of up to $36 \times 36$ cm$^2$ were used in Schmidt-type telescopes), the ease of handling and the moderate cost. Although the response of the emulsion to light is strongly nonlinear and wavelength dependent (see Figure 16.4) and even varies with the ability in developing and fixing it, an immense amount of quantitative measurements were performed thanks to the photographic plate. Moreover, the capability to store the information for more than 100 years has prompted programs of digitization, which are under way in several observatories in order to preserve such treasure (see the "Notes" section).

Photocathodes were employed for maximum photometric precision. The cathode quantum efficiency is at its best in the blue region, although there are products reaching the near IR (see Figure 16.5). The strict linearity between the current and the light flux insures, with relative ease, photometric precisions better than a few thousandths of magnitudes. Therefore, the most precise astronomical photometry relied for a long time on photomultipliers. Furthermore, the capability to operate in photon counting mode was essential to study rapidly time-varying phenomena such as lunar occultations. The main disadvantage of the cathode is its limitation to single object photometry.

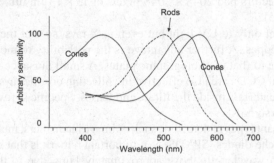

**FIGURE 16.3** The color sensitivity of the eye, arbitrarily normalized to 100. The cones insure the color vision; the rods (dotted curve) insure the night vision.

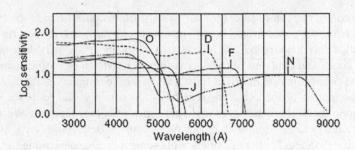

**FIGURE 16.4** Sensitivity of Eastman–Kodak photographic emulsions. Normal emulsions (type O) responded only to blue–green light; the extension to the near IR (types N, I-N), although with lower sensitivity, was obtained with appropriate dyes.

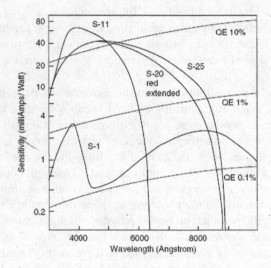

**FIGURE 16.5** Sensitivity of selected photoelectric cathodes.

At the beginning of the '80s of the last century, a linear, high-QE, extended-area, multi-pixel detector became available, namely, the already mentioned CCD, made of a rectangular matrix of pixels on a silicon substrate. Figure 16.6 shows the quantum efficiency (QE) of the two CCDs of the Optical Spectroscopic InfraRed Imaging System (OSIRIS) flown aboard the Rosetta mission to comet 67P/C-G. The detectors had $2048 \times 2048$ pixels of $13 \times 13\,\mu m$ dimension (see Keller et al., 2007; Tubiana et al., 2015).

CCDs are sensitive not only to UV light but even to X-rays, so that they are the chosen detectors in many space telescopes. A further advantage is the possibility to mosaic them to achieve an effective area comparable to that of photographic plates. A small nuisance for astronomical photometric applications is that CCDs are also effective in collecting cosmic rays and particles produced by radioactive decay of materials inside the filters and lenses. Specific software has been developed to remove these spurious signals.

There are other important types of detectors, such as image intensifiers, multichannel plates, CMOS and single-photon avalanche diodes (SPADS). An important remark is that silicon-based devices do not respond to light with wavelength above approximately $1\,\mu m$, due to the intrinsic structure of energy bands. Different materials must be used for the IR, e.g., cadmium telluride; we refer for those to the specialized literature on low-light level astronomical detectors (see the "Notes" section).

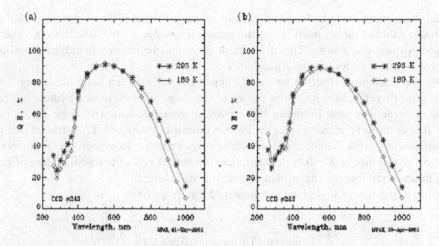

**FIGURE 16.6**   The QE of the two CCD devices of the OSIRIS imaging instrument flown aboard the Rosetta mission. (a) The response of the CCD of the Narrow Angle Camera and (b) that of the Wide Angle Camera. The two curves refer to two operating temperatures. The efficiency reached about 90% at peak sensitivity. The QE in the UV was good down to about 200 nm.

## 16.3   EXTINCTION BY THE EARTH'S ATMOSPHERE

In the visible, the Earth's atmosphere absorbs and diffuses the radiation coming from celestial objects through a number of processes that give rise to both continuous and selective extinctions. For instance, molecular oxygen $O_2$ is very effective to block radiation around 6800 and 7600 Å. The German physicist Josef von Fraunhofer could detect by eye in the far red of the solar spectrum such absorptions that he called B and A, respectively (he examined the spectra from red to blue, and current astronomical practice is from blue to red).

Let us consider the absorption due to a thin layer of atmosphere at height between $h$ and $h + dh$, in the simple model of a plane-parallel atmosphere. The light beam from the star makes an angle $z$ with the zenith, so that the traversed path is $dh/\cos z = \sec z\, dh$.

If $I_\lambda(h)$ is the intensity at the top of the layer, at the exit it will be reduced by the quantity:

$$dI_\lambda = -I_\lambda(h)\cdot k_\lambda(h)\cdot \sec z \cdot dh \tag{16.8}$$

The variable $k_\lambda$ (expressed, e.g., in cm$^{-1}$) represents the absorption per unit length of the atmosphere.

In total, if $I_\lambda(\infty)$ is the intensity outside the atmosphere, at the elevation $h_0$ of the observatory the intensity will be reduced to

$$I_\lambda(h_0) = I_\lambda(\infty)e^{-\sec z \int_{h_0}^{\infty} k_\lambda(h)\cdot dh} = I_\lambda(\infty)e^{-\tau_\lambda(\infty)\cdot \sec z} \tag{16.9}$$

where we have introduced the dimension-less quantity $\tau_\lambda$ called *optical depth*:

$$d\tau_\lambda = k_\lambda(h)dh, \quad \tau_\lambda = \int_{h_0}^{\infty} k_\lambda(h)dh$$

Recalling that the magnitude is defined by the decimal logarithms, we then have

$$m_{\text{ground}} = m_{\text{outside}} - 2.5\, D_\lambda(\infty)\cdot \sec z \tag{16.10}$$

$D_\lambda$ is called the *optical density* of the atmosphere, while the variable $X(z) = \sec z$ is called *air mass*. The minimum value of the air mass is 1 at the zenith; it is $= 2$ at $z = 60°$ (the limit of validity of the present approximate discussion). Therefore, great care must be exerted in calibrating photometric data taken at air masses close to or larger than 2.

Suppose we start observing the star at its upper transit and then keep observing it while its hour angle (and therefore also its zenith distance) increases: we will notice a linear increase of its magnitude in agreement with Equation 16.10, namely, a straight line with slope 2.5 $D_\lambda$ in a graph $(m, \sec z)$. It is common practice to plot the m-axis pointing downward. This straight line is known as the *Bouguer line*, from the name of the 18th-century French astronomer who introduced it. The extrapolation of this line to $X = 0$ (a mathematical absurdity) gives the so-called loss of magnitude at the zenith and therefore the magnitude outside the atmosphere.

According to the formulae derived in Chapters 2 and 3, we have

$$\sec z = \frac{1}{\sin\varphi\sin\delta + \cos\varphi\cos\delta\cos HA} = X(z) \tag{16.11}$$

where $\varphi$ is the latitude of the site, and $\delta$ and $HA$ the coordinates of the star.

For a spherical atmosphere with exponentially decreasing density, the following approximation can be used:

$$X(z) = \sec z\left(1 - \frac{H}{R}\sec^2 z\right) \approx \sec z\left[1 - 0.012\sec^2 z\right]$$

where $H$ is a convenient scale (approximately 8 km) and $R$ the radius of the Earth. Garstang (1989) gives another useful formula:

$$X(z) = \frac{1}{\sqrt{1 - 0.96\sin^2 z}}$$

Table 16.3 shows the continuous extinction of the atmosphere above Mauna Kea, whose elevation above sea level (4200 m) is higher than that of most observatories, so that the transparency of the sky is among the best achievable at ground telescopes.

In the violet–blue region, the transparency quickly drops to zero, due to two major absorbers (see Chapter 11):

1. the ozone $O_3$ molecular absorption in the higher layers;
2. the Rayleigh scattering mostly by the $N_2$ molecule in the troposphere, with a rapid decrease of efficiency with $\lambda^{-4}$ (this is why the clear sky is blue).

**TABLE 16.3**

The Extinction at Mauna Kea

| Wavelength (nm) | Extinction (Magnitude/Air Mass) | Wavelength (nm) | Extinction (Magnitude/Air Mass) |
|---|---|---|---|
| 310 | 1.37 | 500 | 0.13 |
| 320 | 0.82 | 550 | 0.12 |
| 340 | 0.51 | 600 | 0.11 |
| 360 | 0.37 | 650 | 0.11 |
| 380 | 0.30 | 700 | 0.10 |
| 400 | 0.25 | 800 | 0.07 |
| 450 | 0.17 | 900 | 0.05 |

At the other end of the spectrum, ignoring the contribution of dust or smokes in the air, the Rayleigh-dominated transparency is reasonably good until about 0.6 μm. From 0.6 to about 1.0 μm, the water vapor $H_2O$ and the molecular oxygen $O_2$ produce absorptions (recall the B and A bands seen by Fraunhofer) plainly visible in the stellar spectra. The heavy $H_2O$ absorptions around 8200 and 9000 Å are called $\rho$ and $\gamma$ bands, respectively. From 1.0 μm, and especially above 2.5 μm, until approximately 30 μm, water vapor, carbon dioxide $CO_2$ and methane $CH_4$ heavily absorb the light. Spectral bands around 5, 10 and 20 μm have acceptable transparency; therefore, spectrophotometry from the ground is feasible from high-altitude sites and with great care.

Summarizing the previous considerations, astronomical observations with electromagnetic waves can be performed from the ground in the spectral range from $\approx 3200$ Å to $\approx 2.5$ μm. In the IR from 5 to 20 μm, there are three windows accessible from high-elevation sites. The atmosphere becomes again transparent from the millimeter region up to kilometric radio waves, where again the ionosphere blocks the transparency. Astronomical observations in the Gamma, X, UV, far-IR and very long radio waves require stratospheric airplanes, high-flying balloons, suborbital rockets and, finally, satellites flying above 300 km. The lunar surface in the future will offer important observational advantages.

To complete these considerations about the influence of the atmosphere on the photometry and the spectroscopy of the celestial bodies, we must add that the atmosphere itself contributes radiation, both by spontaneous emission and by scattering of natural and artificial lights. If the observatory is close to populated areas, bright emission lines of mercury and sodium from street lamps are observed: Hg at $\lambda\lambda$ 4046.6, 4358.3, 5461.0, 5769.5, 5790.7; Na at 5 $\lambda\lambda$ 683.5, 5890/96 (the yellow D-doublet), 6154.6; Ne at $\lambda$ 6506, and so on. Nowadays, LED and OLED lamps are increasingly used in public and private illumination, with appreciable contamination of the night sky, in particularly in the blue region.

Natural lines come from the atomic oxygen in forbidden transitions (designated with [OI], see Chapter 17 for an explanation of this terminology) at $\lambda\lambda$ 5577.4, 6300 and 6367 and from the molecular radical OH which provides a wealth of spectral lines and bands filling the near-IR region above 6800 Å. The OH comes from the dissociation of the water vapor molecule under the action of the solar UV radiation. Therefore, the atmosphere is a diffuse source of radiation, whose intensity strongly depends on the observatory site. To set an indicative value in the visual band, a luminosity equivalent to one star of the 20th mag per $arcsec^2$ can be assumed.

Figure 16.7 shows the night sky of the Roque de los Muchachos (La Palma, Canary Islands), where the artificial light contribution is fairly small. See for contrast Figure 17.3 which shows the heavily polluted night sky of the Asiago Observatory.

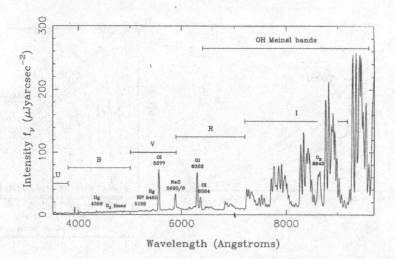

**FIGURE 16.7**  The night sky of the Roque de los Muchachos (La Palma, Canary Islands). (Figure adapted from the ING Technical Note 115: *La Palma night-sky brightness* by Benn and Ellison, 1998; Courtesy C. Benn).

## 16.4   THE BLACK BODY

A very useful concept is that of the black body, of such importance in the development of all physics that we assume it well known to the reader. Only some basic facts, useful for astronomical applications will be recalled here. The black body is an ideal body of uniform temperature $T$ (measured in Kelvin, K), which absorbs all incident radiation and emits, isotropically, unpolarized thermal radiation according to Planck's law:

$$\pi B_\lambda(T)\mathrm{d}\lambda = \frac{2\pi hc^2}{\lambda^5}\frac{1}{e^{hc/k\lambda T}-1}\mathrm{d}\lambda\left(\mathrm{erg/cm^2 s}\right) \tag{16.12}$$

where $h = 6.626 \times 10^{-27}$ erg·s is Planck's constant and $k = 1.381 \times 10^{-16}$ erg/K is Boltzmann's constant. Expressing $\lambda$ in cm,

$$2\pi hc^2 = c_1 = 3.742 \times 10^{-2}\,\mathrm{erg/cm^2\,s}, \quad hc/k = c_2 = 1.439\,\mathrm{cm\,K}$$

where $c_1$ and $\underline{c}_2$ are called the *radiation constants*.

Therefore, the quantity $\pi B_\lambda(T)\cdot\mathrm{d}\lambda$ represents the energy emitted in the interval $(\lambda, \lambda + \mathrm{d}\lambda)$ by the unit surface in the unit time in the outward hemisphere. The function $\pi B_\lambda(T)$ is also called the *emittance* of the black body. Notice that the factor 2 in (16.12) is not connected to the total solid angle. The reason for having $\pi$, and not $2\pi$ in (16.12), and the following formulae resides in the projection effect of the surface at different angles, as will be demonstrated in a following section.

The function $B_\lambda(T)$, namely, the energy emitted into the unit solid angle per unit bandwidth, is called *intensity* or *surface brightness* (erg/cm²·cm·s·sr) of the black body.

Figure 16.8 gives two representations of $B_\lambda(T)$.

All curves go to zero at zero wavelength, and they never cross: the unit area of a hotter black body emits at all wavelengths more energy than a cooler one, even in the IR. The emissivity of the solar surface is approximately that of a black body at 5800 K and the emissivity of the Earth that of a black body at 290 K. Using frequencies instead of wavelengths, the expressions of the emittance and surface brightness become, respectively,

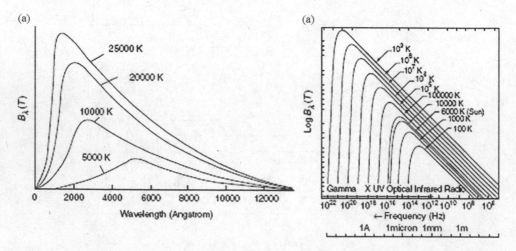

**FIGURE 16.8**   The black body curve for temperatures of astrophysical interest, on linear axes (a) and logarithmic axes (b). Notice that at temperatures higher than 9000 K or lower than 4000 K, the peak of emissivity is outside the visible range.

$$\pi B_v(T)\mathrm{d}v = \frac{2\pi h v^3}{c^2}\frac{1}{e^{hv/kT}-1}\mathrm{d}v \ \left(\mathrm{erg/cm^2\,s}\right)$$

$$(16.13)$$

$$B_v(T) = \frac{2\pi h v^3}{c^2}\frac{1}{e^{hv/kT}-1}\mathrm{d}v\left(\mathrm{erg/cm^2\,Hz\,s\,sr}\right)$$

Planck's law has two limiting cases:

1. If $\left(hc/k\lambda T\right) \ll 1$, then

$$\pi B_\lambda(T) = 2\pi kc T\lambda^{-4} = \frac{c_1}{c_2}\,T\lambda^{-4}\ \left(\mathrm{in\ \ frequencies}\ \pi B_v(T) = 2\pi c^{-2}kTv^2\right)$$

which is the so-called *Rayleigh–Jeans* approximation valid in the IR and radio domains.
2. If $\left(hc/k\lambda T\right) \gg 1$, then

$$\pi B_\lambda(T) = 2\pi hc^2\lambda^{-5}e^{-hc/k\lambda T} = c_1\lambda^{-5}e^{-c_2/\lambda T}\ \left(\mathrm{in\ frequencies}\ \pi B_v(T) = 2\pi hc^2 v^3 e^{-hv/kT}\right)$$

which is the so-called *Wien approximation*, apt to describe the spectral distribution in the visible band of radiation from the surfaces of stars hotter than about 4000 K.

By integration of the monochromatic emittance over all wavelengths, or frequencies, we obtain the *Stefan–Boltzmann* law:

$$\pi\int_0^\infty B_\lambda(T)\mathrm{d}\lambda = \pi\int_0^\infty B_v(T)\mathrm{d}v = \sigma T^4$$

$$(16.14)$$

where $\sigma = 2\pi^5\cdot k^4/15\cdot h^3\cdot c^2 = 5.6696\times 10^{-5}$ (erg/cm$^2$·s·K$^4$) is the Stefan–Boltzmann constant. Or else

$$B = \frac{\sigma}{\pi}T^4 = 1.8047\times 10^{-5}\,T^4\ \left(\mathrm{erg/cm^2\,s\,sr}\right)$$

From Equation 16.12, the Wien law (or displacement law) is easily seen: the wavelength of the maximum value of $B_\lambda(T)$ is inversely proportional to the temperature:

$$\lambda_{max}\cdot T = 0.28979\ \left(\mathrm{cm\,K}\right)$$

$$(16.15)$$

so that the peak of emissivity of a star behaving like a black body of $T = 1\times 10^4$ K (for instance, a star such as Vega or Sirius) is situated at $\lambda_{max} = 2899$ Å, unobservable from the ground; the same applies to hotter stars. At the other extreme, a star having $T = 1\times 10^3$ K has its maximum emissivity in the IR; the emissivity peak of cooler bodies, such as the Earth, is in the far IR or even in the radio domain.

Care must be taken when considering the power emitted per unit frequency and not per unit wavelength. According to d$v$ Wien's displacement law, the peak of emissivity is at

$$v_{max} = \alpha v_{max} = \alpha^h kT \approx \left(5.879\times 10^{10}\ \mathrm{Hz/K}\right)\cdot TkT\,/\,h = 5.88\times 10^{10}\,T\,\mathrm{(Hz)}$$

where $\alpha \approx 2.821$ is a constant resulting from the numerical solution of the maximization equation, $k$ is Boltzmann's constant, $h$ is Planck's constant and $T$ is the temperature. Therefore, the wavelength $\lambda$ corresponding to the frequency $v_{max}$ of maximum $B_v$ is

$$\lambda = c/\nu_{max} = 0.510/T \,(cm)$$

Consequently, the peak of emission per unit frequency corresponds to a wavelength 1.7 times longer than the peak per unit wavelength. For instance, the effective temperature of the Sun is 5778 K. Using Wien's law, the wavelength of the peak emissivity per unit wavelength is approximately 500 nm, in the green portion of the spectrum near the peak sensitivity of the human eye (see Figure 16.3). In terms of power per unit frequency, the emission peaks at 343 THz, namely, at a wavelength around 883 nm in the near IR.

The emittance per unit wavelength at the peak of emissivity is

$$\pi B\left(\lambda_{max}\right) = 1.2865 \times 10^{-4} T^5 \,(erg/cm^2\,cm\,s)$$

which rises extremely rapidly with the temperature. As an example, expressing $\lambda$ in $\mu$m, at $T = 10^4$ K, namely, a temperature roughly corresponding to that of Sirius or Vega, the emittance is

$$\pi B\left(\lambda_{max}\right) = 1.2865 \times 10^{12} \left(erg/cm^2\,\mu m\,s\right)$$

The gradient of the curve of emissivity is given by

$$\Phi_\lambda\left(T\right) = \frac{c_2}{T}\left(1 - e^{c_2/\lambda T}\right) \tag{16.16}$$

which can be measured and used to evaluate the so-called *color temperature* of a star. Because no real star radiates exactly as a black body, this quantity can change according to the spectral range chosen for determining the gradient.

It is also useful to express Planck's distributions in terms of photon emittance per unit wavelength and unit frequency, respectively:

$$\pi N_\lambda\left(T\right) = \frac{2\pi c}{\lambda^4}\frac{1}{e^{hc/k\lambda T} - 1}\,d\lambda\left(ph/s\,cm^2\,cm\right)$$
$$\pi N_\nu\left(T\right) = \frac{2\pi \nu^2}{c^2}\frac{1}{e^{h\nu/kT} - 1}\,d\nu\left(ph/s\,cm^2\,Hz\right) \tag{16.17}$$

Notice the exponents, 4 and not 5, and 2 and not 3, respectively, because each photon has energy $E = h\nu$.

The wavelength of maximum photon emissivity and the corresponding photon flux are

$$\lambda_m T = 0.25506 c_2 = 0.36698\left(cm\,K\right)$$

$$N\left(\lambda_m\right) = 2.1008 \times 10^{11} T^4 \,(ph/cm\,cm^2\,s)$$

The Stefan–Boltzmann law becomes

$$N = 1.52033 \times 10^{11} T^3 \,(ph/cm^2\,s)$$

A "universal black body curve" can be constructed, using as variable the product $\lambda T$. The examination of this curve shows some general results:

*   the energy radiated below $\lambda_{max}$ is one quarter of the total;

- a whole 98% of the total is radiated between 0.5 and 8 $\lambda_{max}$, the rest being equally divided below and above those two limits;
- the decrease of emissivity is much steeper below $\lambda_{max}$ than above it; below 0.17 $\lambda_{max}$, a fraction of only $10^{-10}$ is radiated.

Some useful numbers

- at the normal laboratory temperature of 290 K (17°C), $\lambda_{max} = 10\,\mu m$ and the total emissivity is 401 W/m², corresponding to approximately $1.2 \times 10^{18}$ ph/cm²·s·sr, mostly emitted at $\lambda_{max} = 31.1\,\mu m$ (energy of about $4 \times 10^{-2} eV$);
- at ten times higher temperature, corresponding to the temperature of cool stars, $\lambda_{max} \approx 1\,\mu m$, the emissivity is several MW/m², approximately 1% of the energy is emitted below 5000 Å or above 8 $\mu m$ and about 10% in the visible band;
- raising the temperature by another factor of 10, namely, reaching the surface temperature of a hot star, the emissivity rises by four orders of magnitude, the maximum moves to the far UV and a very small percentage of energy is radiated in the visible band.

## 16.5   COLOR INDICES AND TWO-COLOR DIAGRAMS

Suppose we observe a star in two spectral bands, measuring $m(\lambda_1) = m_0(\lambda_1) - 2.5 \log i(\lambda_1)$ and $m(\lambda_2) = m_0(\lambda_2) - 2.5 \log i(\lambda_2)$, respectively. We define *color index* of the star the difference:

$$c_{12} = m(\lambda_1) - m(\lambda_2) = m_0(\lambda_1) - m_0(\lambda_2) - 2.5 \log \frac{i(\lambda_1)}{i(\lambda_2)}, \quad (\lambda_1 < \lambda_2) \qquad (16.18)$$

The name of this quantity comes from the visual sensation of the color of a luminous source. We also have

$$i(\lambda_2) = i_0 i(\lambda_1) \cdot 10^{0.4 c_{12}}$$

where the constant $i_0$ depends on the two wavelengths and the instrumental sensitivity, according to Equation 16.5.

If the star radiates as a black body, from Equation 16.12 we can easily derive the following expression:

$$c_{12} = m(\lambda_1) - m(\lambda_2) = c_0 + \frac{1.56}{T}\left(\frac{1}{\lambda_1} - \frac{1}{\lambda_2}\right) + f(T)$$

where $f(T)$ would be zero in the Wien approximation and is always a small quantity for $T > 4000$ K. Therefore, for not too cool black bodies we have the relationship:

$$c_{12} = m(\lambda_1) - m(\lambda_2) \approx A_{12} + \frac{B_{12}}{T} \qquad (16.19)$$

For instance,

$$m_{blue} - m_{vis} \approx -0.64 + \frac{7200}{T}$$

which is one of several useful formulae named after H.N. Russell.

However, real stars do not radiate as black bodies. Following Johnson and Morgan, it has been agreed that the zero point of the color indices is defined by a set of stars having surface temperatures similar to that of Vega, namely, approximately 10,000 K. Those stars have the spectroscopic

**TABLE 16.4**

Color Indices of the Black Body in the UBV Photometric System

| T | U–B | B–V | T | U–B | B–V |
|---|---|---|---|---|---|
| 4,000 | +0.37 | +1.13 | 20,000 | −1.01 | −0.16 |
| 6,000 | −0.25 | +0.62 | 25,000 | −1.06 | −0.15 |
| 10,000 | −0.69 | +0.14 | 40,000 | −1.14 | −0.29 |
| 15,000 | −0.91 | −0.07 | ∞ | −1.28 | −0.44 |

designation A0-V (Arabic zero, Roman 5), as explained in Chapter 17. For those standard stars, $c_{12} = 0$ *by definition*, no matter how the two wavelengths might have been selected:

$$U - B = B - V = V - R = R - I = \cdots = 0$$

It is plainly evident that color index equal to zero does not mean equal fluxes in the different spectral bands.

Let us take the black body radiation law and the sensitivity of the UBV photometric system. Formula (16.5) will allow to calculate the (U–B, B–V) color indices of the black body as a function of the temperature, as shown in Table 16.4.

The numbers in this table are only indicative values; they change by few hundredths of magnitude according to the particular realization of the UBV system (see, for instance, Matthews and Sandage, 1963). Two points are worth noticing:

1. in this particular photometric system, the color indices saturate with increasing temperature, as foreseen by Equation 16.19. Therefore, it is easier to estimate the temperature of stars cooler than $10^4$ K than of hotter ones. The reason can be easily understood from Figure 16.8, which shows that in the blue-visual bands, the black body curves for $T > 10^4$ K are almost parallel to each other, so that the gradients will be essentially equal;
2. the color indices of the $10^4$ K black body are not zero, because real stars do not radiate exactly as black bodies. A more detailed explanation is given later on and in Chapter 17.

Let us obtain the color indices of a large number of stars, randomly chosen over the celestial sphere, and plot their position in the (U–B, B–V) plane. The result is given in Figure 16.9, where all stars brighter than magnitude +6.5 have been used.

Most stars are located on a main sequence running from the top left (the hottest stars) to the bottom right (the coolest stars). Other stars occupy the region to the right of this main sequence, towards a boundary defined by the black body locus. A number of stars are also found below the main sequence. The precise interpretation of this two-color indices diagram (usually called in the abbreviated form two-color diagram) must wait until Chapter 17; some preliminary notions are in order now:

- the A0 stars at (U–B, B–V) = (0, 0) are actually the most distant ones from the black body locus. In Chapter 17, we shall see that the reason for this distance is the presence of hydrogen absorption lines (Balmer series), especially affecting the U filter;
- in the upper part of the diagram, the stars having colors very similar to those of the black body are the so-called white dwarfs, briefly mentioned in Chapter 13;
- below the A0 stars, there is a short heavily filled region where stars of large radii and intermediate temperatures (giant stars) are located;

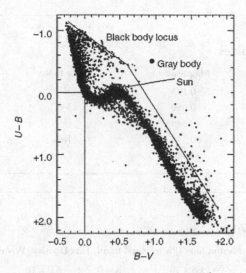

**FIGURE 16.9**  The (U–B, B–V) diagram for bright stars. In stellar photometry, it is customary to draw the U–B axis positive downward. The black body locus, the gray body (equal emissivity in all bands), the A0 position and Sun are also indicated.

- some points in the intermediate region between the main sequence and the black body line are simply unresolved binaries, with stars having different colors (say a hot plus a cool star seen as a single source). Unresolved normal galaxies, which in this context behave like a mixture of stars, are also found in the same region;
- there are points to the right of the black body line. Sometimes, the star simply varied in brightness during the observations; in other much more interesting cases, these peculiar color indices reveal the presence of non-thermal emission mechanisms (e.g., strong emission lines or continuous synchrotron mechanism). These cases occur in planetary nebulae or active galactic nuclei and quasars, where strong emission lines are superimposed to the continuous spectrum;
- the colors of some stars are not the true ones; they are affected by interstellar absorption, as explained in a later section. This effect is particularly evident in the upper part of the diagram, between the main sequence and the black body locus, because the stars in that region are usually very luminous, very distant and close to the Milky Way plane, with a large quantity of intervening matter.

## 16.6  CALIBRATION OF THE APPARENT MAGNITUDES IN PHYSICAL UNITS

Until now, the system of astronomical magnitudes has not been connected to any particular physical unit. This necessary calibration can be performed using the Sun, the Moon or a set of convenient stars such as Vega. Regarding the Sun, the following lines and Table 16.5 report some useful values of the solar flux at 1 AU and outside the terrestrial atmosphere:

- $f(\lambda_v) = 1.83 \times 10^5$ erg/cm²·s integrated over the visual band;
- the bolometric correction is $Cv = -0.08$ mag, and the bolometric flux is $f_{bol} = 1.36 \times 10^6$ erg/cm²·s. Using units of the photovisual system, the solar light flux corresponds to $1.27 \times 10^5$ lux. This quantity is also called *solar constant*; it is equivalent to $1.36$ kW/m², or to $1.95$ cal/cm²·min⁻¹. In the present epoch, the solar constant has only small oscillations, by no more than few thousandths.

**TABLE 16.5**
**Solar Color Indices**

| Color Index | Magnitude | Color Index | Magnitude |
|---|---|---|---|
| U–B | +0.195 | V–I | +0.88 |
| B–V | +0.650 | J–H | +0.310 |
| V–R | +0.540 | H–K | +0.060 |
| R–I | +0.340 | K–L | +0.034 |
| V–K | +1.486 | L–M | −0.053 |

**TABLE 16.6**
**Calibration Factors for Vega**

| Band | Flux Density (W/m²·μ) | Photon Flux (Ph/m² s·μ) | Band | Flux Density (W/m²·μ) | Photon Flux (ph/m²·s·μ) |
|---|---|---|---|---|---|
| R = 0 | $7.8 \times 10^{-8}$ | $1.66 \times 10^{11}$ | K = 0 | $4.0 \times 10^{-10}$ | $4.5 \times 10^{9}$ |
| I = 0 | $8.3 \times 10^{-9}$ | $1.39 \times 10^{11}$ | L = 0 | $8.1 \times 10^{-11}$ | $1.2 \times 10^{9}$ |
| J = 0 | $3.3 \times 10^{-9}$ | $2.0 \times 10^{10}$ | M = 0 | $2.1 \times 10^{-11}$ | $5.1 \times 10^{8}$ |
| H = 0 | $3.9 \times 10^{-10}$ | $9.6 \times 10^{9}$ | N = 0 | $1.0 \times 10^{-12}$ | $5.1 \times 10^{7}$ |

The direct utilization of the Sun to calibrate the normal stellar observations is very difficult; thus, a number of stars with characteristics as close as possible to solar have been selected. The following stars are among the solar analogs: 16 Cyg B, VB64, HD 10,5590 and HR 2290.

From the calibration of Vega, we derive the following correspondences between magnitudes and fluxes (outside the terrestrial atmosphere):

$$U = 0 : f(37000A) = 4.4 \times 10^{-9} \left(\text{erg/cm}^2\, \text{s}\, A\right)$$

$$B = 0 : f(4450A) = 7.2 \times 10^{-9}\, (\text{erg/cm}^2\, \text{s}\, A)$$

$$V = 0 : f(5556A) = 3.44 \times 10^{-9} \left(\text{erg/cm}^2\, \text{s}\, A\right) = 3.44 \times 10^{-8} \left(\text{W/m}^2\, \mu\right)$$

For a generic star of magnitudes U, B, V, the flux is given by

$$f(\lambda_U) = 10^{-0.4(U+20.90)}, \quad f(\lambda_B) = 10^{-0.4(B+20.36)}, \quad f(\lambda_V) = 10^{-0.4(V+21.08)},$$

In photon fluxes, recalling that the energy of a photon is $E = 1.98610 \times 10^{-8}/\lambda(A)$ erg, we have

$$N_{\text{ph}}(U) \approx 800 \times 10^{-0.4U}, \quad N_{\text{ph}}(B) \approx 1500 \times 10^{-0.4B},$$

$$N_{\text{ph}}(V) \approx 1000 \times 10^{-0.4V} \left(\text{ph/cm}^2\, \text{s}\, A\right)$$

Other useful conversion factors from Vega (approximate within probably ±5%) are given in Table 16.6.

## 16.7　APPARENT DIAMETERS AND ABSOLUTE MAGNITUDES OF THE STARS

Let us assume that the star is a spherical body of radius $R$, emitting as a black body of temperature $T$, so that the luminous energy radiated over the whole sphere in the bandwidth $d\lambda$ is $dF_\lambda = 4\pi R^2 B_\lambda(T)$ $d\lambda$. Furthermore, suppose that the star is at rest with respect to an observer located at distance $d$

from it and that space is perfectly transparent. The luminous flux through the sphere of radius $d$ will then be

$$df_\lambda = 4\pi R^2 B_\lambda(T) d\lambda / 4\pi d^2 = \left(\frac{R}{d}\right)^2 B_\lambda(T) d\lambda \qquad (16.20)$$

Therefore, the flux received from such star decreases according to the inverse square of the distance. Integrating over all wavelengths, the received bolometric flux is

$$f_{bol} = f_0 \left(\frac{R}{d}\right)^2 T^4$$

The ratio $(R/d)$ is proportional to the apparent diameter $\theta$, while its square is proportional to the solid angle $\omega$ of the star as seen by the observer:

$$\frac{R}{d} = \tan\frac{\theta}{2} \approx \frac{\theta}{2}(\text{rad}), \quad \omega = \pi\left(\frac{R}{d}\right)^2 \approx \pi\left(\frac{\theta}{2}\right)^2 \ (\text{sr})$$

so that

$$df_\lambda = f_0 B_\lambda(T)\theta^2 d\lambda, \quad F_{bol} = F_0 T^4 \theta^2$$

The apparent magnitudes are then given by

$$m_\lambda = m_0 - 2.5\log B_\lambda(T) - 5\log\theta,$$
$$m_{bol} = m_{0bol} - 2.5\log T^4 - 5\log\theta \qquad (16.21)$$

From Equation 16.21, we see that the bolometric correction does not depend on the distance nor on the radius of the star:

$$C_V = m_{bol} - m_{5500} = m_0' + 2.5\log B_{5500}(T) - 10\log T$$

The zero point of $C_V$ was originally chosen to be zero for temperatures of 6500 K, so that it becomes negative for both higher and lower temperatures. For black bodies, the bolometric corrections are approximately $-5.4$ at $T = 6 \times 10^4$ K, $-0.4$ at $T = 10^4$ K, $-0.08$ for the Sun, $-1.0$ at $T = 4 \times 10^3$ K and $-4.0$ at $T = 2.5 \times 10^3$ K. For real stars, the bolometric corrections were very uncertain for the hotter and the cooler stars, until the space age permitted observations in the far UV and far IR.

We explicitly mention that the quantity $f_0 B(T) = f_\lambda/\theta^2$, namely, a quantity proportional to the surface brightness of the star, is constant along the line of sight (provided the space is transparent) and, in particular, it equals the emissivity from the surface of the star.

Now, let us introduce the concept of *absolute magnitude*, by considering a particular distance $d = 10$ pc as standard distance. The absolute magnitude $M$ is the apparent magnitude a star would have if located at the standard distance. From Equation 16.21,

$$M_\lambda = m_{0\lambda} - 2.5\log\left(\frac{R}{10}\right)^2 - 2.5\log B_\lambda(T),$$

$$M_\lambda - m_\lambda = -2.5\log\left(\frac{d}{10}\right)^2 \qquad (16.22)$$

and similar for the bolometric magnitude. The quantity $y_\lambda$

$$y_\lambda = m_\lambda - M_\lambda = 5\log\left(\frac{d}{10}\right) = -5\log 10\pi \qquad (16.23)$$

where $d$ is in parsec, and $\pi$ is the annual parallax of the star in arcsec, is called *distance modulus* of the star. Therefore,

$$m_\lambda = M_\lambda + y_\lambda = M_\lambda + 5\log d - 5 = M_\lambda - 5\log\pi - 5,$$

$$M_\lambda = m_\lambda - y_\lambda = m_\lambda - 5\log d + 5 = m_\lambda + 5\log\pi + 5$$

From the definition given by Equation 16.23, it would seem that the distance modulus is independent of the wavelength. However, the interstellar space might not be transparent in the direction of that object, so that the distance derived from inverting Equation 16.23 may well depend in a systematic way on the wavelength. This is why we have explicitly kept the wavelength in the notation of the distance modulus $y_\lambda$.

A word of clarification about the notation: apparent magnitudes are indicated with small letters, $m_\lambda$, absolute magnitudes with capital letters, $M_\lambda$. However, the apparent magnitudes in Johnson's photometric system and its extensions (UBVRIYJHKLMNQ) are indicated with capital letters, U, B, V, etc.; therefore, the absolute magnitude in Johnson's V band will be indicated with $M_V$, and similar for the other bands.

Regarding color indices, it is obvious that they do not depend on the distance (always in a transparent space), so that

$$c_{12} = m(\lambda_1) - m(\lambda_2) = M(\lambda_1) - M(\lambda_2), \quad C_V = m_{bol} - m(\lambda) = M_{bol} - M(\lambda)$$

It is easy to extend the results found in Equation 16.20 and following, to cases where the star does not radiate as a black body, by substituting the appropriate spectral distribution $I_\lambda$ in place of the Planck function; of course, the simple relation with the temperature would be lost. Furthermore, the star is a gaseous sphere, so that its diameter can depend to some extent on the wavelength, through the optical depth of the atmosphere. Therefore, even the apparent diameter will be a function of $\lambda$.

Let us consider in particular the Sun. Its distance is 1/206,265 pc, so that its absolute magnitudes are

$$M_\odot = m_\odot + 5 + 5\log 20,6265 = m_\odot + 31.57,$$

$$M_\odot(V) = -26.74 + 31.57 = +4.83, \quad M_\odot(bol) = -26.82 + 31.57 = +4.75$$

The average values of its radius, apparent diameter and apparent solid angle are, respectively,

$$R_\odot = 6.96 \times 10^{10}\,\text{cm}, \quad \theta_\odot = 33' = 0.0095\,\text{rad}, \quad \omega_\odot = 6.81 \times 10^{-5}\,\text{sr}$$

From these values and from the bolometric flux observed outside the Earth's atmosphere, we derive the following quantities:

- bolometric flux through the unit area of the solar surface: $F_\odot = 6.29 \times 10^{10}$ erg/cm$^2$ s;
- bolometric surface brightness: $I_\odot = F_\odot/\pi = 2.04 \times 10^{10}$ erg/cm$^2$ s sr;
- total luminosity: $L_\odot = 4\pi R_\odot^2 F_\odot = 3.83 \times 10^{33}$ erg/s;
- from the Stefan–Boltzmann law, the effective temperature $T_\odot$, namely, the temperature of a black body of same radius which would emit the same energy is $T_\odot = 5777$ K.

From the previous considerations, we understand that these values will be only approximately correct for the real Sun. We shall not be surprised if the measured temperature depends on the wavelength or the position over the solar disk (higher at the center, lower at the limb, the so-called limb-darkening) because of the different optical depths along the line of sight.

Using the Sun as reference for radii and luminosities, we obtain for a generic star:

$$\log L = -3.147 + 2\log R + 4\log T_{\text{eff}},$$

$$M_{\text{bol}} = 4.75 - 2.5\log\frac{L}{L_\odot} = 42.36 - 10\log T_{\text{eff}} - 5\log\frac{R}{R_\odot},$$

$$C_V = -42.54 + 10\log T_{\text{eff}} + \frac{29{,}000}{T_{\text{eff}}},$$ 

$$\log T_{\text{eff}} = 4.221 - 0.1(V + C_V) - 0.5\log\theta$$

(16.24)

where $\theta$ is expressed in milliarcsec (mas). Equation 16.24 is the basis for the photometric determination of stellar diameters. For example, consider a $V = 10$ hot star, having $T_{\text{eff}} = 2.7 \times 10^4$ K and $C_v = -2.9$. From Equation 16.24, we derive an apparent diameter $\theta = 0.01$ mas. A cooler star, with $T_{\text{eff}} = 4 \times 10^3$ K and the same apparent magnitude, would have a diameter approximately ten times higher. Recalling the considerations made in Chapter 11 about the angular extent of the diffraction figure of the telescope and the atmospheric seeing, these values of $\theta$ clearly demonstrate the difficulty of the direct measurement of the apparent diameter of stars.

Other useful formulae, associated with the name of H.N. Russell and valid for sufficiently hot stars (namely, when Wien's approximations to a black body is acceptable) are

$$\log F(4400) = -0.4M_B + 9.11 - 2.5\log\frac{R}{R_\odot},$$

$$\log F(5550) = -0.4M_V + 8.85 - 2.5\log\frac{R}{R_\odot}$$

(both in erg/cm²·s A);

$$M_b = -0.72 + \frac{36{,}700}{T} - 5\log\frac{R}{R_\odot}, \quad M_v = -0.08 + \frac{29{,}500}{T} - 5\log\frac{R}{R_\odot}$$

$$M_b - M_v = m_b - m_v = -0.64 + \frac{7200}{T}$$

The monochromatic magnitudes in the blue and visual bands $m_b$, $m_v$ used in the above relations are slightly different from the heterochromatic B, V magnitudes of Johnson's system.

## 16.8   THE HERTZSPRUNG–RUSSELL DIAGRAM

Around 1913, E. Hertzsprung and H.N. Russell independently discovered a most important relationship between the absolute magnitudes of normal stars and their temperatures (or equivalently, their color indices or their spectral types, as we shall see in Chapter 17). From the Stefan–Boltzmann law, we expect that stars can have any combination of radius and temperature, so that the distribution of stars in the plane ($\log L$, $\log T$) or in any of the equivalent planes ($M_{\text{bol}}$, $\log T$), ($M_{\text{bol}}$, B–V), ($M_V$, B–V) should be a uniform one. We shall refer in the following to any of those planes as a Hertzsprung–Russell (H–R) diagram. Observations prove instead that uniform distribution is not the case: stars are preferentially found in well-defined regions of the H–R diagram. Given the temperature, the radius assumes specific values only, for reasons that must be intrinsic to the equilibrium structure of the star. This behavior is represented in Figure 16.10. Figure 16.10a shows the diagram obtained by using data of the Hipparcos astrometric satellite; this sample of stars is therefore within approximately 50 parsec of the Sun. GAIA has obtained data of higher precision on a larger number of stars, but on a different photometric system. Therefore, for consistency with the previous discussion, we retain here the Hipparcos data, as sufficiently representative of the basic considerations. See "Notes" section for a further discussion of the GAIA data. Figure 16.10b shows the theoretical H–R diagram, which can be obtained from the Stefan–Boltzmann law; the diagonal lines are lines of equal radii in units of the solar radius. Stars do not uniformly fill the plane, and they preferentially populate a main sequence, two high radii sequences (giants and supergiants, at 100 and 1000 solar radii) and a low radius sequence (white dwarfs, at 0.01 solar radius).

The observational data represented by the two-color and H–R diagrams are at the very basis of all theories of stellar structure and evolution. Especially important is the examination of the H–R diagram for well-selected samples of stars, for instance, those belonging to galactic open clusters (e.g., the Hyades), to galactic globular clusters (e.g., M3), to small galaxies in the local group (e.g., Fornax). The study of these stars is important not only because all stars have the same distance from the Sun and the same interstellar reddening, but also because they presumably had a common origin in time and space and an identical initial chemical composition. The very different H–R diagrams derived from these diverse samples led Baade, around 1955, to propose the concept of different stellar populations, namely, Population I represented by open galactic clusters and Population II represented by globular clusters. The classical work on stellar populations is contained in the volume edited by O'Connell (1957, Bibliography section). Following Baade's seminal ideas, the concept of stellar population has been extended to cover the earlier phases of the evolution of the Universe.

The density of points in the observed H–R diagrams is not representative of the true density of stars in space, namely, of the so-called *luminosity function*, a statistical function which represents the distribution of stars among the luminosities. Faint stars greatly outnumber the bright ones, so that the commonly held concept of the Sun being a normal star is somewhat misleading: no more than approximately 5% of all stars are brighter than the Sun, and most are members of binary or multiple systems. The main sequence is observed extending to at least $M_v = +18$, where true stars give place to *brown dwarfs* (no nuclear reactions in their interiors), and further down to planets.

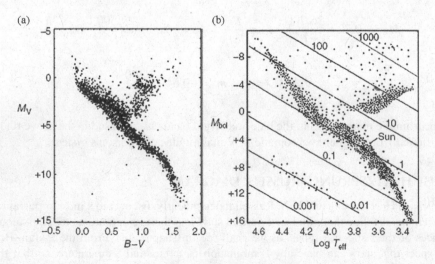

**FIGURE 16.10**    (a) The position of nearby stars in the $(M_v, B–V)$ plane, from Hipparcos data. (b) Observational data superimposed to the theoretical H–R diagram; the diagonal lines are lines of equal radii in units of the solar radius.

Today, planets around other stars are found with relative ease, thanks especially to very precise photometric transit and radial velocity measurements; at the time of writing (Dec, 2019), their number was around 4000 and constantly increasing (see Chapter 15). These fascinating topics cannot be discussed here; given the richness of these fields of research, see, for instance, Barbieri (2019, Bibliography section).

## 16.9   INTERSTELLAR ABSORPTION AND POLARIZATION

The assumption that the interstellar space is perfectly transparent is not correct, as discovered by Trumpler and others around 1930. The Milky Way contains a quantity of diffuse matter in the form of interstellar clouds of gases and dust, which absorb and diffuse the radiation in a manner strongly dependent on the wavelength. Only the effects of the continuous absorption produced by the interstellar matter on the colors of the stars and census of galaxies will be treated here. For the extremely important presence of discrete spectral features of the interstellar and intergalactic medium, we refer to specialized literature, e.g., Savage and Sembach (1996, Bibliography section).

The diffuse interstellar medium is preferentially distributed along the galactic plane, so that the absorption varies with the galactic coordinates of the line of sight.

Let $m(\lambda)$ be the observed magnitude and $d$ the distance in parsec of a given star; if the interstellar matter absorbs the amount $A_\lambda$ (in magnitudes), then the true magnitude is $m_0(\lambda) = m(\lambda) - A(\lambda)$, and the absolute magnitude is

$$M_0(\lambda) = m(\lambda) - A(\lambda) + 5\log(d/10)$$

As already remarked, until allowance is made for interstellar absorption, the measured distance modulus can depend on the wavelength.

In order to derive the function $A(\lambda)$, we make recourse to two sets of stars having the same spectral type as judged from the absorption lines (see Chapter 17): the first set of nearby and presumably unabsorbed stars and the second of much more distant stars. Stebbins and Whitford around 1950 performed a classical work along these lines of reasoning, with observations extending from 3200 to 10,000 Å (see Whitford, 1951). They demonstrated that the absorption is much more important in the blue than in the red. Therefore, a heavily absorbed star will also show a color much redder than the true one (that expected from its spectral type). The color dependence of interstellar absorption justifies the utilization of interstellar *"reddening"* to indicate the same effect of absorption and scattering. Such identification of absorption with reddening would not be correct if the interstellar matter also produced a gray absorption, which however has not been detected.

In the visible band, the interstellar reddening is very different from that caused by the terrestrial atmosphere. The latter depends essentially on $\lambda^{-4}$ (*Rayleigh scattering*), as expected from scattering centers with dimensions much smaller than the wavelength (namely, molecules). The former has a dependence approximately as $\lambda^{-1}$ (*Mie scattering*), as expected from particles having the same dimensions of the wavelength (namely dust). The explanation in terms of dust was first suggested by Lindblad around 1938 and then by Oort and van de Hulst. Observations by satellites have extended the knowledge of $A(\lambda)$ to the UV and to the IR (see, for instance, Cardelli et al. (1989), Draine (2003, Bibliography section)).

Restricting our considerations to the UBV bands, let us introduce the so-called *color excesses* $E(U–B)$ and $E(B–V)$, namely, the difference between the observed reddened color indexes and the unreddened ones:

$$E(U - B) = (U - B) - (U - B)_0, \quad E(B - V) = (B - V) - (B - V)_0$$

Let $R$ be the ratio between the absorption in the V band, $A(V)$ and $E(B-V)$, and same for the B band. The observations provide the following values:

$$R = \frac{A(V)}{E(B-V)} = 3.2 \pm 0.1, \quad A(B) = \frac{4}{3}A(V),$$

$$E(U-B) \approx 0.72 \times E(B-V)$$

(16.25)

along any direction in the Milky Way (apart from some notable localized, exceptions, e.g., toward the Orion Nebula). Therefore, the parameter $Q = (U-B) - 0.72(B-V)$ is essentially independent from the total absorption. In other words, Q is a *reddening-free* parameter, function only of the spectral type (SpT, see Chapter 17) of the star. However, the relationship Q(SpT) is not a single valued one.

Now, let us consider an intrinsically hot (blue) and very luminous star, located in the upper left part of the two-color diagram (Figure 16.10; these stars have spectral type SpT indicated with the letters O and B). Suppose that an intervening dust cloud absorbs its light by $A(V)$ magnitude. Its color indices will be modified by the quantities $E(B-V) = A(V)/R$, $E(U-B) \approx 0.72E(B-V)$, so that the star will move in the H–R diagram downwards to the right, along a line of slope 0.72 by the corresponding amounts, ending in the region between the main sequence and the black body locus. Notice that in the upper left region of the diagram, the slope of the main sequence is slightly steeper than the reddening line; otherwise, we could not distinguish between a temperature and an absorption effect from the color indices alone. This separation of effects is instead impossible (with the UBV system) for stars such as the Sun or cooler. Therefore, the UBV system is not the most suitable one to study the interstellar absorption. However, it provided the initial useful reddening criteria for the bluest and most luminous stars, which sample the Milky Way plane to large distances.

Regarding the different amounts of extinction in the different galactic directions, the interstellar matter is distinctively concentrated at low galactic latitudes, where both the high luminosity and temperature O-B stars and the gaseous hydrogen clouds emitting the 21-cm (1420 MHz) spectral line are most abundant. Making an analogy with a plane-parallel atmosphere (thus ignoring the dependence on the galactic longitude $l$), suppose we are located in the symmetry plane of the Milky Way and observe a galaxy beyond all absorbing matter at galactic latitude $b$. In such simple model, the galactic absorption is identical in the Northern and Southern galactic hemispheres, so that the modulus of $b$ is employed in the following formulae. If $\tau_{90}$ is the total optical depth of the absorbing matter toward the galactic pole, the corresponding value in direction $b$ will be

$$\tau_{|b|} = \frac{\tau_{90}}{\sin|b|} = \tau_{90}\text{cosec}|b|$$

(16.26)

so that in magnitudes,

$$m_{90} = m_0 + A_{90}, \quad m_b = m_0 + A_{90}\text{cosec}|b|$$

(16.27)

The practical meaning of Equation 16.27 is as follows: suppose we count galaxies, well distributed over the celestial sphere, at increasingly fainter magnitudes. In a transparent Euclidean universe, their number increases with the volume, while their apparent luminosity $f$ decreases with the square of the distance, so that the cumulative number of galaxies brighter than magnitude $m$ increases with the power 3/2, namely,

$$N(f) = N_0 f^{3/2}, \quad \log N(m) = \log N_0 + 0.6m$$

(16.28)

The galactic absorption will produce a decrease of this expected number equal to $A_{90}$ cosec $|b|$; therefore, from the counts, we ought to be able to derive $A_{90}$, a procedure analogous to Bouguer's line for the air mass at the zenith. Hubble (1936, Bibliography section) was indeed able to derive the classic values of absorption at the galactic poles, $A_{90}(V) = 0.18$ mag, $A_{90}(B) = 0.25$ mag.

Hubble's original values have been revised several times, thanks to better data and taking into account the expansion of the Universe, that decreases the slope of the counts below 0.6 according to the distance of the galaxies. The total galactic extinction and reddening in a given direction have been calibrated using the brightness temperature $T$ of the 21-cm spectral line, which is proportional to the total number of H atoms, $N_H$, in that direction. Knapp and Kerr (1974) found the following relationships for extragalactic objects:

$$N_H = 5.8 \times 10^{21} E(B - V) \text{atoms cm}^{-2}, \quad A(V) = 0.6 \times 10^{-21} N_H \text{mag}$$

Therefore, accurate maps of $T(l, b)$ provide one of the best ways to correct the magnitudes and colors of extragalactic objects for interstellar reddening (see, e.g., Burstein and Heiles, 1984).

As a final remark on interstellar matter, the scattering of radiation by a cloud of interstellar particles will *polarize* the light of stars beyond the cloud. The classical studies of interstellar polarization were performed by Hiltner (1952), showing a fair regularity of the polarization according to the direction in the Milky Way. These observations provided the first evidence of a weak interstellar magnetic field (intensity of few millionths of gauss), orienting in preferential directions the dust particles. See also Cøyne (1987, Bibliography section).

Actually, the polarization of cosmic bodies is not only of interstellar origin. More generally, polarization can be detected from Gamma rays to radio frequencies, from a variety of galactic and extragalactic objects, such as planetary surfaces and minor bodies, binary systems, pulsars, AGN and quasars. For a recent review of this challenging subject, requiring sophisticated technologies and theories, see Mignani et al. (2019, Bibliography section). The case of Solar System bodies will be examined in more depth in the next section.

## 16.10   EXTENSION TO THE BODIES OF THE SOLAR SYSTEM

The previous considerations are not entirely suitable for Solar System objects, which reflect the solar light. Here, we shall not consider the far-IR (thermal) radiation coming from the bodies nor the fluorescent light from comets (see Chapter 17). Furthermore, the AU, not the parsec, is the suitable distance unit; moreover, the distance of the objects from the Sun and from the Earth is greatly variable with their ephemerides.

Let us indicate with $r$ the distance of the body from the Sun and with $\Delta$ its distance from the Earth, both in AU. The absolute magnitude $H(1,1)$ is defined as the observed one, when the body is at unit distance from Sun and Earth (the observer) at the opposition; the quantity $H(1,1)$ can be derived from the apparent magnitude at the opposition:

$$H(1,1) = m(r, \Delta) - 5 \log r \cdot \Delta \tag{16.29}$$

Or else, if the body is a fully illuminated perfectly reflecting circular disk of diameter $2s$ and $m_\odot$ is the apparent magnitude of the Sun ($m_\odot = -26.74$ in the visual band),

$$H(1,1) = m_\odot - 5 \log s(\text{AU}) = 14.14 - 5 \log s(\text{km}) \tag{16.30}$$

For instance, the largest body between Mars and Jupiter, namely, Ceres, has a diameter of 945 km and an average distance from the Sun of 2.75 AU. According to Equation 16.30, it would have $H(1,1) \approx 0.80$. However, a body will reflect only a fraction of the light coming from the Sun, according to its surface characteristics (oceans, clouds, rocky or icy terrain, dust-covered land, etc.). Indeed, Ceres has $H(1,1) = +3.4$, consistent with a reflectivity of 10%.

More precise considerations are therefore in order. For instance, the body, as seen from the Earth, might be not entirely illuminated. The observed magnitude will therefore depend on both the surface properties of the body and the observing geometry. The *terminator* is the line dividing the sunlit hemisphere from the shadowed regions. Let $\alpha$ be the so-called *phase angle*, namely, the angle between the Sun and the Earth as seen from the body ($\alpha = 0°$ at opposition, $180°$ at conjunction) and $\varphi(\alpha)$ the *phase function*, describing the fraction of the illuminated disk normalized to unity at zero phase.

If the observations are carried out from space, e.g., from Rosetta, $\alpha$ becomes the angle between the Sun and the spacecraft as seen from the observed object and it can assume values larger than those observable by the terrestrial observer.

For instance, if $\varphi(\alpha)$ is simply proportional to the area, then

$$\varphi(\alpha) = \frac{1 + \cos\alpha}{2} \tag{16.31}$$

All real bodies differ from this simple relationship. In particular, the observations prove that the Moon and most asteroids become decidedly brighter at zero phase angle than the value extrapolated from larger angles (the so-called *opposition effect*). This effect is partly due to the complex scattering law from the terrain (coherent backscattering) and partly to the influence of superficial roughness. The shadows indeed influence the surface luminosity, and they disappear at zero phase angle. To describe in more detail the photometric properties of illuminated bodies, we introduce two definitions of the *albedo*:

- the (bolometric) *geometric* albedo $p$ is the ratio between the total brightness of the body at $\alpha = 0°$ and that of a perfectly diffusing *disk* having the same diameter of the body and at the same distance from the Sun. Inserting the phase angle $\alpha$ and the albedo $p$, Equations 16.29 and 16.30 become, at opposition (null phase angle),

$$m(r, \Delta, 0) = m_\odot + 5\log r \cdot \Delta - 5\log s(\text{AU}) - 2.5\log p$$

(notice the coefficient 2.5 in front of $\log p$) and then

$$\log 2s(\text{km}) = -0.2H(1,1) - 0.5\log p + 3.12$$

At a generic phase angle,

$$m(r, \Delta, \alpha) = H(1,1) + 5\log r \cdot \Delta - 2.5\log p\varphi(\alpha)$$

For the Moon, $p = 0.12$; for Venus, $p = 0.46$; for Jupiter and Io, $p = 0.45$; for the carbonaceous-type asteroids (C-type), $p = 0.05$; for the silicate-rich asteroids (S-type), $p = 0.20$, and so on;
- the (bolometric) *Bond* albedo. Consider the integral of $\varphi(\alpha)$ over the entire solid angle, the so-called phase integral is

$$q = \frac{1}{\pi} \int\limits_{4\pi}^{0} \varphi(\alpha)\,d\omega = 2\int\limits_{0}^{\pi/2} \varphi(\alpha)\sin\alpha\,d\alpha$$

For a perfectly diffusing disk, $q = 1$, for a perfectly diffusing sphere, $q = 1.5$ (Lambert's law), for a metallic sphere, $q = 4$ and so on. The product $A_B = pq$ is called Bond albedo. It expresses the ratio between the total light diffused from the sphere and the total light illuminating it. For the Moon, $A_B = 0.07$, for Ceres $A_B = 0.035$, for Venus and Jupiter $A_B = 0.7$ and so on.

Furthermore, the reflectivity of the different bodies is wavelength dependent in different manners. For instance, the colors of the asteroids are usually redder than the solar ones; in addition, the reflected light is polarized. Figure 16.11 shows as example the photometry of the asteroid 21 Lutetia, as observed by the OSIRIS camera aboard Rosetta from phase angle 0° to 160° (a, adapted from Masoumzadeh et al., 2015) and its polarization from phase angle 0° to 30° from ground observations (b, adapted from Gil-Hutton, 2007).

Although the UBV photometric system has found wide application for asteroidal studies, more suitable band passes have been identified, for instance, the eight-band ECAS photometry (see the "Notes" section).

As a final remark, asteroids rotate, and thus, the viewing geometry from Earth can appreciably change at different oppositions; furthermore, several objects are unresolved binary systems, with a further complication of the light curve.

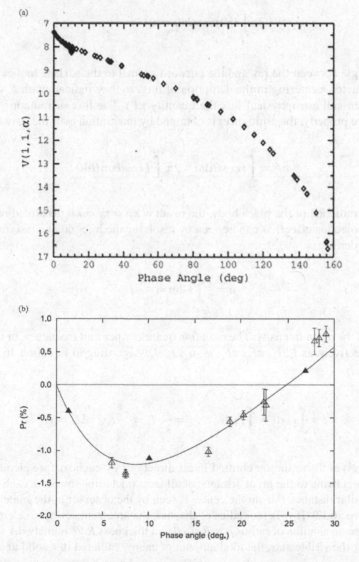

**FIGURE 16.11** Photometry and polarimetry of asteroid 21 Lutetia. (a) Photometry observed by the OSIRIS camera from phase angle 0° to 160° (Adapted from Masoumzadeh et al., 2015, with permission by Elsevier); (b) polarization from phase angle 0° to 30° from ground observations (Adapted from Gil-Hutton, 2007, reproduced with permission @ ESO).

## 16.11   RADIATION QUANTITIES

This section presents a more formal definition of the photometric quantities used until now. The specific intensity, or simply intensity, $I$ is the flux of radiation at a given point in a given direction across the unit surface normal to that direction per unit time and per unit solid angle. In general, the intensity depends on two angular variables $(\theta, \varphi)$, with $0 \leq \theta \leq \pi$, $0 \leq \varphi \leq 2\pi$. For simplicity, we shall assume azimuthal symmetry around $\varphi$. Integrating over the solid angle $d\omega = \sin\theta\, d\theta\, d\varphi$, we obtain the flux of radiation through the unit surface or flux density $\pi F$:

$$\pi F = \int_{4\pi}^{0} I\cos\theta\, d\omega = 2\pi \int_{0}^{\pi} I(\theta)\cos\theta\sin\theta\, d\theta$$

$$(16.32)$$

$$= 2\pi \int_{-1}^{1} I(\theta)\cos\theta\, d(\cos\theta)$$

where $\theta$ is the angle between the ray and the outward normal to the surface (notice several authors do not have the factor $\pi$ entering in the definition of flux, so they indicate with $F$ what is here $\pi F$; some authors even call astrophysical flux the quantity $\pi F$). The flux of radiation emitted from a unit surface (more properly, the emittance) is obtained by integration over the outward hemisphere:

$$\pi F = \int_{2\pi}^{0} I\cos\theta\, d\omega = 2\pi \int_{0}^{\pi/2} I\cos\theta\sin\theta\, d\theta \qquad (16.33)$$

For the isotropic radiation of the black body, the result is $\pi F = \pi I = \pi B$ (as already stated, $\pi$ not $2\pi$, because of the projection effect, as can be seen by resolving the integral at constant $I$).

The radiation density is

$$u = \frac{1}{c} \int_{4\pi}^{0} I\, d\omega = \frac{4\pi}{c} J \qquad (16.34)$$

where $J$ indicates the mean intensity. The radiation quantities per unit frequency, or unit wavelength, are written, respectively, as $I_\nu$, $I_\lambda$, $\pi F_\nu$, $\pi F_\lambda$, $u_\nu$, $u_\lambda$, $J_\nu$, $J_\lambda$. According to Equation 16.1, the following relations apply:

$$I = \int_{0}^{\infty} I_\nu\, d\nu = \int_{0}^{\infty} I_\lambda\, d\lambda, \quad I_\lambda = \frac{c}{\lambda^2} I_\nu = \frac{\nu^2}{c} I_\nu, \quad \lambda I_\lambda = \nu I_\nu$$

The observer receives the radiation emitted in all directions by each surface element of the hemisphere facing him. Owing to the great distance of all stars (including the Sun), each surface element located at an angular distance $\theta$ from the center is seen by the observer at the same angle $\theta$ from its normal (see Figure 16.12). If $R$ is the radius of the star, the area connecting all elements having the same $\theta$ is therefore an annulus of radius $\rho = R\sin\theta$ and thickness $R d\theta$, namely $dA = 2\pi R\sin\theta \cdot R d\theta$.

Summing over the visible star, the total amount of energy radiated in a solid angle $d\omega$ is

$$E_\lambda = 2\pi R^2 d\omega \int_{0}^{\pi/2} I_\lambda(\theta)\cos\theta\sin\theta\, d\theta$$

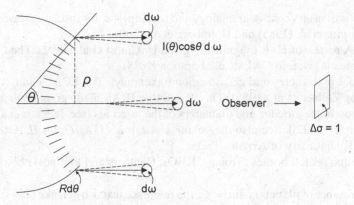

**FIGURE 16.12**  The stellar radiating hemisphere. The observer at infinity receives the flux coming from all directions $\theta$ between 0 and $\pi/2$; the dashed region indicates the area with $\theta$ between $(\theta, \theta + d\theta)$.

Let us take, as solid angle $d\omega$, a unit area $\Delta\sigma = 1$ at the observer, distant $d$ from the star. Therefore, $d\omega = 1/d^2$. If the space is perfectly transparent and the emission isotropic, all the energy $E_\lambda$ will flow through $\Delta\sigma$, so that the observed flux $f_\lambda$ will be

$$f_\lambda = I_\lambda \frac{\pi R^2}{d^2} = \omega I_\lambda \qquad (16.35)$$

In other words, the observed spectrum gives the energy distribution of the outwards flux; the star behaves like a disk of surface $\pi R^2$ subtending a solid angle $\omega$ (for instance, the Sun viewed from the Earth subtends a solid angle $\omega = 6.8 \times 10^{-5}$ sr). Equation 16.35 is valid even if the emissivity is non-isotropic, provided $I_\lambda$ is meant as an average intensity.

These last considerations, in particular Equation 16.35, have shown a remarkable degree of symmetry between source and observer, namely, the quantity $\sigma_\lambda = f_\lambda/\omega$, that we can call surface brightness, stays constant and equal to $I_\lambda$, independent of the distance. The surface brightness of the Sun is the same at Mercury, at the Earth and at Pluto. This result, of distance-independent surface brightness, also applies to diffuse nebulae or even to distant galaxies, provided the interstellar absorption and the metrics of the expanding Universe for objects at cosmologic distance are properly taken into account.

## NOTES

- Programs to digitize the archives of photographic plates are carried out in several observatories, see Web Sites section. The References section quote two papers: Omizzolo et al. (2005) and Nesci et al. (2007), based on digitized plates of the Vatican and Asiago Observatories archives, respectively.
- References to a wealth of photometric systems are given by Mermilliod et al. (see the Web Sites section). See also Moro and Munari (2000) and Fiorucci and Munari (2003).
- For a general discussion of light detectors, see Rieke (2002, Bibliography section). SPADs are described by Cova et al. (2004).
- The extinction coefficients for Mauna Kea are taken from the Gemini website: https://www.gemini.edu/sciops/telescopes-and-sites/observing-condition-constraints/extinction
- For the air mass, night sky, artificial lights and zodiacal light, see, for instance, Kasten and Young (1989) and Garstang (1989 and 1991). See also the site of the International Dark Sky association: http://www.darksky.org/index.html.
- The great percentage of dust in the zodiacal cloud originates from comets of the Jupiter family (Nesvorný et al., 2010). For early measurements from space of zodiacal light, see Torr et al. (1979); Reach (1997, Bibliography section).

- For the calibration of Vega, solar analogs and the Sun, see, for instance, Hayes and Latham (1975), Campins et al. (1985) and Holmberg et al. (2006).
- A pre-GAIA review of H–R diagrams is given by Chiosi et al. (1992). The GAIA data on H–R diagrams is given in GAIA Collaboration (2018).
- For asteroidal photometry and classification (taxonomy). The ECAS eight-color system is described by Zellner et al. (1985). The IR satellite IRAS made great contributions to the determination of the albedos and diameters of the asteroids (see Tedesco et al. (1992), and Bus and Binzel (2002)). See also the volumes: *Asteroids I, Asteroids II, Asteroids III* and *Asteroids IV*, University of Arizona Press.
- For comets and related bodies (Trojans, KBOs, Centaurs and Plutinos) photometry: Jewitt (2015).
- For the reflectance of planetary surfaces, the reference text is by Hapke (1993, Bibliography section). Hapke's methods are the basis of data interpretation for illuminated bodies.

## EXERCISES

1. Calculate the (U–B, B–V) color indices $c$ of a gray body, namely, of a body having $F_U = F_B = F_V$. By definition:

$$U - B = c_{UB}^0 - 2.5 \log \frac{F_U}{F_B}, \quad B - V = c_{BV}^0 - 2.5 \log \frac{F_B}{F_U}$$

Using Vega's calibration:

$$F_U/F_B = 4.4/7.2, \quad F_B/F_V = 7.2/3.7, \quad c_{UB}^0 = -0.53, \quad c_{BV}^0 = +0.72,$$

which are therefore the color indices of a flat (gray) emission spectrum.

2. Calculate the photon flux collected in the visual band by a CCD at the Cassegrain focus of a 2-m telescope from a star of apparent magnitude $V$ at the zenith.

   The effective area of a telescope with a primary mirror of radius $r$ and area occultation $\varepsilon^2$ by the secondary is $A_{eff} = \pi r^2 (1 - \varepsilon^2)$. Usually, $\varepsilon^2$ is of the order of 10%, so that the effective area of the 2-m telescope is approximately $28,000 \, cm^2$. The overall reflectivity (two mirrors) in the visual band is around $0.9^2 \approx 0.8$; the CCD quantum efficiency for a good device is around 0.8; the sky transparency at the zenith is around 0.8; the filter transmissivity is about 0.9.

   If the bandpass of the filter is $800 \, \text{Å}$, we finally obtain the detection rate:

$$N_{ph} \approx 1080 \times 28,000 \times 0.8 \times 0.8 \times 0.9 \times 800 \times 10^{-0.4V}$$

$$\approx 1.39 \times 10^{10} \times 10^{-0.4V} \text{ counts/s}$$

Integrating for $\Delta t$ seconds, the total number of detected photons is

$$N_{ph} \cdot \Delta t \approx 1.39 \times 10^{10} \times 10^{-0.4V} \Delta t.$$

For an ideal (noise-free) detector, the signal-to-noise (S/N) ratio is limited by Poisson's statistics, according to which the noise is simply the square root of the signal:

$$\frac{S}{N} = \frac{N_{ph} \cdot \Delta t}{\sqrt{N_{ph} \cdot \Delta t}} = \sqrt{N_{ph}} \cdot \sqrt{\Delta t}$$

The value of S/N must be at least 5 for a sure detection and larger than 50 for a good photometry. Notice that in this ideal case, the value of S/N increases with the square root of the area of the telescope and of the exposure time. In practice, however, the detector is not noise free; moreover, the atmospheric sky brightness contributes an additional signal, which is an added noise and which varies from site to site. In a dark observatory, a typical value of the sky brightness is 1 star of $V = +21.0$ per $arcsec^2$.

The background brightness is present even in observations from space, because there are always a number of unresolved stars and galaxies inside each pixel. To give some numbers, at low galactic latitudes the sky background is equivalent to 250 stars of apparent magnitude $V = +10$ per square degree, at high galactic latitudes to 35 such stars. In different units, this galactic background is equivalent to one star of $V = +21.8/arcsec^2$ and of $V = +23.9/arcsec^2$, or to $N_{ph} \approx 2.0 \times 10^{-2} m^{-2} s^{-1} A^{-1} arcsec^{-2}$ and to $N_{ph} \approx 2.8 \times 10^{-3} m^{-2} s^{-1} A^{-1} arcsec^{-2}$, respectively. Furthermore, at low ecliptic latitude there is a background of solar light scattered by the interplanetary dust known as zodiacal light (and more specifically, *gegenschein* at 180° from the Sun).

3. Calculate the magnitude and color indices of an unresolved binary star.

    Suppose the two stars have the following characteristics:

    star A,   $V_A = +4.40$,   $(U - B)_A = +0.05$,   $(B - V)_A = +0.58$

    star B,   $V_B = +5.90$,   $(U - B)_B = +0.47$,   $(B - V)_B = +0.89$

    From the general relation $f_{A+B} = f_A \left(1 + 10^{-0.4(m_B - m_A)}\right)$, we have

    $V_{A+B} = +4.40 - 0.24 = +4.16$,   $B_{A+B} = +4.98 - 0.19 = +4.79$,

    $U_{A+B} = +5.03 - 0.13 = +4.90$,   $(U - B)_{A+B} = +0.11$,   $(B - V)_{A+B} = +0.63$

    This calculation can be extended to an arbitrary number of unresolved stars, e.g., a distant globular cluster or a distant galaxy.

4. Given that the interstellar absorption is larger in U than in B and in V, discuss why $E(U-B)$ is only 72% of $E(B-V)$.

5. Calculate the equilibrium temperature of a planet of radius $R$ at a distance $d$ from the Sun. The total energy radiated from the Sun is:

    $$E_\odot = \sigma \cdot T_\odot^4 \cdot 4\pi R_\odot^2$$

    the flux at $d$ (AU) is

    $$f = \frac{\sigma \cdot T_\odot^4 4\pi R_\odot^2}{4\pi \cdot d^2} = \sigma T_\odot^4 (R_\odot/d)^2$$

    and the cross section of the planet is $\pi R^2$. If $A$ is the Bond albedo, the fraction $(1 - A)$ $f\pi R^2$ will be absorbed to heat the planet. Furthermore, we assume that the planet rotates fast enough to distribute the heat over its entire body, so that the temperature is the same over the entire surface. Therefore, we have

    $$\sigma \cdot T^4 = \frac{\sigma \cdot T_\odot^4}{4\pi \cdot d^2} \frac{4\pi \cdot R_\odot^2}{4\pi \cdot R^2} \cdot \pi \cdot R^2 (1 - A),$$

    $$T = T_\odot \left[\left(\frac{R_\odot}{d}\right)^2 \frac{1 - A}{4}\right]^{\frac{1}{4}} = 4.1 \times 10^3 (1 - A)^{1/4} \sqrt{\frac{R_\odot}{d}}$$

    In conclusion, the equilibrium temperature of a planet does not depend on its radius, but only (weakly) on its albedo. For instance, for the Earth,

$\sqrt{R_\odot/d} \approx 0.068$, $\quad A = 0.39$, $\quad (1-A)^{1/4} = 0.88$, $\quad T = 248\,\text{K}$. With the same reasoning, we would find for Jupiter, $T = 111$ K.

For both the Earth and Jupiter, these values are lower than the measured ones, hinting at more complex physical situations. Internal heat sources are required in addition to the solar one. The Earth's internal heat comes from a hot core and radioactive decay of elements in the crust and that of Jupiter, most likely from phase transition in the metallic $H_2$ core. All other gaseous planets have considerable internal sources.

# 17 Elements of Astronomical Spectroscopy

Spectroscopy is a most powerful way to study the heavenly bodies, from planets to comets, asteroids, stars, nebulae and distant extragalactic objects. Among the many steps leading to contemporary spectroscopy, we recall the first observations made by Newton of the dispersive power of prisms. In 1814, the German physicist Joseph von Fraunhofer discovered dark bands in the solar spectrum. In the second half of the 19th century, Foucault, Kirchhoff, Bunsen and others identified some dark bands with those produced by known chemical elements. In 1868, Angstrom produced a very accurate solar spectrum. Father A. Secchi, S.J. of the Specola Vaticana, around 1870 pioneered the classification of stellar spectra by their colors and presence of dark and bright bands. By visual observations made in 1864, the Italian astronomer G. Donati observed for the first time three emission bands in comet Tempel 1864 II. Few years later, Huggins associated the bands seen by Donati to emissions from the molecule of carbon ($C_2$, those bands are known as *Swan bands*); in 1881, he obtained the first photographic spectrum of a comet. It was then possible to identify other prominent emissions, due to other transitions of the $C_2$ molecule, to the molecular radical CN in the violet (388.3 nm), to atomic sodium Na in the yellow (588.6 nm), and so on. In 1882, Rowland obtained diffraction gratings of excellent quality, providing a spectral resolution and precision in wavelength much superior to those given by prisms.

From the spectrum, one obtains the temperature, pressure, chemical composition, radial velocity, rotational status, possible multiplicity, radiation emission or scattering mechanisms, and the presence of electric and magnetic fields. By means of suitable calibration, fundamental stellar quantities such as total luminosity, mass, radius and age can be determined. Different theoretical approaches and techniques will be needed for the different objects and the different spectral bands. In this chapter, we examine in some detail the stellar case in the extended visible band from the near UV to the near IR. Spectra of other celestial bodies (planets, asteroids, nebulae, active galaxies, etc.) will be shown and discussed. The extended visible spectral region is very rich in absorption lines and bands, thus providing good classification criteria. The spectral regions accessible from space, namely the gamma, X and UV bands, and the IR from about 5 μm to the sub-mm, also provide very useful information. Moreover, for Solar System bodies, satellites give the possibility to perform mass spectroscopy and even geochemical analysis of their matter. These topics cannot be covered here, if not for brief mentions. The Web Sites section provides information on a selection of such space instruments.

In addition to observational facts, simple theoretical arguments will be expounded to help with the interpretation and classification of spectra of normal and peculiar stars, assuming basic knowledge of atomic and molecular spectroscopy. Atomic lines are produced by radiative transitions among the quantized internal energy levels; each line has a frequency corresponding to the difference between the energies of the two levels. For molecular lines and bands, the situation is more complex: in addition to electronic energy levels, the molecule has quantized degrees of freedom associated with its vibrations and rotations. As a matter of specific notations, astronomers indicate the ionization stages of atoms with a Roman numeral +1; for instance, H I is neutral hydrogen, H II then indicates the proton, neutral helium is indicated with He I, singly ionized helium with He II, two-times ionized carbon with C III, and so on. Astronomical spectra often display lines of very high ionization stages, corresponding to temperatures nearly impossible to reach in the laboratory, for instance, Fe XV or Ni XVI in the solar corona, whose temperature exceeds millions of degrees. For molecules, the notation with "+" is used; for example, ionized water seen in comets is indicated with $H_2O^+$. Forbidden lines (namely, those arising from metastable states via magnetic dipole or electric quadrupole transitions) are indicated by the wavelength of that transition followed by the symbol of the element within square brackets, e.g., $\lambda5577$ [O I], $\lambda5007$ [O III]. Intermediate cases, named semi-forbidden transitions, are also encountered and denoted with an open square bracket at the end, [. These transitions, possible in astrophysical conditions, are often impossible to reproduce in the laboratory. Indeed, forbidden lines, first observed in celestial sources such as planetary nebulae, were initially associated with an unknown element called *nebulium*. Only afterwards, the lines were assigned to the correct element (see the "Notes" section).

## 17.1  SPECTROSCOPIC TECHNIQUES

In the visible, near UV and near IR, spectra are generally obtained by means of spectrographs having as basic dispersive elements either prisms or gratings (or a combination of the two, the so-called *grisms*). Very schematically, the light focused on the focal plane of the telescope is intercepted by an aperture, usually by a rectangular one, therefore a *slit* (in other cases, a round aperture may be more convenient). The function of the slit is to limit the solid angle of the accepted beam. In addition, its jaws provide a convenient reflective surface on which the star and the surrounding field can be seen. As discussed in Chapter 11, the Earth's atmosphere produces a dispersion of the stellar light in the vertical plane; therefore, it is advisable to align the slit of the spectrograph in that plane, to avoid systematic chromatic errors. After the slit, the light goes through a collimator, namely, a lens system or a mirror, which produces a parallel beam, which in its turn illuminates the dispersive element (or elements). Finally, another complex lens system (camera) produces on the detector a series of quasi-monochromatic images of the illuminated portion of the slit (see an example in Figure 17.1).

In some cases, the dispersive element is placed directly in parallel light before the entrance pupil of the telescope; this is, for instance, the case of the objective prisms used with wide-field astrographs or Schmidt telescopes. In this manner, all stars in the field produce simultaneously a low-resolution spectrum. In other cases, the focal plane of the telescope is equipped with a number of adjacent movable slits or optical fibers, in order to obtain simultaneous spectra of dozens of stars. Finally, the dispersion of light can be realized by entirely different means, e.g., by Fourier spectroscopy in the near IR.

In contemporary astronomical practice, the spectrum is examined and displayed with the wavelength increasing to the right. In addition to the covered spectral range, two characteristics of the spectrum are usually given, namely, the *reciprocal dispersion D* (in A/mm, often improperly called dispersion) and the *spectral resolution* $R = \lambda/\Delta\lambda$, which are determined by the slit width, the dispersive element, the optical configuration and the detector pixel size. If the dispersive device is a prism, both $D$ and $R$ are wavelength dependent. In astronomical applications, two contrasting elements often need some kind of compromise: it would be desirable to obtain the maximum theoretical resolution by keeping the slit width at a minimum; however, the seeing conditions often require widening the slit to collect as much light as possible. The dimensions of the slit on the telescope focal plane (the height, namely, the dimension perpendicular to the width) are usually given in

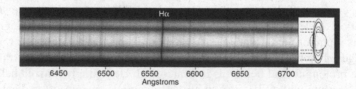

**FIGURE 17.1**   In this spectrum of Saturn covering the Hα region, the slit went through the disk and the rings, as indicated on the right side; the inclination of the spectral lines is due to the rotation of the disk and of the rings. (CCD spectrum obtained with the 1.2m Galileo telescope at Asiago Observatory).

arcseconds. The dimensions of the slit on the detector are determined by the ratio of the focal length of the collimator to that of the camera (briefly, the *collimator/camera ratio*) and, if the case, by the anamorphic magnification of the diffraction grating.

In the past, when the detector was the photographic emulsion, the height of the spectrum used to be enlarged by moving the star up and down along the slit during the exposure time. This technique facilitated the examination of the spectral features against the grain of the emulsion, but it considerably increased the telescope time. Nor it could be applied to extended objects such as comets, planets, nebulae and galaxies; otherwise, all the spatial information along the slit would have been lost. Solid-state detectors have largely overcome this problem. In all cases, the spectral lines are the quasi-monochromatic images of the illuminated slit in the different wavelengths.

Restricting our discussion to the extended visible region, the spectra of normal stars have two components, i.e. a continuum, which can be approximated by a black body emission curve, with over-imposed atomic lines in absorption. Figure 17.2 shows an example, namely, the intensity calibrated spectrum of the blue (hot) star Vega (α Lyr), characterized by prominent absorption lines of the Balmer series of H. The crowding of lines continues until the 10,000 K black body like continuum is abruptly interrupted around 3640 Å by the Balmer discontinuity.

The photosphere of the Sun displays a continuum at a temperature of approximately 5800 K plus atomic and molecular absorption features. In cooler stars, molecular absorption bands are more prominent than atomic features. The variety of cases will be shown by several figures in the following paragraphs. Notice that, even considering normal stars, the behavior of the continuum and the atomic and molecular features present in their spectra depend not only on the spectral band, but also on the spatial region entering the spectrograph, e.g. in the solar case, the sunspot, or the chromosphere, or the corona.

There are important cases of celestial bodies with atomic emission lines (peculiar stars, diffuse and planetary nebulae, active galaxies, quasars, etc.) and/or molecular emission bands, e.g., cometary comae. Figure 17.3 shows three photographic prismatic spectra of Halley's periodic comet observed at three dates in 1985 and 1986. Several molecular bands of CN, $C_3$, $C_2$ (Swan's band), $NH_2$ and atomic lines are visible in emission. The continuum is solar light scattered by cometary dust particles. Notice that the strength of the molecular bands decreases on the blue side, with sharp red edges. In other cases, the shape of the molecular bands decreases to the red, with sharp blue

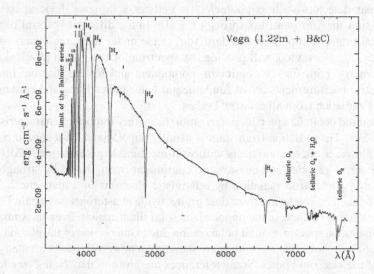

**FIGURE 17.2** Intensity calibrated CCD spectrum of the blue star Vega (α Lyr), from the Balmer discontinuity in the near UV to the near IR. The strongest telluric absorption features in the near IR are also indicated. (Spectrum Asiago Observatory, courtesy P. Ochner and A. Siviero).

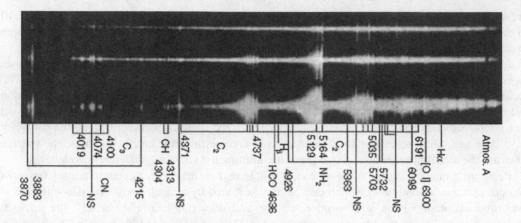

**FIGURE 17.3** Photographic prismatic spectra of Comet Halley observed at three dates in 1985 and 1986, reproduced in positive (emissions bright, absorptions dark). The strongest emission lines in the night sky of artificial origin are indicated with NS. (Spectra by CB, Asiago Observatory).

edges. As in Figure 17.1, the spatial resolution of the cometary spectra in Figure 17.3 is perpendicular to the dispersion.

A word of clarification: on the photographic plates, the light darkens the grains of the emulsion, so that the bright continuum appears dark and the absorption lines appear clear on the originals (negative); the situation reverses on positive prints such as in Figure 17.3. On solid-state devices, the absorptions appear as dips of the continuum (the absorptions remove energy from the continuum), and the emissions are bright (they add energy to the continuum), as in Figure 17.2.

In other objects, only the continuum is visible, for instance, in hot white dwarfs, while at the other extreme, the continuum is so faint that only emission lines are readily visible, e.g., in some planetary nebulae or active galactic nuclei. As an example, Figure 17.4 shows the spectrum of the planetary nebula M57 (Ring Nebula). The ordinate is in log scale, in order to bring out the faintest features.

Utilizing linear detectors such as photoelectric cells or solid-state devices, one can measure with great precision the luminous flux through the slit, in the different spectral bands. The measurement of a calibrated comparison standard star, made under the same observation conditions (sky transparency, air mass, etc.), will provide the spectrum of the star in physical units, namely, its spectrophotometry, both for the continuum component and for the discrete lines and bands. Spectrophotometric measurements are of fundamental value in verifying and calibrating the theoretical models of radiation from all celestial bodies.

As already pointed out in Chapter 16, polarization is another important characteristic of the light from cosmic bodies. The radiation from stars is usually unpolarized, as it should be for the black body emission. However, there are objects with well-measurable polarization, of different origins. For instance, in active galaxies and quasars the continuum is originated by strongly non-thermal processes, namely, synchrotron radiation by relativistic electrons in a magnetic field. In comets, polarization is induced by scattering from dust grains while in asteroids (see again Figure 16.11) by reflection from a slanted surface, of the unpolarized solar illumination. Even the sunspots, observed in high spatial resolution spectra, exhibit polarization due to interaction of the plasma with localized magnetic fields. Spectropolarimetric studies are therefore a powerful means for studying the physical conditions of the celestial bodies. Some references are given in the "Notes" section, in addition to those of Chapter 16.

Figure 17.5 shows the spectra of two extragalactic objects with strong emission lines. Left panel shows the uncalibrated spectrum of the active galaxy Markarian 259, dominated by the strong

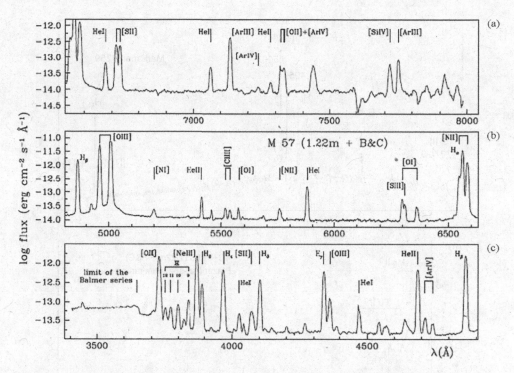

**FIGURE 17.4** Intensity calibrated CCD spectrum of the planetary nebula M57 (Ring Nebula) from the Balmer limit (c) to the near IR (a). The ordinate is in log scale, in order to bring out the faintest features. (Spectrum Asiago Observatory, courtesy P. Ochner and A. Siviero).

emission lines of the HI Balmer series and of the forbidden transitions of [O II], [O III] and [N II]. Right panel shows the intensity calibrated spectrum of the blazar 4C 31.63 (namely, a blazing quasistellar object), with conspicuous emissions of ionized iron and oxygen.

## 17.2   THE ANALYSIS OF THE SPECTRAL LINES

The first step in the analysis of spectral lines (here assumed in absorption) is the measurement of their wavelengths. Then, for each line one proceeds to the identification of the responsible element, of its excitation and ionization stage, and to its assignment to a given series of energy transitions. This procedure is not always possible, even today several lines of the solar spectrum defy identification. Two most useful quantities can be obtained for each line, namely, the *equivalent area* and the *equivalent width*. Suppose we have a well-calibrated spectrum, with reciprocal dispersion high enough to discern strong and weak absorption lines. We can measure with precision the line profile $F_\lambda$, namely, the intensity along the curve extending from $\lambda_1$ to $\lambda_2$, $\lambda_0$ being the wavelength at the center (see Figure 17.6). The continuum has been arbitrarily normalized to 1.

The total strength of the line is specified by the so-called *equivalent area* $A_{\lambda 0}$, defined as

$$A_{\lambda 0} = \int_{\lambda_1}^{\lambda_2} (1 - F_\lambda)\, d\lambda \qquad (17.1)$$

a quantity which is a measure of the energy subtracted by the line to the continuum. The *equivalent width* $W_\lambda$ of the line is the width in angstrom of a rectangle having the same total integrated energy as the true line. Notice that the equivalent width alone does not discriminate between a shallow wide line and a narrow deep one, so that it is advisable to indicate also the full

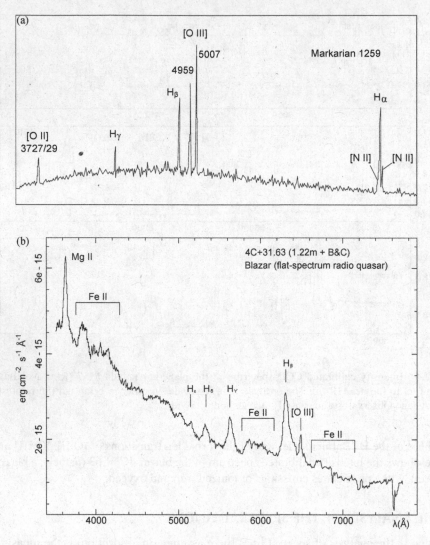

**FIGURE 17.5**   (a) Uncalibrated spectrum of the active galaxy Markarian 1259. (b) Intensity calibrated spectrum of the blazar 4C 31.63. (Spectra Asiago Observatory by CB and courtesy of P. Ochner and A. Siviero).

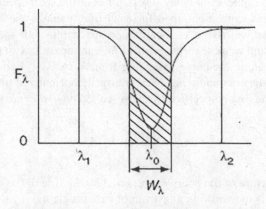

**FIGURE 17.6**   The line profile, equivalent area, and equivalent width. On the line profile, we distinguish between the center (core) and the wings.

width half maximum (FWHM), namely, the width (usually in Angstrom) of the line at half its intensity. The value of $W_\lambda$ and FWHM can critically depend on the choice of the continuum, because sometimes its position can be uncertain. This difficulty is especially important in very high-resolution spectra, where faint lines may be present in the extended wings of the profile of the strong line under investigation. Therefore, great caution is needed in the quantitative interpretation of the spectra.

On the theoretical side, the equivalent width depends on the number of atoms $N$ in the atmosphere capable of absorbing that transition (number of atoms in the present context means the total number of those atoms in the column of unit cross section through the stellar atmosphere). Therefore, it depends on the temperature and density (or pressure) inside the gas and on the chemical composition of the star. Furthermore, the strength of a given line depends on the probability of that transition, expressed by a so-called *f-factor*, or *oscillator strength*, described in a later section. The relationship between the line intensity and the number of atoms, or better the product $N \cdot f$, is far from linear. In addition, it depends on the broadening mechanisms affecting the shape and width of the line.

Suppose, for instance, that the only broadening mechanism is the natural width of the line, namely, the width dictated by Heisenberg's uncertainty principle. The absorptivity of the gas will be high at the center of the line, while it will drop to almost zero at a small distance from it. Therefore, a small number of atoms will produce an absorption at the center of the line, and almost no wings. By adding increasing numbers of atoms, the center of the line will deepen and finally saturate, while the wings will rapidly grow. As a consequence, the equivalent width regime will pass from a linear dependence on $N \cdot f$ at low numbers, to a much slower one, essentially $W = \sqrt{N \cdot f}$, at higher column densities.

Collisions between atoms and molecules will also broaden the line, because the internal energy levels of the particle and so the wavelengths of the corresponding transitions will be altered. In most stellar types, the broadening due to collisions is much more important than that due to the natural width; however, the resulting profile is the same as natural broadening (except for H and He, as explained later). The H and K lines of Ca II in the solar atmosphere provide a good example of collisionally broadened line profile.

Another important cause of broadening is thermal agitation, namely, the superposition of the Doppler shifts due to the different velocities of each atom or molecule. In this case, at low gas densities the line is broad and not very deep. By adding successively more atoms, the linear regime continues for a longer interval than in the case of collisional broadening, until eventually the intensity saturates; no matter how many more atoms are added, $W_\lambda$ will stay constant.

The superposition of natural, collisional and Doppler broadenings determines the overall behavior of $W_\lambda$ with $N \cdot f$, namely, the so-called *curve of growth*. The previous oversimplified discussion has shown that this curve will be composed of three parts (see Figure 17.7):

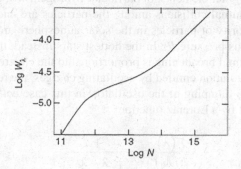

**FIGURE 17.7** Idealized behavior of the curve of growth in a log–log diagram.

1. at low $N \cdot f$, the intensity will be proportional to $N \cdot f$;
2. for intermediate values of $N \cdot f$, the center of the line is deep, but the wings are relatively unimportant; the intensity will remain almost constant for a range of $N \cdot f$;
3. for very large values of $N \cdot f$, the intensity will increase with $\sqrt{N \cdot f}$.

The crossover between the three regimes depends on the relative importance of the three mechanisms; Doppler broadening is sensitive only to temperature, but collisions are also sensitive to density (or pressure). In reality, other broadening mechanisms must be considered, for instance, the Stark effect, which is so important for H and He lines that the previous considerations on the curve of growth are incapable of explaining the observed profiles. Similar difficult problems are caused by the Zeeman effect in sunspots or magnetic stars (see later for an explanation of the two effects). On the other hand, stellar rotation causes a characteristic broadening of the profiles, but does not affect the curve of growth.

Therefore, the precise theory has considerable complexity, due to the several factors that contribute to the width of the line. Heisenberg's uncertainty principle prevents a line from being strictly monochromatic; moreover, the hyperfine structure adds to the natural width, but these factors usually do not contribute more than a few mÅ in the visible. Macroscopic effects are much more important. Let us consider in more detail the thermal Doppler broadening. In a gas in thermal equilibrium with temperature $T$, each particle of mass $m$ acquires a kinetic energy given by

$$\frac{1}{2}mv^2 = \frac{3}{2}kT \tag{17.2}$$

where v is the velocity vector modulus. Maxwell's function (see also Equation 12.27) gives the velocity distribution function of particles of mass $m$ according to the temperature $T$:

$$dN(v) = \frac{N}{\sqrt{\pi}} e^{-\left(v^2/2kT\right)} \frac{dv}{v} = 4\pi N \left(\frac{m}{2\pi kT}\right)^{3/2} v^2 e^{-mv^2/2kT} dv \tag{17.3}$$

Maxwell's distribution is not always appropriate to astrophysical conditions; for instance, highly collimated jets can be described as gases with three different temperatures along the three principal axes. By virtue of the Doppler effect, each particle of mass $m$ absorbs at larger or smaller wavelengths according to its velocity, so that the line profile will be described as a Gaussian function with FWHM given by

$$\text{FWHM} = 7.16 \times 10^{-7} \lambda \sqrt{\frac{T}{m}} \tag{17.4}$$

Therefore, the lines of the lighter elements, in particular hydrogen, will be the widest.

As already discussed, mutual collisions among the particles are another cause of broadening. For instance, the great majority of particles in the solar atmosphere are HI atoms, whose density is proportional to the gaseous pressure $P_g$. In the hottest stars instead, the fastest particles are free electrons, so that the collisional broadening is proportional to the electron pressure $P_e$.

In the classic theory of radiation emitted by oscillating charges, the resulting alteration of wavelength can be interpreted as damping of the oscillator, in this case *collisional damping*. The line profile can be approximated by a Lorentz function:

$$\frac{F(\lambda)}{F(\lambda_0)} = \frac{1}{1 + \left(\dfrac{\lambda - \lambda_0}{\text{FWHM}}\right)^2} \tag{17.5}$$

While the Doppler broadening is more efficient at the center of the line, the collisional damping contributes more to the wings.

A third effect, the *Stark broadening*, is due to the microscopic electric fields caused by free charges passing near the radiating atom or ion. These short-lived fields of random direction and amplitude are sufficient to perturb the energy levels (especially the outer ones) of the atom or ion. The effect is particularly important for hydrogen; it contributes to the different aspect of the lines between stars of different radii and atmospheric densities (namely, between dwarfs, giants and supergiants, as defined later). In dwarfs, the density is higher, so that the Stark effect is stronger. In giants and supergiants, the density is decidedly lower, and the broadening is essentially Doppler. The broadening of the H I lines increases as $n^2$ (where $n$ is the principal quantum number): the higher lines of the Balmer series become progressively broader until they finally merge with one another. The last visible line on a high-resolution spectrum is related to the electron density $N_e$ ($cm^{-3}$) by the Inglis–Teller formula:

$$\log N_e = 22.0 - 7.0 n_{last}$$

The lines of helium and other heavier elements exhibit a more complex behavior, showing the so-called *quadratic* Stark effect. Detailed calculations have been given, for instance, by Vidal et al. (1973).

In the Sun, high spatial resolution allows the detection of macroscopic electric fields well-ordered over lengths of several hundred kilometers, although of short temporal duration. Such fields are present in particular, in the proximity of dark spots when a magnetic polarity inversion produces a flare. The spectral lines then separate into several components by virtue of this macroscopic Stark effect; the different components are differently polarized.

A macroscopic magnetic field produces the *Zeeman effect*, well observable in the Sun especially in the proximity of dark spots. The lines of particular metals are easily seen resolved in different components having different polarization. In a generic star, the varying direction of the magnetic field along the line of sight usually prevents the resolution of the different components, so that the Zeeman effect is seen as a broadening of the line. However, in peculiar stars, white dwarfs, pulsars and so on, the magnetic field can be very strong, reaching intensities of several hundred Megagauss and even Gigagauss, allowing the separation of the individual components.

Finally, the broadening can be due to ordered movements of the atmosphere (currents), to turbulence, or more simply to the rotation of the entire star around an axis with arbitrary inclination $i$ to the line of sight. The generally unknown inclination allows the measurement of $v_{rot} \sin i$. Several blue stars have rotational velocities exceeding 300 km/s, while stars cooler than the spectral type F0, for instance, the Sun, are much slower rotators. The slow rotation of the Sun implies that the total angular momentum of the Solar System is dominated by the planetary revolutions, while its mass is dominated by that of the Sun.

Around 1960, a pattern of radial velocity shifts was discovered in the solar atmosphere, with small amplitude and quasi-periodic variations. The strongest wave has a period of about $5^m$, but there is a whole spectrum of such oscillations. It was soon realized that this pattern of waves was due to oscillations of the entire solar structure. This interpretation gave rise to a new method to investigate the inner solar structure by means of waves analogous to geoseismology, namely, to *helioseismology*. This refined method of spectrophotometric analysis was soon extended to other stars, a discipline named *asteroseismology*. For helioseismology, see, for instance, Thompson et al. (2003); for asteroseismology, see, for instance, Aerts et al. (2010), both in Bibliography section.

## 17.3 DETAILED BALANCE AND THE BOLTZMANN EQUATION

In a closed system, the so-called *detailed balance* condition may be fulfilled: every transition is balanced by the opposite process; for example, the emission of a photon of energy $hv$ by a jump from state B to state A is balanced by the absorption of a photon of the same energy in a transition from A to B. The medium is then in a condition of thermodynamic equilibrium, at least locally, a condition referred to as local thermodynamic equilibrium (LTE): the population of the energy states depends only on the temperature, while the density of the medium affects only the processes rates. Therefore, from Kirchhoff's first law, at each place the ratio between volume emissivity and linear absorption coefficient, namely, the so-called *source function*, is a function of temperature only:

$$S_v = j_v/k_v \tag{17.6}$$

In the following, $S_v(T)$ will be identified with Planck's radiation function. In reality, there are astrophysical situations where this identification is not even approximately legitimate, but we do not examine them here.

Consider the detailed balance between two states, $n$ (upper) and $m$ (lower), separated by an energy $\Delta E_{nm} = h\nu$. Downward transitions (emissions) can take place by either spontaneous or stimulated emission. A radiation density $u_v$ (see Equation 16.34) produces stimulated emission at a rate $N_n \cdot u_v \cdot B_{nm}$, where $N_n$ is the number of atoms per cubic centimeter in the $n$th level (the photons produced by stimulated emissions are coherent with the incident electromagnetic field; this is the basic process of masers and lasers). Spontaneous emissions occur at a rate $N_n \cdot A_{nm}$ even in the absence of the radiation field. Upward transitions (absorptions), stimulated by the incident radiation field, occur at a rate $N_m \cdot u_v \cdot B_{mn}$. The coefficients $A_{nm}$, $B_{mn}$ are determined by the detailed balance condition:

$$N_n \left( A_{nm} + u_v B_{nm} \right) = N_m \cdot B_{mn} \cdot u_v$$

In many astronomical applications, stimulated emission is negligible, so that

$$N_n \cdot A_{nm} \approx N_m \cdot B_{mn} \cdot u_v$$

In a gas composed of a given element (either neutral or ionized), in thermodynamic equilibrium, the relative population of two energy levels $n, m$ separated by energy $\Delta E_{mn}$ is given by the Boltzmann equation:

$$\frac{N_n}{N_m} = \frac{g_n}{g_m} e^{-\Delta E_{mn}/kT} = \frac{g_n}{g_m} e^{-h\nu/kT} \tag{17.7}$$

where the terms $g_n$ and $g_m$ are the *statistical weights* of the two levels, namely, the number of indistinguishable states having the same energy in each level, or else the number of electrons that can occupy that level without violating the Pauli exclusion principle. Einstein showed that, if the populations are given by Equation 17.7, then the radiation density $u_v$ must be

$$u_v = \frac{4\pi}{c} B_v = \frac{8\pi h\nu^3}{c^3} \frac{1}{e^{h\nu/kT} - 1}$$

where $B_v$ is the Planck function. Furthermore, the transition probabilities are related by

$$A_{nm} = \frac{8\pi h\nu^3}{c^3} B_{nm} = \frac{8\pi h\nu^3}{c^3} \frac{g_m}{g_n} B_{mn}, \qquad g_m B_{mn} = g_n B_{nm} = \frac{8\pi^3}{3h^2} S_{mn} \tag{17.8}$$

where $S_{mn} = S_{nm}$ is the so-called *line-strength* (electric dipole for permitted lines, or magnetic dipole or electric quadrupole for forbidden lines). In the astronomical literature, instead of $S_{nm}$ the quantity $g \cdot f$ (named weighted *oscillator strength*) is usually found:

$$gf = g_m f_{mn} = -g_n f_{nm} = \frac{8\pi^2 m_e \nu}{3he^2} S_{mn}, \qquad A_{nm} = \text{const} \cdot \frac{gf}{g_n \lambda^2}$$

where $m_e$ and $e$ are, respectively, mass and electric charge of the electron; the value of the numerical constant depends on the adopted unit system.

Let us consider again the Boltzmann formula 17.7 applied to the case of H I. Each energy level $i$ is referred to the ground level ($n = 1$):

$$\frac{N_i}{N_1} = \frac{g_i}{g_1} e^{-E_i/kT}, \qquad i = 2,3,\dots$$

where $E_i$ is the excitation potential of the $ith$ level above the ground state. By summing over all $i$'s, we obtain the total population of that atom:

$$N = \sum N_i = \frac{N_1}{g_1}\left[\frac{g_2}{g_1} e^{-E_2/kT} + \frac{g_3}{g_1} e^{-E_3/kT} + \cdots\right] = \frac{N_1}{g_1} U(T) \tag{17.9}$$

where $U(T)$ is called the *partition function*. Owing to the facts that $E_i$ tends to a limit and that $g_i \propto i^2$, this series seems to be unbound. However, this apparent mathematical divergence is physically removed by the collisions with the nearby atoms (Bohr's radii increase with $i^2$). The correction for the highest terms, which depends on pressure and temperature, could be applied by artificially lowering the ionization potential by a small amount.

With appropriate methods, the correct partition function $U(T)$ can be calculated for H I and for all astrophysically important atoms and ions (see, for instance, Gray, 1976).

Table 17.1 gives approximate values of $U(T)$ for a few elements, for two temperature values, both neutral and first ionized. In most cases, $U \approx g_1$ independent of $T$, but there are exceptions, such as Na I, Ca I and Ca II, Fe I and Fe II.

The Boltzmann formula can be written in a more practical way. By expressing $\Delta E_{mn}$ in electronvolts (eV), letting $\Delta E_{mn}(\text{eV}) = \chi_{mn}$, using powers of 10 and introducing the usual notation in the astronomical literature $\Theta = 5040/T$, the law can be written as

$$\frac{N_n}{N_m} = \frac{g_n}{g_m} 10^{-\Delta E_{mn}(\text{eV})\, 5040/T}, \qquad \log\frac{N_n}{N_m} = \log\frac{g_n}{g_m} - \chi_{mn} \cdot \Theta$$

## TABLE 17.1
## Partition Functions of Selected Ions

| Element | $g_1$ | Neutral | | | $g_1$ | First Ionization | |
| | | $U$ | | | | $U$ | |
| | | $T = 5000$ | $T = 10,000$ | | | $T = 5000$ | $T = 10,000$ |
|---------|-------|-----------|-------------|---|-------|-----------|-------------|
| H  | 2  | 1.995  | 1.995  | 1  | 1.000  | 1.000  |
| He | 1  | 1.000  | 1.000  | 2  | 1.995  | 1.995  |
| C  | 9  | 9.913  | 10.000 | 6  | 6.026  | 6.026  |
| N  | 4  | 4.074  | 4.571  | 9  | 8.913  | 9.332  |
| O  | 9  | 8.710  | 9.333  | 4  | 3.981  | 4.074  |
| Ne | 1  | 1.000  | 1.000  | 6  | 5.370  | 5.623  |
| Na | 2  | 2.042  | 3.981  | 1  | 1.00   | 1.00   |
| Ca | 1  | 1.174  | 3.548  | 2  | 2.188  | 3.467  |
| Fe | 25 | 26.915 | 54.954 | 30 | 42.658 | 63.096 |

## 17.4   THE SAHA EQUATION

The ratio between the partition functions of ion plus electron and of neutral atom, in other words the ionization balance in thermal equilibrium, is given by Saha's equation:

$$\frac{N_{i+1}}{N_i} N_e = \frac{2u_{i+1}}{u_i} \frac{(2\pi m)^{3/2} (kT)^{3/2}}{h^3} e^{-E_i/kT} = 4.83 \times 10^{15} T^{3/2} \frac{g_{i+1}}{g_i} e^{-E_i/kT} \qquad (17.10)$$

where $E_i$ is the ionization potential of the ground state of the $i$th ion.

The term $u_e = 2(2\pi mkT)^{3/2}/h^3$ represents the partition function of the free electron, namely, its density of states in phase space; the factor 2 accounts for the two possible spin states of the electron.

The Saha equation does not have the same degree of general validity as does the Boltzmann equation, because ionization can also be caused by non-LTE phenomena, e.g., by collision with a stream of particles having a preferred direction. Therefore, in specific cases Saha's equation does not represent the true ionization degree of the plasma.

Table 17.2 shows the ionization potentials $\chi_i$ (expressed in eV) for the same elements of Table 17.1.

Furthermore, let us call $U$ the partition function of the ion and use the variable $\Theta = 5040/T$. In logarithmic form (base 10), Saha equation becomes

$$\log\left(\frac{N_{i+1}}{N_i} N_e\right) = -\chi_i \cdot \Theta - \frac{3}{2}\log\Theta + 20.9366 + \log\frac{2U_{i+1}}{U_i} \qquad (17.11)$$

Making use of the perfect gas law applied to the electronic component $P_e = N_e kT$, the equation can be written as

$$\log\left(\frac{N_{i+1}}{N_i} P_e\right) = -\chi_i \cdot \Theta + \frac{5}{2}\log T - 0.4772 + \log\frac{2U_{i+1}}{U_i} \qquad (17.12)$$

The pressure is expressed in dyn/cm² (1 dyn/cm² = $10^{-6}$ bar = 0.1 Pa).

An example of the Saha equation is given in Figure 17.8, which refers to conditions of temperature and pressure appropriate to the solar photosphere and corona. In the photosphere, all calcium atoms remain neutral until the temperature reaches 3000 K; above such value of $T$, a fraction begins to be singly ionized. At 4000 K, Ca I and Ca II are equally abundant; at 5000 K, all calcium becomes singly ionized; at 7000 K, Ca II and Ca III are in the same proportion, and so on. In the corona, the temperature rises above millions of K, the pressure is extremely low, and iron and other heavy elements become multiply ionized.

---

### TABLE 17.2
### Ionization Potentials (eV) of Selected Elements

| Element | I | II | III | IV | V | VI | VII | VIII | IX |
|---------|-------|-------|-------|-------|--------|--------|--------|--------|--------|
| H       | 13.60 |       |       |       |        |        |        |        |        |
| He      | 24.59 | 54.42 |       |       |        |        |        |        |        |
| C       | 11.26 | 24.38 | 47.89 | 64.49 | 392.08 | 489.98 |        |        |        |
| N       | 14.53 | 29.60 | 47.45 | 77.47 | 97.89  | 552.06 | 667.03 |        |        |
| O       | 13.62 | 35.12 | 54.93 | 77.41 | 113.90 | 138.12 | 739.32 | 871.30 |        |
| Ne      | 21.56 | 40.96 | 63.45 | 97.11 | 126.21 | 157.93 | 207.26 | 239.09 | 1195.8 |
| Ca      | 6.11  | 11.87 | 50.91 | 67.15 | 84.43  | 108.78 | 127.70 | 147.40 | 188.70 |
| Fe      | 7.87  | 16.16 | 30.65 | 54.80 | 75.50  | 100.00 | 128.30 | 151.12 | 235.00 |

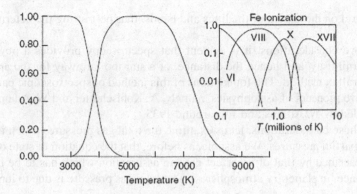

**FIGURE 17.8**   Ionization equilibrium between neutral and ionized calcium at solar photospheric conditions of temperature and pressure. The inset shows the situation of iron in the solar corona, where the temperature exceeds $1 \times 10^6$ K.

The Saha equation shows that at constant temperature, the degree of ionization will depend on the electron pressure $P_e$. Therefore, let us consider two stars having equal masses and effective temperatures, but very different radii, for instance, a dwarf and a giant. The ionization will be lower in the dwarf. Conversely, the dwarf will need a higher temperature to reach the same ionization as the giant. This fact introduced a severe complication in the earlier interpretation of the stellar spectra, which was clarified by the work of Russell, Fowler and Milne.

Table 17.3 shows the electron pressure $P_e$ (in barye) in the atmospheres of dwarf, giant and supergiant stars.

From the above discussion regarding the Boltzmann and Saha equations, we have concluded that the intensities of the absorption lines depend on two fundamental variables, namely, temperature and electron pressure (all other variables, such as chemical composition, magnetic fields and rotation, are ignored in this context). This fact implies that the lines of several elements, having different ionization potentials, must be simultaneously taken into account for proper classification of a stellar spectrum.

Instead of electron pressure, we can use the intrinsic luminosity (or the absolute magnitude) as second variable. The stellar mass is indeed a fairly stable parameter for stars of the same temperature; therefore, two stars having the same temperature but different radii will have very different densities and surface pressures and different mechanisms of interaction between atoms and molecules. The consequence is that the spectral lines and bands of a giant star have a different appearance from that of a star of the same temperature but smaller radius, a difference that can be easily noticed even on spectra of moderate spectral resolution. For instance, the interatomic Stark effect produces, in the hotter stars, Balmer lines that are narrower in giants than in dwarfs; in the cooler stars, the blue band of CN is stronger in giants than in dwarfs. Therefore, a proper classification of stellar spectra must show the influence of both temperature and luminosity (or equivalently of the radius, surface

**TABLE 17.3**

**Typical Stellar Electron Pressures $P_e$**

| | | Electron Pressure $P_e$ (barye) | | |
|---|---|---|---|---|
| $T$ (K) | $\Theta$ | Dwarfs | Giants | Supergiants |
| 10,080 | 0.5 | 320 | 100 | 33 |
| 5040 | 1.0 | 4 | 0.8 | 0.2 |
| 3360 | 1.5 | 0.57 | 0.02 | 0.004 |

gravity or pressure) on the aspect of the lines and bands, thus permitting the determination of both variables.

From the above considerations, it is evident that spectroscopy provides a powerful means to determine the luminosity, and hence the distance, of a star too far away for the application of the trigonometric parallax method. The foundations of this method of spectroscopic parallaxes go back to the works of two pioneers of astrophysics, namely, A. Kohlschütter and W. Adams in 1914. Their work was expanded by W. Adams and Joy, around 1935.

To complete these considerations, let us examine the total gas pressure, which is the sum of the ion and electron partial pressures. We assume, as before, that the equation of state of a stellar atmosphere can be described by that of a perfect gas (an assumption which has to be taken with great caution, for instance, in planetary atmospheres). The total gas pressure is due to ions and electrons:

$$P_{gas} = (N_i + N_e)kT \tag{17.13}$$

where $N_i$ is summed over all ionization stages (including neutral). In a stellar atmosphere, all matter is essentially composed of hydrogen and helium, with traces of other elements collectively named "metals" (even if their chemical properties are in general not metallic) and indicated with Z. Although their density is low, metals can contribute an appreciable number of free electrons, so that

$$N_i = N(\mathrm{H\,I}) + N(\mathrm{H\,II}) + N(\mathrm{He\,I}) + N(\mathrm{He\,II}) + N(\mathrm{Z\,I} + \mathrm{Z\,II} + \mathrm{Z\,III} + \cdots)$$

$$N_e = N(\mathrm{H\,II}) + N(\mathrm{He\,II}) + N(\mathrm{Z\,II} + \mathrm{Z\,III} + \cdots)$$

summed over all metals ionization states.

Two limiting regimes can be considered in this respect:

- in hot stars ($T > 10^4$ K), hydrogen is totally ionized, so that $N_e \approx N_i$, $P_e/P_{gas} \approx 1/2$, independent of the precise chemical composition of the atmosphere;
- in cool stars ($T < 5 \times 10^3$ K), hydrogen is neutral, but metals are ionized, so that $N_e = N$ (Z II+ Z III+ ...); the ratio $P_e/P_{gas} = N$ (Z II+ Z III+...)/[$N$ (H I)+$N$(He I)+$N$(Z I)+$N$(Z II+ Z III+...)] is much smaller than 1 and sensitive to the chemical composition. In precise atmospheric models, the contribution of molecules must be included.

Therefore, the Sun is in an intermediate situation. The degree of ionization of an average stellar atmosphere (an average between a metal rich and a metal poor composition) is given in Table 17.4, relating gas pressure, electron pressure and temperature. As expected, at constant electron pressure $P_e$ the gas pressure $P_g$ varies by many orders of magnitude with $T$.

## TABLE 17.4
## Ionization Degree of an Average Stellar Atmosphere

| | $\Theta = 0.2$ | $\Theta = 0.6$ | $\Theta = 1.0$ | $\Theta = 1.4$ |
|---|---|---|---|---|
| $\mathrm{Log}_{10} P_e$ | ($T = 25{,}200$) | ($T = 8400$) | ($T = 5040$) | ($T = 3600$) |
| −2 | −1.8 | −1.67 | +0.78 | +2.4 |
| 0 | +0.29 | +0.35 | 3.9 | 5.3 |
| 2 | 2.30 | 2.98 | 6.7 | 8.5 |
| 4 | 4.31 | 6.84 | 10.0 | 12.4 |

The general result, namely, that the ionization increases with the decrease of density, applies also to the interstellar medium (ISM). Lines of highly ionized elements (e.g., C IV) are observed in the ISM even if the temperature of the gas is extremely low; the reason for this is that, once an element is ionized by the absorption of a UV photon, the probability of recombination is essentially nil because of the very low density.

Saha's equation can be extended to molecules. Consider for simplicity a diatomic molecule AB, composed by two atoms A and B. The dissociation rate is governed by a similar equation:

$$N_A N_B / N_{AB} = \frac{U_A U_B}{Q_{AB}} \left( \frac{2\pi m_{AB} kT}{h^2} \right)^{3/2} e^{-D/kT} \tag{17.14}$$

where $m_{AB} = m_A m_B / (m_A + m_B)$ is the reduced mass of the molecule, $U_A$ and $U_B$ are the partition functions of the two atoms, $D$ is the dissociation energy and $Q_{AB} = Q_{rot} \cdot Q_{vib} \cdot Q_{el}$ is the molecular partition function. Data for some astrophysically relevant molecules are provided in Tables 17.5 and 17.6.

Numerically, expressing $m$ in atomic mass units (amu), $D$ in eV, $N$ in cm$^{-3}$:

$$\log \frac{N_A N_B}{N_{AB}} = 20.2735 + \frac{3}{2} \log m_{AB} + \frac{3}{2} \log T - D/\Theta + \log \frac{U_A U_B}{Q_{AB}} \tag{17.15}$$

Equation 17.14 can be considered a particular case of the general law of mass action found by Guldberg and Wage.

## TABLE 17.5
### Diatomic Molecules

| Name | Dissociation Potential (eV) | Ionization Potential (eV) | Equilibrium Distance (Å) |
|------|------|------|------|
| $H_2$ | 4.5 | 15.4 | 0.74 |
| $C_2$ | 6.3 | 12.2 | 1.24 |
| CH | 3.5 | 10.6 | 1.12 |
| CO | 11.1 | 14.0 | 1.13 |
| CN | 7.8 | 14.2 | 1.17 |
| $O_2$ | 5.1 | 12.1 | 1.21 |
| OH | 4.4 | 12.9 | 0.97 |

## TABLE 17.6
### Polyatomic Molecules

| Name | Dissociation Potential (eV) | Ionization Potential (eV) | Equilibrium Distance (Å) |
|------|------|------|------|
| $H_2O$ | 5.1 | 12.6 | 3.5 |
| $N_2O$ | 1.7 | 12.9 | 4.0 |
| $CO_2$ | 5.3 | 13.8 | 3.8 |
| $NH_3$ | 4.3 | 10.2 | 3.0 |
| $CH_4$ | 4.4 | 13.0 | 3.5 |

## 17.5   SPECTRAL CLASSIFICATION OF STARS AND THE ABUNDANCE OF THE ELEMENTS

In order to devise a coherent scheme of spectral classification of stellar spectra, namely, to assign stars to a particular spectral type SpT, the generally adopted criteria are the behavior of the continuum with the wavelength (in other words, the color of the star as a measure of its surface temperature) and the relative intensities of absorption lines or bands. Each scheme is strongly dependent on the spectral resolution or on the spectral response of the available equipment. As it occurred for laboratory spectroscopy, many classification schemes were devised well in advance of the physical understanding of the stellar atmospheres and interiors.

In each classification, stars having well-known properties are used as a template, against which the spectrum of each particular star can be examined. Then, the empirical classification is calibrated in terms of the fundamental physical quantities temperature $T$ and luminosity $L$ (by virtue of the Stefan–Boltzmann law, luminosity and radius play essentially the same role). Other variables can be included in the scheme, e.g., the chemical composition $\mu$, the magnetic field $\mathbf{H}$ and the rotation, but in order to do so, finer observations are needed. The book by Jaschek and Jaschek (see Bibliography section) provides additional details.

Therefore, each particular classification scheme is at least bi-dimensional: $SpT = SpT(T, L) = SpT(T, R)$. Let us examine the influence of the temperature, $T$, which ranges for normal stars from approximately 50,000 down to 1000 K, by ordering the stars as a function of their color, from the bluest to the reddest. We easily see that the appearance of the absorption lines and bands changes appreciably. For example, consider the behavior of the Balmer series of hydrogen going from the hotter to the cooler stars: the intensity of the lines increases with the decrease of $T$ from 50,000 K to reach a maximum at 10,000 K and decreases again to almost disappear around 3500 K. This behavior is a temperature, not a chemical composition, effect.

Indeed, all stars have essentially the same composition. Hydrogen and helium are by far the dominant elements; if we put conventionally equal to 1 the number of oxygen atoms per unit volume, then the standard mix $\mu$ is

$$\mu = 1600\,\mathrm{H} + 160\,\mathrm{He} + 1\,\mathrm{O} + 0.5\,\mathrm{N} + 0.3\,\mathrm{C}$$

$$+ 0.2\,\mathrm{Ne} + 0.1\,\mathrm{Fe} + 0.06\,\mathrm{Mg} + \cdots$$

Compositional differences from star to star are essentially due to the heavier elements, to the point that all elements other than H and He are called "metals", as already pointed out. The composition can be given by number of atoms, but also by mass (see Table 17.7). In theoretical applications, the percentages by mass are indicated by $X$ (hydrogen), $Y$ (helium) and $Z$ (all the rest), with $X + Y + Z = 1$, as in Table 17.8.

It is plainly evident that the standard stellar composition is very different from that of the Earth, Mercury, the Moon or Mars. Indeed, considering that the average density of the Sun is so much

**TABLE 17.7**

**Schematic Standard Mix of Chemical Elements, by Number and by Mass**

| Element | Number | Mass | Electrons |
|---|---|---|---|
| H | 100 | 100 | 100 |
| He | 9.8 | 39 | 20 |
| C, N, O, Ne | 0.15 | 2.2 | 1.1 |
| All the rest | 0.01 | 0.4 | 0.21 |

**TABLE 17.8**

**Theoretical Notation of the Standard Mix by Mass**

$X$ (mass fraction of H) = 0.71

$Y$ (mass fraction of He) = 0.27

$Z$ (mass fraction of "metals") = 0.02

lower than that of the Earth, namely, 1.4 against 5.4 g/cm³, we conclude that the Sun's composition must be dominated by the lighter elements and the Earth's by the heavier ones. Jupiter, Saturn and the other giant planets have densities comparable to the solar one. On the other hand, stars dominate the mass of the visible Universe, so that hydrogen and helium are the main universal constituents.

Indeed, following the initial event usually named "Big Bang", when the expanding Universe cooled down and reached temperatures of billions of degrees Kelvin, nucleosynthesis occurred. The rapid cooling allowed the formation of the first elements through fusion reactions involving protons $p$ and neutrons $n$ present in the environment. The main processes were the $p$–$p$ reaction producing deuterium (D) and the D–D reaction producing helium (He3 and He4). In addition, minor channels such as D+He3, $p$+He3 and $n$+He3 were also present. The light elements lithium (Li6 and Li7) and beryllium (Be7) were produced in minute quantities by subsequent processes involving H, He3 and He4. The primordial nucleosynthesis essentially stopped at Li7, because there are no stable elements with atomic mass $A = 5$ or $A = 8$. In essence, around 75% of the mass of primordial ordinary matter (in physical terms, *baryonic* matter) was H, and 25% was He4, with tiny fractions of D, He3, Li7 and Be7. See, for instance, De Angelis and Pimenta (2018, Bibliography section). A first generation of stars produced heavier elements through nuclear fusion processes in their interiors, where temperatures ranged from $10^8$ to $10^6$ K, with heavier stars being hotter. Soon after, the formation of galaxies, their clusters, stars and perhaps planets, was already underway. Such an evolution of structures was accompanied by fast chemical evolution. The explosions of first-generation stars as supernovae (SN) enriched the ISM of heavy elements, so that successive generations of stars had those elements and more complex nuclear reactions at their disposal. To be more precise, there are two basic types of SN:

- Type I SN occur in binary systems with at least one star of approximately one solar mass, a surface temperature of around $10^4$ K and a radius comparable to that of Earth, namely, a so-called white dwarf, the prototype being the faint companion of Sirius A mentioned in Chapter 12. A distinctive feature of Type I SN is the absence of hydrogen lines in their spectra. Type I SN belonging to the subclass "Ia" are the best standard candles with which to measure the expansion of the Universe and provide firm evidence of the existence of the dark energy accelerating the expansion.
- Type II SN show hydrogen lines in their spectra and result from the rapid collapse of a star of eight to fifty solar masses. Type II SN are categorized based on the resulting light curve following the explosion. Type II-L SN show a steady decline of the light curve, whereas Type II-P SN display a period of slower decline (a plateau) in their light curve followed by a normal decay.

The details of the chemical evolution of the Universe are very complex and depend not only on the cosmological epoch but also on the environment where the star or the nebula was located. In a drastic approximation, we can say that after 1 Gy all main chemical elements were present, both in the stars and in the interstellar nebulae. Our knowledge of the overall abundance of the elements in our Solar System originates from three quite different sources: the solar photospheric spectrum, the pristine meteorites, in particular the so-called CI chondrites, and cometary dust. Meteoritic abundances, including their isotopic content, can be measured to exquisite accuracy in the laboratory, with the caveat that elements such as H, He, C, N, O and neon (Ne) are all volatile and hence depleted in meteorites. Table 17.9 (adapted from Asplund et al., 2009, Bibliography section) shows the abundance of selected elements, in a $\log_{10}$ scale, conventionally normalized to 12 for H. The composition is dominated by H, He, C, N, O and so on in decreasing order until the strong peaks of Fe and nickel (Ni) are reached. Cometary dust analyzed in the laboratory (thanks, for instance, to the mass samples retrieved and brought back to Earth by the Stardust mission) and through mass spectroscopy in dedicated *in situ* space mission was proved to be closer to solar abundances being less depleted in light elements with respect to CI chondrites.

As already pointed out, the detailed composition varies with cosmic epoch and ambient, but three general problems are worth quoting. First, there is the problem of deuterium (D), the only stable isotope of H. The Big Bang produced a tiny amount of D, as confirmed, for instance, by the very low D/H ratio in the solar nebula (approximately $10^{-5}$). Nuclear processes in the interior of stars do not produce new D, because it appears only as an intermediate product. However, observations show that the values of D/H in the ISM, planetary atmospheres, terrestrial water, comets, etc. span a factor of more than 50, raising in some cases the value of D/H well above the standard terrestrial value of $1.5 \times 10^{-4}$ for the ocean water (the so-called Vienna model VSMOW). Therefore, enrichment or fractionation processes are required to explain the wide range.

Another puzzle is the under-abundance of the light elements Li, Be and boron (B) with respect to H and He on the one side, and C, N and O on the other. Although Li, Be and B are formed by nuclear fusion in the deep interior of stars, they are also destroyed by other reactions close to the surface.

## TABLE 17.9
## Solar Photospheric Abundance (by Number) of Selected Chemical Elements

| Symbol | Atomic Number | Atomic Weight | Log Abundance |
|--------|--------------|---------------|---------------|
| H | 1 | 1.008 | 12.00 |
| He | 2 | 4.003 | 10.9 |
| Li | 3 | 6.941 | 1.1 |
| Be | 4 | 9.012 | 1.4 |
| B | 5 | 10.811 | 2.7 |
| C | 6 | 12.011 | 8.4 |
| N | 7 | 14.007 | 7.8 |
| O | 8 | 15.999 | 8.7 |
| F | 9 | 18.998 | 4.6 |
| Ne | 10 | 20.179 | 7.9 |
| Na | 11 | 22.9898 | 6.2 |
| Mg | 12 | 24.305 | 7.6 |
| Al | 13 | 26.9815 | 6.5 |
| Si | 14 | 28.086 | 7.5 |
| P | 15 | 30.974 | 5.4 |
| S | 16 | 32.06 | 7.1 |
| Ar | 18 | 39.948 | 6.4 |
| Ca | 20 | 40.08 | 6.3 |
| Mn | 25 | 54.9380 | 5.4 |
| Fe | 26 | 55.847 | 7.5 |
| Ni | 28 | 58.71 | 6.2 |
| Mo | 42 | 114.82 | 1.9 |
| Xe | 54 | 131.30 | 2.2 |

An additional formation mechanism is the bombardment of heavy atoms by high-energy particles called "cosmic rays", capable of splitting them into lighter ones in a process called "spallation".

Until recently, much heavier elements represented another puzzle, because their formation requires the presence of many free neutrons in the ambient. In 2017, the detection of short-duration gamma-ray bursts accompanied by a gravitational wave event, with a transient optical-infrared counterpart, provided an additional formation mechanism: the merging of two neutron stars produced a kilonova (or meganova), whose spectra showed the presence of such heavy elements (see Covino et al., 2017; Pian et al., 2017).

We now describe two widely used classification schemes, devised at Harvard and Yerkes, respectively.

## 17.6   THE HARVARD AND THE MK CLASSIFICATION SCHEMES

At the beginning of the 20th century, E.C. Pickering and coworkers produced the Harvard classification scheme by examining more than 250,000 photographic spectra, obtained with objective prisms from both hemispheres. The stars were brighter than approximately the ninth magnitude. The spectra, covering the blue range from 3300 to 4900 Å, were ordered in a discrete sequence of seven classes having decreasing temperatures. The classes were designated with the following capital letters: O, B, A, F, G, K and M (this succession of letters reflects the many steps used to consolidate the final ordering). The planetary nebulae were indicated with P and the novae with Q, both objects being characterized by strong emission lines. In 1922, the IAU recommended some modifications to the original notation, which are still in use today. For instance, the letter "e" indicates the presence of emission lines, "k" the presence of interstellar lines and "p" a peculiarity. The Harvard catalog, known as HD (Harvard Durchmunsterung, but later also Henry Draper Catalog), was published starting in 1918 (see Cannon and Pickering, Bibliography section). The HD scheme has been thoroughly revised by Houck (1975–1988, Bibliography section).

The HD classification was followed by that produced at Yerkes Observatory by Morgan, Keenan and Kellman (1943, Bibliography section; see also Garrison, 1984), known as MKK, often simply as MK classification. As in the HD system, the Yerkes photographic spectra were obtained with a prismatic dispersion of approximately 100 A/mm at H$\gamma$. The dominant variable temperature $T$ was estimated inside seven intervals, designated with the same capital letters O, B, A, F, G, K and M used in the HD system. Each interval was subdivided into ten finer subclasses with Arabic numbers from 0 to 9, always in the sense of decreasing temperature:

$$\text{MKK SpT} : O4,\ldots,O9; \quad B0,B1,\ldots,B9; \quad A0,A2,\ldots,A9;$$

$$G0,G2,\ldots,G9; \quad K0,K1,\ldots,K9; \quad M0,M1,\ldots,M9$$

Not all subclasses were actually used; for instance, A1 or G1 was lacking, in the sense that the template did not contain any star of that subtype. For other subtypes, a fractional classification was needed, e.g., O9.5 or B0.5.

With respect to Harvard's scheme, Yerkes classification takes into account the absolute luminosity, indicated with the five Roman numerals I, II, III, IV and V. These five luminosity classes are I: supergiants (subdivided into Ia, Ib and Iab), II: bright giants, III: giants, IV: subgiants and V: dwarfs (also called *main sequence* or *zero age sequence* in theoretical works). Unfortunately, the designation "dwarf" can be misleading: do not confuse dwarfs with white dwarfs, which actually are not part of the MKK scheme of luminosities. Indeed, luminosity classes VI and VII, used by several authors to indicate subdwarfs and white dwarfs, do not belong to the original Yerkes scheme.

Therefore, examples of complete SpT designations in the MKK classification are B2 Ia, A0 V, F5 III and so on. The Sun belongs to the MKK SpT G2 V, while in the original Harvard classification, it was a G0 star.

Examples of photographic and charge-coupled device (CCD) spectra of stars classified in the MKK system can be found on the websites indicated in the "Notes" section. Here, we give some of the main criteria of classification.

*Type O* (old Harvard types Pd, Oe, Oe5): blue color, $50,000 \leq T \leq 25,000$ K; strong absorption lines of the Pickering series of He II, whose maximum intensity is reached at O5; a significant percentage of the stars hotter than O4 also display emission lines (see the paragraph on peculiar stars). From O4 to O9, the relative intensities of the absorptions of He II, He I, H, Si IV are used to fix the SpT. He I increases from O4 to O9. At O7, the lines of N III reach their maximum, and C III $\lambda 4647$ begins to appear. These are the rarest of all main-sequence stars.

Typical stars: O5 ζ Pup; O6 λ Cep; O7 S Mon; O8 λ Ori; O9 10 Lac.

*Type B*: blue–white color, $25,000 \leq T \leq 12,000$ K. He II disappears, He I reaches its maximum intensity, and H I gradually increases and is a good criterion of luminosity (as was shown by Abetti, Lindblad and Schalen). The ratios Si III/Si IV and Si II/He I are good indicators of SpT; for instance, the ratio $\lambda 4552$ Si III/$\lambda 4089$ Si IV is <1 at B0 and >1 at B1. As luminosity criteria, the ratio $\lambda 4089$ Si IV/$\lambda 4009$ He I can be used in addition to the Balmer lines. Among the B stars, the appearance of Balmer lines in emission, often inside the broader absorptions, is not infrequent; those stars are indicated as Be.

Typical stars: B0 Ia ε Ori; B0 V δ Sco; B2 III γ Ori; B3 V η UMa; B8 Iab β Ori; B7 V α Leo.

*Type A*: white color, $12,000 \leq T \leq 8000$ K. The transition from B to A for main-sequence stars is estimated by the disappearance of He I lines. The Balmer lines, usually quite broad, reach their maximum in this class. On low- and medium-resolution spectra, the K-line of Ca II at $\lambda 3933$ is clearly visible, while the H-line of Ca II at $\lambda 3970$ is masked by the $H_\varepsilon$ line of the Balmer series, which can be resolved up to the 11th line. The increasing intensity of the metal lines can be used as a temperature indicator; for instance, the Mn II line at $\lambda 4030/34$ starts to be seen at A2. The Balmer lines are indicators of luminosity as they are for the B stars, and so are the Fe II lines, whose intensity increases with $L$. Less than 1% of the main-sequence stars in the solar neighborhood are A-type stars.

Typical stars: A0 III α Dra; A0 IV γ Gem; A0 V α Lyr (see again Figure 17.2); A2 Ia α Cyg; A3 III β Tri; A7 III γ Lyr.

*Type F*: white–yellow color, $8000 \leq T \leq 6000$ K. The spectrum is similar to A-type, although the Balmer lines are weaker and the H, K of Ca II and other metal lines are stronger. The molecular band at $\lambda 4310$ (the so-called G-band, a blend of CH and metal lines) is clearly visible. Several lines of ionized metals are good indicators of luminosity, e.g., Sr $\lambda 4077$ and $\lambda 4215$, Ti II $\lambda 4161$ and $\lambda 4399$, Fe II $\lambda 4233$, which are stronger on the giants. There is a noticeable luminosity effect also in the violet continuum, which is bluer in the dwarfs than in the giants; the effect increases on going to cooler stars and becomes very evident in the K type. About 3% of the main-sequence stars in the solar neighborhood are F-type stars.

Typical stars: F0 Ib α Lep; F0 V γ Vir; F2 IV β Cas; F5 Ib γ Cyg; F5 V β Vir.

*Type G*: yellow color, $6000 \leq T \leq 5000$ K. Spectrum similar to the solar one. Strong metal lines; in particular, H, K lines of Ca II are stronger than the Balmer lines. The ratio Fe II $\lambda 4325$/Hγ is a good indicator of $T$. The CN violet bands at $\lambda 4216$ and $\lambda 4144$ are good indicators of luminosity, being much stronger in giants than in dwarfs. A similar but weaker effect is shown by the G-band. Class G main-sequence stars make up about 8% of the main-sequence stars in the solar neighborhood.

Typical stars: G0 Ib α Aqr; G0 II α Sag; G0 IV η Boo; G2 V the Sun and solar analogs like 16 Cyg and BD +00-2717, see Figure 17.9; G5 IV μ Her; G8 I ζ Cyg I; G8 III δ Boo; G8 IV β Aql; G8 V ζ Boo A.

*Type K*: yellow–orange color, $5000 \leq T \leq 4000$ K. The blue continuum becomes weaker and weaker, and the spectrum resembles that of the sunspots. The H, K lines of Ca II reach their maximum at K0, while the intensity of the Ca I lines at $\lambda 4227$ and $\lambda 4454$ increases along the sequence. The $T$ criteria are the same used for the G-type; the CN band at $\lambda 4216$ can still be used for $L$. K-type stars make up about 12% of the main-sequence stars in the solar neighborhood.

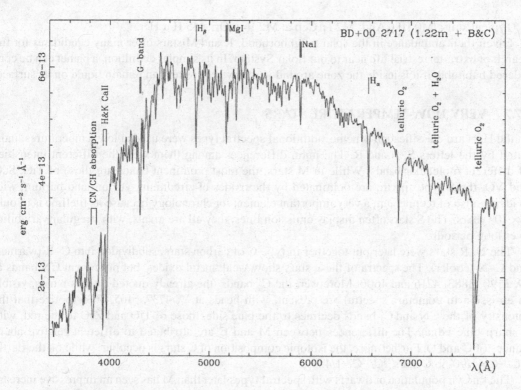

**FIGURE 17.9** The flux calibrated solar analog BD +00-2717. (CCD spectrum taken at the Asiago Observatory, courtesy P. Ochner and A. Siviero).

Typical stars: K0 III ε Cyg; K0 IV η Cep; K1 IV γ Cep; K2 Ib ε Peg; K2 III α Ari; K2 V ε Eri; K3 Ib η Per; K3 Iγ Λql I; K3 III δ And; K5 II ζ Cyg; K5 III α Tau; K5 V 61 Cyg A.

*Type M*: reddish color, $4000 \leq T \leq 2500$ K. The spectrum is dominated by the appearance of molecular bands, typically those of TiO, ScO and VO (see Figure 17.10). The temperature is estimated by the strength of these bands or of the Ca I line λ4227. The Balmer series can still be used for the determination of luminosity. Class M stars are by far the most common. About 76% of the main-sequence stars in the solar neighborhood are class M stars. However, they have such low luminosities that none is bright enough to be seen with the unaided eye (being the brightest known M-class main-sequence star M0V Lacaille 8760, with visual magnitude +6.6).

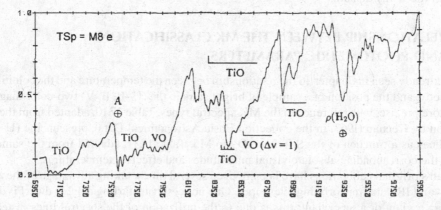

**FIGURE 17.10** Uncalibrated spectrum of a very cold M8e star, with typical oxide bands. The telluric A-band due to $O_2$ and $\rho$-band due to $H_2O$ are also indicated. (Spectrum by CB, Asiago Observatory).

Typical stars: M0 III β And; M2 I μ Cep a; M2 Ib α Ori; M5 II α Her.

Given their abundance in the solar neighborhood, K and M stars offer many candidates for the search of extra-terrestrial life near to our Solar System. In its simpler definition, a planet can be considered habitable if it is inside the zone around its star where water can remain liquid on its surface.

## 17.7   VERY LOW-TEMPERATURE STARS

In the Harvard classification scheme, additional spectral types were used at low temperatures, indicated by the letters S, N and R. The main differences among them were the different intensities of different molecular bands. While in M stars, the most prominent bands are those of TiO, ScO and VO, the S-type spectra are dominated by the oxides of zirconium, yttrium and barium, with evidence also of technetium, a very important element for chronology because its lifetime is of only $2 \times 10^6$ years. The S stars often display emission lines; they all are giants, with irregular variability over long periods.

The N, R stars were later put together in type C of carbon stars, subdivided into C–R (warmer) and C–N (cooler). The spectra of these stars show weak metal oxides, but prominent CN bands at λλ 3590, 3883, 4216 and 4606. Moreover, the $C_2$ bands (the already quoted Swan's bands, visible in emission in cometary spectra) are present, with heads at λλ 4737, 5165, 5635 (notice that the intensity of the CN and $C_2$ bands declines to the blue side, those of TiO and ZrO to the red, with a sharp blue edge). The differences between M and C are attributed to different relative abundances of C and O. Furthermore, the isotopic composition of C stars is peculiar: while on the Earth, $^{13}C/^{12}C \approx 1/90$, in C stars $^{13}C/^{12}C \approx 1/4$.

The known population of dwarfs with spectral types later than M has seen an impressive increase in recent times, thanks to the availability of detectors with good sensitivity in the IR, located in high elevation Observatories like the Japanese 8m Subaru at Mauna Kea or in space telescopes. These very cool and very red dwarfs, intrinsically very faint at optical wavelengths, fall into two new categories: L dwarfs, whose near-IR spectra feature carbon exclusively in the form of CO; and even cooler T dwarfs, whose near-IR spectra show carbon predominantly in methane ($CH_4$). See, for instance, Basri et al. (2000) and Geballe et al. (2002). The L and T spectral types thus make the first substantial addition to the original Harvard and MKK schemes, completing a continuous spectral sequence from the hottest O stars ($T_{eff} \approx 50,000$ K) to the coolest known brown dwarfs ($T_{eff} \approx 750$ K).

In general, the majority of cool stars are variable; a notable example is Mira Ceti (o Cet), the prototype of the Mira Ceti variables, classified as M9e, because for a great part of its light cycle the Balmer series and the lines of Fe II appear in emission. The presence of emission lines in cool stars is also indicative of chromospheric activity, which can be observed not only at optical wavelengths but also in the X-ray domain.

## 17.8   RELATIONSHIP BETWEEN THE MK CLASSIFICATION AND PHOTOMETRIC PARAMETERS

We have already seen in Chapter 16 the relationship between the temperature and the color indices of a black body, and the position of a sample of bright stars in the (U–B, B–V) two-color diagram. We can be more precise, with reference to the MK spectral types. Table 17.10, adapted from the original calibration by Keenan (1963, in the collection Basic Astronomical Data), presents the (U–B, B–V) color indices as a function of the SpT from O9 to M5. Table 17.11, adapted from the same source, provides the corresponding absolute visual magnitudes and effective temperatures.

Keenan's original calibrations have been revised several times, taking advantage of the availability of UV and IR data from orbiting telescopes. One notices that, starting at F2, a dwarf is decidedly hotter than a giant or a supergiant; this is due to the utilization of the spectral lines instead of the continuum as the main criterion for evaluating temperature, as discussed before.

**TABLE 17.10**

**Values of Color Indices (U–B, B–V) as Function of SpT for the Different Luminosity Classes**

| SpT | U–B | B–V | SpT | V U–B | V B–V | III U–B | III B–V | Ib U–B | Ib B–V |
|---|---|---|---|---|---|---|---|---|---|
| O5–V | −1.19 | −0.33 | F0 | +0.03 | +0.30 | | | +0.15 | +0.17 |
| O9–V | −1.12 | −0.31 | F2 | +0.00 | +0.35 | | | +0.18 | +0.23 |
| B0–V | −1.08 | −0.30 | F8 | +0.02 | +0.52 | | | +0.41 | +0.56 |
| B2–V | −0.84 | −0.24 | G2 | +0.12 | +0.63 | | | +0.63 | +0.87 |
| B5–V | −0.58 | −0.17 | G5 | +0.20 | +0.68 | +0.56 | +0.86 | +0.83 | +1.02 |
| B8–V | −0.34 | −0.11 | K0 | +0.45 | +0.81 | +0.84 | +1.00 | +1.17 | +1.25 |
| A0–V | 0.00 | 0.00 | K5 | +1.08 | +1.15 | +1.81 | +1.50 | +1.80 | +1.60 |
| A2–V | +0.05 | +0.05 | M2 | +1.18 | +1.49 | +1.89 | +1.60 | +1.95 | +1.71 |
| A5–V | +0.10 | +0.15 | M5 | +1.24 | +1.64 | +1.58 | +1.63 | +1.60 | +1.80 |

**TABLE 17.11**

**Absolute Visual Magnitudes $M_V$ and Effective Temperatures $T_{eff}$ of the Different Spectral Types and Luminosity Classes**

| TSp | V $M_V$ | V $T_{eff}$ | I $M_V$ | I $T_{eff}$ | TSp | V $M_V$ | V $T_{eff}$ | III $M_V$ | III $T_{eff}$ | I $M_V$ | I $T_{eff}$ |
|---|---|---|---|---|---|---|---|---|---|---|---|
| O5 | −5.7 | 42,000 | | | F2 | +3.5 | 7000 | | | −6.5 | 7000 |
| O9 | −4.5 | 34,000 | −6.5 | 32,000 | F8 | +4.0 | 6300 | | | −6.5 | 5750 |
| B0 | −4.1 | 30,000 | | | G0 | +4.4 | 5900 | | | −6.4 | 5400 |
| B2 | −2.5 | 21,000 | −6.4 | 17,600 | G2 | +4.7 | 5800 | | | −6.3 | 5000 |
| B5 | −1.2 | 15,000 | −6.2 | 13,600 | G5 | +5.2 | 5560 | +0.9 | 5000 | −6.2 | 4900 |
| B8 | −0.3 | 11,500 | −6.2 | 11,000 | K0 | +5.9 | 5150 | +0.7 | 4700 | −6.0 | 4500 |
| A0 | +0.6 | 9,800 | −6.3 | 10,000 | K5 | +7.4 | 4400 | −0.3 | 4100 | −5.8 | 4000 |
| A2 | +1.4 | 9,000 | −6.5 | 9,400 | M0 | +8.8 | 3850 | −0.4 | 3700 | −5.5 | 3600 |
| A5 | +2.0 | 8,200 | −6.6 | 8,600 | M2 | +10.0 | 3550 | −0.6 | 3500 | −5.5 | 3400 |
| F0 | +2.6 | 7,300 | −6.6 | 7,500 | M5 | +12.3 | 3200 | −0.3 | 3400 | −5.5 | 2900 |

## 17.9 SPECTRA OF PECULIAR STARS

Several stars defy a simple classification, requiring additional parameters to describe their spectra properly. For instance, stars with strong lines of Ba, Sr and rare earths are found among the G and K giants. At higher temperatures, between F5 and B5, there are the so-called peculiar A stars (Ap). These stars have a strong and rapidly varying magnetic field, with an associated variability of the spectral lines of metals such as Mn, Eu, Cr, Sr and Si. On the other hand, in the same temperature range, metallic A stars (Am) are found, with no strong magnetic field, under-abundance of Ca and great overabundance of heavy elements and rare earths.

A good percentage of the hotter stars show emission lines of H I, Fe II, Si II and Ti II inside the corresponding absorptions. The explanation is the presence of a thin, rotating disk of matter around the star. The appearance of the spectrum also depends on the inclination of such a disk to the line of sight. The hot star P Cyg displays a remarkable feature, namely, strong narrow emissions accompanied on the blue side by wide absorptions. The stars with these characteristics, obviously named P-Cyg stars, must be surrounded by a large thick expanding envelope. P Cyg itself has shown large variations in brightness over the past centuries; today, it is around the fifth magnitude.

Among the hotter stars, we find the so-called Wolf–Rayet (W stars, the old Harvard types Oa, Ob, Oc), which have $T > 50,000$ K. W stars are very rare in space, but extremely luminous ($M \approx -9$), so that they can be observed even in external galaxies. W stars are also found at the center of planetary nebulae. They are presumably binaries, very old and very massive. Their spectra are characterized by broad emission lines of He I and He II; carbon from C II to C IV; nitrogen from N II to IV; and oxygen from O II to O VI. The W type is subdivided into two subclasses, namely, the WN and the WC. The WN display strong lines of N, and the WC have strong lines of C and O. The lines of N on one hand, and those of C and O on the other, seem to be mutually exclusive.

White dwarfs are indicated by the prefix D in front of the spectral type, e.g., DB and DA. The spectrum of the DB shows essentially only He lines and that of the DA only the Balmer lines. The difference is too large to be attributable only to a temperature effect; a chemical diversity must also play an important role. Furthermore, the masses are significantly different: those of the DBs being around 0.3 and those of the DAs around 0.7 solar masses. There are white dwarfs with essentially continuous, featureless spectra (designates as DCs) and others, where the magnetic field is so large, several Megagauss, that the spectral lines are exceedingly large due to the Zeeman effect.

## 17.10 SPECTRA OF SOLAR SYSTEM OBJECTS

Solar System bodies such as the Moon, Mercury, Mars, the asteroids, the dust component of the cometary comae and tails, the zodiacal light, in the visible region of the spectrum reflect and scatter the solar light, with only slight modifications due to the reflectance from the solid materials of their surfaces.

Madden and Kaltenegger (2018) have published a catalog of reference spectra and geometric albedos for 19 objects representative of the different types of Solar System bodies, from 0.45 to 2.5 μm.

Figure 17.11 shows as examples the reflectance spectra from the blue to the near IR of two so-called near-earth asteroids, NEAs. For clarity, the category near-earth objects (NEOs) comprises both NEAs and near-earth comets (NECs), at moment numbering approximately 22,000 and 200, respectively. The two spectra have been normalized to the solar spectrum by using the intermediate

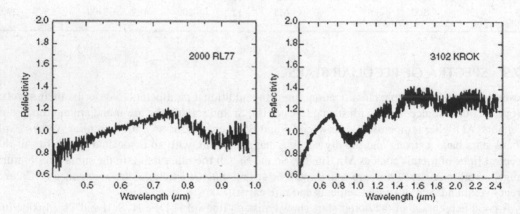

**FIGURE 17.11** Reflectance spectra of two Near Earth Asteroids. (Adapted from Lazzarin et al., 2004, with permission by Elsevier).

of solar analog stars. Notice how different the two spectra are, an indication of different surface properties of the two bodies.

Figure 17.12 shows the spectra of two distant trans-Neptunian object (TNO), namely, 2400 DW on the left panel (Figure 17.12a) and 486958 Arrokoth (2014 MU69) on the right one (Figure 17.12b). At their distance, the Sun is an almost point-like source. The first object has semi-major axis $a = 39.5$ AU, eccentricity $e = 0.218$, inclination $i = 20°.6$ and period $P = 248$ years (notice how similar these orbital elements are to those of Pluto). Its estimated diameter is 1500 km. The spectrum shows a neutral and nearly featureless visible continuum and water ices absorption bands at 1.5 and 2.0 μm (Fornasier et al., 2004). The second object, dubbed Ultima Thule, is the most distant Solar System object visited so far by a spacecraft (the NASA New Horizons mission). Its orbital values are semi-major axis $a = 44.6$ AU, eccentricity $e = 0.05$, inclination $i = 2°.4$ and period $P = 298$ years. Its estimated diameter is about 30 km. The spectrum (Stern et al., 2019) shows tentative absorption bands due to water ice (at 1.5 and 2.0 μm) and methanol (at 2.3 μm).

Venus has the densest atmosphere of the four terrestrial planets, consisting of more than 96% carbon dioxide, 4% molecular hydrogen and less than 0.01% water vapor. The atmospheric pressure at the planet's surface is 92 times that of Earth. Venus is by far the hottest planet in the Solar System, with a mean surface temperature of 735 K. The planet is shrouded by an opaque layer of highly reflective clouds of sulfuric acid ($H_2SO_4$), preventing its surface from being seen in visible light. Noticeable $CO_2$ absorption bands are seen in the near-IR around 1.0, 1.2 and 1.4 μm, while the $H_2O$ bands around 0.9, 1.1 and 1.4 μm so conspicuous in the Earth's reflectance spectra are totally absent.

In low-resolution spectra, Mars and Mercury have a nearly featureless spectrum. Mercury though shows a thin transient escaping atmosphere, well visible from the ground in the light of the Na yellow doublet (see, e.g., Mangano et al., 2009; Leblanc et al., 2008).

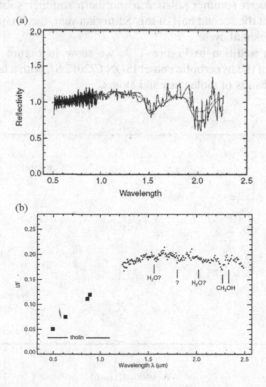

FIGURE 17.12 (a) Reflectance spectrum of the TNO 2004 DW, obtained with the TNG. (Adapted from Fornasier et al., 2004, reproduced with permission @ ESO). (b) Reflectance spectrum of the TNO 486958 Arrokoth (Courtesy A. Stern and W. Gully).

Jupiter and the other giant planets have a very different situation, as their atmospheres produce important absorption bands in the visible region. Molecular bands of ammonia ices ($NH_3$) dominate the spectra of Jupiter and Saturn, while the bands of methane ($CH_4$) ices are the most conspicuous in the spectra of Uranus and Neptune. These methane bands produce a greenish color of Uranus and Neptune when observed by eye with a telescope of modest aperture.

The James Webb Space Telescope (JWST), with its very wide response from the visible to the mid-IR, will be particularly valuable for investigating species that cannot be observed from the ground due to telluric absorption. Thus, great advances in the study of the giant planets are likely to come in the near future. As an example, Figure 17.13, adapted from Noorwood et al. (2015), shows synthetic spectra in the near-IR from 0.8 to 5 μm of the four giant planets, in preparation for observations with the JWST.

In this spectral region, the opacity of Uranus and Neptune atmospheres is dominated by the opacity of methane, which varies over several orders of magnitude with the depth. Also observable in these planets are stratospheric emission bands from $CO_2$, CO, HCN that may have an external origin. Possible sources of these emissions include infalling ring particles, material from satellites, impacts from comets or Kuiper Belt objects.

The operational lifetime of the JWST will also play an important role to cover a fraction of orbital periods of Uranus (84 years) and Neptune (165 years). Uranus will approach solstice in 2030 (northern summer) and Neptune the equinox in 2046 (northern spring), so that important signs of climatic variations may be detectable.

The large angular sizes of Jupiter and Saturn will allow very detailed studies of atmospheric dynamics on these planets. Reflectivity observations that make use of the variable opacity of methane and center-to-limb variations will enable to determine the altitude of the cloud features in storms and vortices. Furthermore, the Cassini mission explored Saturn over half of a Saturnian year, roughly from southern summer solstice to northern summer solstice; JWST's anticipated lifespan will cover most of the second half of this Saturnian year, the two providing together a more complete look at a full seasonal cycle.

Regarding comets, in addition to Figure 17.4, we show in Figure 17.14 two versions of a low-resolution spectrum of the hyperbolic comet ISON C/2012 S1, with a faint solar-type continuum and prominent emission bands of molecules and atoms.

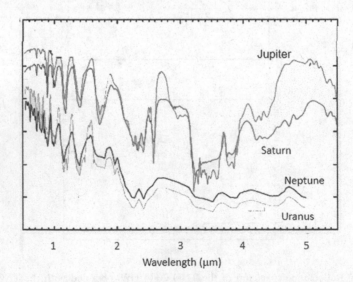

**FIGURE 17.13** Near-infrared model spectra of the four giant planets from 0.8 to 5 μm. (Adapted from Noorwood et al., 2015, reproduced with IOP permission).

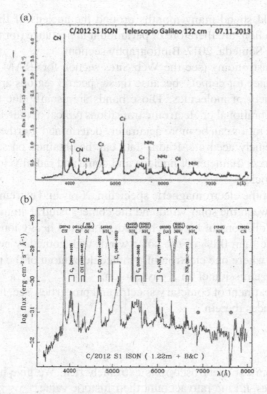

**FIGURE 17.14** Flux-calibrated CCD spectrum of Comet ISON C/2012 S1 in the visible region. (a) Linear ordinate; (b) logarithmic ordinate to bring out faint features. (Spectrum Asiago Observatory, courtesy of P. Ochner and A. Siviero).

The interpretation of cometary spectra would deserve a specific section for their peculiar chemical and physical conditions, rapidly varying with the distance from the Sun. In essence, the emission bands are due to a mechanism of fluorescence, namely, to absorption of the solar UV light by the internal energy levels of the different molecules, followed by re-emission in other wavelengths. Actually, the solar spectrum in the UV band is not adequately approximated by a Planck's law at a temperature around 5900 K, as instead is in the visible band. The many solar absorptions and emissions in the UV spectrum influence the fluorescence mechanism, and so does the variable radial velocity of the comet with respect to Sun. Therefore, the energy levels are not populated by collisions as in equilibrium conditions, but by a rapidly varying radiation field. Several other factors enter in the interpretation of the spectra in addition to the distance, such as the rotation of the nucleus and the orientation of the spin axis with respect to the orbital plane, the chemical composition, the percentage of volatile components versus solid particles, the interaction with magnetic fields and particles of solar origin (the solar wind). As an example of such difficulties, consider the behavior of the CO molecular emission compared to that of the $CO_2$, which at first sight seems to contradict the values of excitation, ionization and dissociation in Tables 17.5 and 17.6, because in several comets the CO coma is seen at a larger solar distances than the $CO_2$ coma. Here, in addition to the relative abundance of the molecules (the $CO_2$ seemingly more abundant than the CO), the sublimation temperature has an important role, 25 K for CO versus 72 K for $CO_2$. See the detailed paper by Womack et al. (2017).

The distance from the Earth enters also in the interpretation of the data, because the finite spatial and temporal resolution of a given instrument severely limits the comprehension of the various processes occurring inside the coma in the proximity of the nucleus. Space missions such as the

ESA Rosetta, which could spend many months around the nucleus of the comet 67P/C-G, both before and after the perihelion, are therefore of paramount importance (for the Rosetta mission, see, for instance, Barbieri and Someda, 2017, Bibliography section).

Millimeter and radio astronomy (see the Web Sites section for IRAM and ALMA telescopes) have a particular importance for comets, because these spectral regions are very rich of molecular bands due to a great variety of molecules. Those bands are usually due to rotational transitions, simpler to interpret that vibrational or electronic transitions typical of the visual and UV bands; spin temperatures and isotopic ratios can be more accurately determined. Furthermore, the measurement of radial velocity is extremely accurate. Radar data can be obtained on sufficiently close comets, measuring not only distance, diameter of nucleus and coma and radial velocity, but also properties and velocities of dust particles.

At the other extreme of the electromagnetic spectrum, X-ray and extreme UV (EUV) emissions, due to the interaction between the solar wind and the comet's neutral atmosphere, are a fundamental property of comets. Collisions of highly charged solar wind heavy ions (in particular C, O, N, which constitute not more than 0.1% of the solar wind) with cometary neutral species result in a charge-exchange process where one electron of a molecule or atom of the neutral coma is captured by the heavy ion, with the emission of an X-ray photon.

For a more detailed treatment of cometary spectra and properties, see, for instance, Barbieri and Bertini (2017) and references therein.

## NOTES

The literature on astronomical spectroscopy is extremely rich. We give here and in Bibliography section only few references, taking into account their historic value.

- The books by Herzberg (1944 and 1950) and White (1944) are classical reference textbooks for atomic and molecular spectroscopy. Moore (1945) published a fundamental catalog of reference spectral lines of astrophysical interest. Meinel et al. (1968) provided a very useful catalog of emission lines from celestial sources, including those due to the night sky. Another reference handbook was compiled by Drake (1996). For another recent introduction on atomic spectroscopy, see Martin and Wiese *Atomic Spectroscopy*, http://physics. nist.gov/Pubs/AtSpec/AtSpec.PDF. More specific for stellar astrophysics are the textbooks by Gray (1976), Böhm-Vitense (1989), Schatzman and Praderie (1990). For gaseous nebulae Osterbrock (1974). For solar astrophysics, Zirin (1988).
- Regarding the forbidden lines, Bowen (1927) was able to identify the so-called *nebulium* lines, so prominent in planetary and diffuse nebulae, as forbidden transitions of O II, O III and N II. The forbidden lines in the solar corona were identified by Edlén (1943). See also Osterbrock (1974) and Garstang (1995).
- The works of Struve and Elvey (ca. 1930) were central to evaluating the influence of rotation. See Gray (1999, Bibliography section).
- Two important papers for the interpretation of cometary spectra are those by Biermann (1951) and Swings (1965) cited in References section.
- Reproductions of photographic stellar spectra useful for the MK classification are given by Abt et al. (1968, Bibliography section) and Jacoby et al. (1984). A digital reproduction of the original MK catalog of CCD spectra, obtained by R.O. Gray, together with other interesting material on stellar spectroscopy, can be found at: http://nedwww.ipac.caltech. edu/level5/spec_stars.html.
- Spectrophotometry of stars can be found for instance in Breger (1976) and Hayes et al., (1979)
- For spectropolarimetry, in addition to Coyne ed., (1987 in Bibliography section), see, for instance, di Serego et al. (1996)

- A most important spectral library is ELODIE (see the Web Sites section) by Ph. Prugniel and C. Soubiran, which includes 1962 high-resolution spectra of 1388 stars obtained with the ELODIE spectrograph at the Observatoire de Haute–Provence 193 cm telescope in the wavelength range 400–680 nm. It provides a large coverage of atmospheric parameters: $T_{eff}$ from 3000 to 60,000 K, log g from −0.3 to 5.9 and [Fe/H] from −3.2 to +1.4.
- Although this chapter is mostly devoted to spectra in the extended visible region, we provide in Web Sites section references to UV and IR satellites, such as IUE, HST, ISO, IRAS, Herschel and Spitzer. In particular, the study of the UV spectrum down to approximately the 1216A Lyman-α hydrogen line was greatly extended by the long-life satellite International Ultraviolet Explorer (IUE), which operated for 18 years from 1986 until 1994, observing more than 10,000 stars. The HST, operating since more than 30 years, has returned an exceptional amount of high-quality data. For instance, Carpenter et al. (2018, References section) provide reference spectra for evolved M stars. For a discussion of the hottest stars and associated stellar wind, see, for instance, Nota et al. (1996, 1997). A program is underway as part of the UV Legacy Library of Young Stars as Essential Standards (ULLYSES). This target list includes massive stars in the Large and Small Magellanic Clouds, as well as the sample of T Tauri stars to be observed in nearby Milky Way star-forming regions.
- Examples of IR spectra from IRAS are given by Olnon et al. (1986, References section). The Spitzer Atlas of Stellar Spectra (SASS), which includes 159 stellar spectra in the band from 5 to 32 μm, was published by Ardila et al. (2010).

# Bibliography

We quote here the books and monographies cited in this second edition.
Papers published in scientific journals are reported in **References**.
Useful web sites are quoted in **Web Sites**.

Abt H.A., Meinel A.B., Morgan W.W., Tapscott J. (1968) *An Atlas of Low Dispersion Grating Stellar Spectra*, Kitt Peak National Observatory, Tucson, AZ.

Aerts C., Christensen-Dalsgaard J., Kurtz D. (2010) *Asteroseismology. Astronomy and Astrophysics Library*, Springer, Dordrecht, NY, ISBN 978-1-4020-5803-5.

Aitken R.G. (1935) *The Binary Stars*, Dover, New York.

Allen C.W. (2000) *Astrophysical Quantities*, 4th ed., A.N. Cox, Editor In Chief, AIP, New York, (the fourth edition has been thoroughly revised and enlarged, available also on CD-ROM).

Allen R.H. (1963) *Star Names, Their Lore and Meaning,* Dover, New York.

Andrews L.C. (2004) Field Guide to Atmospheric Optics, SPIE. Vol. FG02; second edition 2019, SPIE Vol. FG41, ISBN: 9781510619371.

Asplund M., Grevesse N., Sauval A.J., Scott P. (2009) The chemical composition of the Sun, *Annual Review of Astronomy & Astrophysics* **47**, Issue 1, pp. 481–522.

Backer C.D., Hellings R.W. (1986) Pulsar timing and general relativity, *Annual Review of Astronomy and Astrophysics* **24**, p. 537.

Bakich M.E. (1995) *The Cambridge Guide to Constellations*, Cambridge University Press, Cambridge.

Barbieri C. (2006) *Fundamentals of Astronomy*, 1st ed., Taylor & Francis Group, Routledge, ISBN 10: 0750308869 / ISBN 13: 9780750308861.

Barbieri C., Bertini I. (2017) Comets, *La Rivista Del Nuovo Cimento*, https://www.sif.it/riviste/sif/ncr/econtents/2017/040/08/article/0.

Barbieri C., Someda C.G. eds. (2017) *From Giotto to Rosetta, 30 Years of Cometary Science from Ground and Space*, Accademia Galileiana di Padova, ISBN 9-78-88-98216-09-55.

Barbieri C. (2019) *A Brief Introduction to the Search for Extra-Terrestrial Life*, CRC Press Focus Book, ISBN 978-0-367-19194-8.

Binney J., Merrifield M. (1998) *Galactic Astronomy,* Princeton Series in Astrophysics, Princeton University Press, Princeton, NJ.

Bohm D. (1965) *The Special Theory of Relativity*, Benjamin, New York.

Böhm-Vitense E. (1989) *Stellar Astrophysics,* Vol. 3, Cambridge University Press, Cambridge.

Cannon A.J., Pickering E.C. (1924) *The Henry Draper Catalogue*, Annals of Harvard College Observatory vols. 91–100.

Carrington, R.C. (1863), *Observations of the Spots on the Sun from November 9, 1853, to March 24, 1861, made at Redhill*, Williams and Norgate, London, England.

Celletti A., Perozzi E. (2010) *Celestial Mechanics: The Waltz of the Planets*, Springer Praxis Books, Berlin, Heidelberg, ISBN-13: 978-0-387-307770.

Celletti A. (2012) *Stability and Chaos in Celestial Mechanics*, Springer Praxis Books, ISBN-10: 3642261566.

Chiosi C., Bertelli G., Bressan A. (1992) New developments in understanding the H-R Diagram, *Annual Review of Astronomy and Astrophysics* **30**, p. 235.

Ciufolini I., Wheeler J.A. (1995) *Gravitation and Inertia*, Princeton University Press, Princeton, NJ.

Copernicus N. (1542) De Revolutionibus Orbium Coelestium, Norimbergae apud Joh. Petreium.

Coyne G.V., Hoskins M.A., Pedersen O. eds. (1983) *Gregorian Reform of the Calendary*, Pontificia Academia Scientiarum, Vatican State.

Coyne G.V. ed. (1987) *Polarized Radiation of Interstellar Origin*, Vatican Observatory Press, Vatican State.

Danby J.M.A. (1998) *Fundamentals of Celestial Mechanics*, Willmann-Bell, Richmond, VA.

Danjon A. (1980) *Astronomie Genérale*, Librairie Blanchard, Paris, second edition 1994.

De Angelis A., Pimenta M. (2018) *Introduction to Particle and Astroparticle Physics Multimessenger Astronomy and its Particle Physics Foundations*, 2nd ed., Springer, ISBN 978-3-319-78180-8; ISBN 978-3-319-78181-5 (eBook), doi: 10.1007/978–3–319–78181–5.

Dick S., McCarthy D., Luzum M. (2000) *Polar Motion: Historical and Scientific Problems, IAU Colloquium 178*, ASP Conference Series, Vol. 208. ed. Steven Dick, Dennis McCarthy, Brian Luzum, ISBN: 1-58381-039.

Draine B.T. (2003) Interstellar dust grains, *Annual Review of Astronomy and Astrophysics* **41**, pp. 241–289, doi: 10.1146/annurev.astro.41.011802.094840.

Drake G.W.F. ed. (1996) *Atomic, Molecular, and Optical Physics Handbook*, AIP Press, Woodbury, NY.

Dravins D. (2019) Intensity interferometry, *Volume 3 of The WSPC Handbook of Astronomical Instrumentation*, D. Burrows ed., 5 Volumes, doi: 10.1142/9446.

Eddington A. (1920) *Space, Time and Gravitation*, Cambridge University Press, Cambridge.

Einstein A. (1905) Zur Elektrodynamik bewegter Koerper, *Annalen der Physik* **17**, p. 891.

Einstein A. (1915) Zur allgemeinen Relativitätstheorie, Sitzungsberichte der Königlich Preußischen Akademie der Wissenschaften, Berlin, pp. 778–786.

Galilei G. (1610) *Sidereus Nuncius*, Venetiis apud Thomas Baglioni, Venice.

Garrison R.F. ed. (1984) *The MKK Process and Spectral Classification*, David Dunlap Observatory, Toronto.

Gray D.F. (1976) *The Observation and Analysis of Stellar Photospheres*, Wiley/Cambridge University Press, New York.

Gray D.F. (1999) *Stellar Rotation and Precise Radial Velocities, IAU Colloquium* Vol 170, 243–254, J.B. Hearnshaw, C.D. Scarfe eds. doi: 10.1017/S0252921100048624.

Green R. (1985) *Spherical Astronomy*, Cambridge University Press, Cambridge.

Hanbury Brown R. (1974) *The Intensity Interferometer; Its Application to Astronomy*, Wiley, ISBN 978-0-470-10797-3. ASIN B000LZQD3C.

Hapke B. (1993) *Theory of Reflectance and Emittance Spectroscopy*, Cambridge University Press, Cambridge, UK; second edition 2009, Online ISBN: 9781139025683.

Heilbron J.L. (1999) *The Sun in the Church*, Harvard University Press, Cambridge, MA.

Herzberg G. (1944) *Atomic Spectra and Atomic Structure*, Dover, New York.

Herzberg G. (1950) *Spectra of Diatomic Molecules*, Van Nostrand, Princeton, NJ.

Hesser E. ed. (1980) Star Clusters, *IAU Symposium 85*, Victoria, B.C., Canada, August 27–30, 1979, Ed. J.E. Hesser, Reidel Publishing Company, Dordrecht, ISBN 90-277-1087-2 (HB), 90-277-1088-0 (PB).

Hirshfeld H.W. (2001) *Parallax: The Race to Measure the Cosmos*, Freeman and Co., New York, ISBN 10: 0716737116 / ISBN 13: 9780716737117.

Hoffleit D., Jascheck C. (1984) *The Bright Stars Catalogue*, Yale University Observatory, New Haven, CT.

Hohenkerk C.Y., Yallop B.D., Smith C.A., Sinclair A.T. (1992) *The Explanatory Supplement to the Astronomical Almanac*, Chapter 3, P.K. Seidelmann ed., University Science Books.

Houck N. (1975–1988) *An Atlas of Objective Prism Spectra*, University of Michigan, Ann Arbor, MI.

Hubble E. (1936) *The Realm of the Nebulae*, Yale University Press, New Haven. ISBN 9780300025002.

Jaschek C., Jaschek M. (1990) *The Classification of Stars*, Cambridge University Press, Cambridge.

Kennan P.C. (1963) Classification of Stellar Spectra. *Basic Astronomical Data: Stars and Stellar Systems*, K. A. Strand, ed., University of Chicago Press, Chicago, IL, 1968, p. 78.

Keenan P.C., McNeil R.C. (1976) *An Atlas of Spectra of the Cooler Stars*, Ohio State University Press, Colombus, OH.

Kovalevsky J., Brumberg V.A. eds. (1986) Relativity in Celestial Mechanics and Astronomy. *I.A.U. Symposium 114*, Leningrad, Russia, USSR, May 29–31, 1985, Eds. J. Kovalevsky & V.A. Brumberg, Reidel Publishing Company, Dordrecht, ISBN 90-277-2189-0 (HB), 90-277-2190-4 (PB).

Kovalevsky J. (2001) *Modern Astrometry*, 2nd ed., Springer, Berlin.

Lagrange J.L. (1772) *Essai sur le Problème des Trois Corps*, Oeuvres de Lagrange, Vol. 6, pp. 229–332, Gauthier-Villars, Paris, 1867–77.

Landau L., Lifschitz E. (1982) *Mechanics*, 4th Ed., MIR, Moscow.

Lissauer J.J., De Pater I. (2019) *Fundamental Planetary Science*, updated edition, Cambridge University Press, Cambridge.

Longair M. (2015) Bending space–time: A commentary on Dyson, Eddington and Davidson (1920) 'A determination of the deflection of light by the Sun's gravitational field', *Philosophical Transactions of the Royal Society A* **373**, Issue 2039, doi: 10.1098/rsta.2014.0287.

Luyten W.J. (1963) Proper motion surveys, *Basic Astronomical Data*, K.A. Strand ed., pp. 46–54, University of Chicago Press, Chicago.

Mashhoon B. (2008) Gravitoelectromagnetism, *The Measurement of Gravitomagnetism: A Challenging Enterprise*, L. Iorio ed., pp. 29–39, Nova Science, New York, abstract http://arXiv.org/abs/gr-qc/0311030v2.

Meeus J. (1991) *Astronomical Algorithms*, Willmann-Bell, Richmond, VA.

Meinel A.B., Aveni A.F., Stockton M.W. (1968) *Catalog of Emission Lines in Astrophysical Objects*, Optical Sciences Center and Steward Observatory, University of Arizona, Tucson, AZ.

Mendillo M., Nagy A., Waite J.H. eds. (2002) *Atmospheres in the Solar System: Comparative Aeronomy*, American Geophysical Union, Washington, DC.

Mignani R., Shearer A., Słowikowska A., Zane S. eds. (2019) *Astronomical Polarisation from the Infrared to Gamma Rays*, Springer Astrophysics and Space Science Library, 460, Springer International Publishing, Cham, Switzerland, ISBN 978-3-030-19714-8.

Minnaert M.G.J. (1993) *Light and Color in the Outdoors*, Springer, New York.

Møller C. (1962) *The Theory of Relativity*, Oxford at Clarendon Press, London.

Moore C.E. (1945) *A Multiplet Table of Astrophysical Interest*, Princeton University Observatory, Princeton, NJ.

Morbidelli A. (2007) *Trans-Neptunian Objects and Comets*, D. Jewitt, A. Morbidelli & H. Rauer eds., Saas-Fee Advanced Course 35, Springer International Publishing, ISBN-10: 3642091091.

Morgan W.W., Keenan P.C., Kellman E. (1943) *An Atlas of Stellar Spectra*, University of Chicago Press, Chicago, IL.

Morgan W.W., Keenan P.C. (1973) Spectral classification, *Annual Review of Astronomy and Astrophysics* **11**, p. 29.

Newcomb S. (1906) *A Compendium of Spherical Astronomy*, Dover, New York.

Newton I. (1687) *Philosophiæ Naturalis Principia Mathematica*, first edition, Londini, iussu Societatis Regiae ac typis Josephi Streater, anno MDCLXXXVII, London, England.

Nieto A. (1972) *The Titius-Bode Law of Planetary Distances: Its History and Theory*, Pergamon Press, New York.

O'Connell D.J.K. ed. (1957) *Stellar Populations*, Vatican Observatory Press, Vatican State.

Oort J. (1965) Stellar Dynamics, *Stars and Stellar Systems*, Vol: 5, Chapter 21, pp. 455–509, University of Chicago Press, Chicago, IL.

Oppolzer von T. (1887) Canon der Finsternisse, Denkschriften der Kaiserlichen Akademie der Wissenschaften in Wien, Math.-naturw. Kl., Band 52.

Osterbrock D.E. (1974) *Astrophysics of Gaseous Nebulae*, W.H. Freeman, San Francisco, CA.

Reach W.T. (1997) The Structured Zodiacal Light: IRAS, COBE, and ISO Observations. Diffuse Infrared Radiation and the Irts 124: 1. Bibcode:1997ASPC..124...33R. SP Conference Series, Vol. 124, eds. H. Okuda; T. Matsumoto; and T. Rollig, p. 33.

Richichi A. (2004) Combining Optical Interferometry with Lunar Occultations, Spectroscopically and Spatially Resolving the Components of the Close Binary Stars, Proceedings of the Workshop held 20–24 October 2003 in Dubrovnik, Croatia, ASP Conference Series, Vol. 318, eds. R.W. Hidlitch, H. Hensberge, K. Pavlovski, pp. 148–156.

Richichi A. (2019) Lunar and planetary occultation, *3rd volume of The WSPC Handbook of Astronomical Instrumentation*, Chapter 15, D. Burrows ed., 5 Volumes, doi: 10.1142/9446.

Rieke G.H. (2002) *Detection of Light: From the Ultraviolet to the Submillimeter*, Cambridge University Press, Cambridge.

Roth G.D. (1994) *Compendium of Practical Astronomy*, Vol. 3, Springer, New York.

Saastamoinen J. (1979) On the Calculation of Refraction in Model Atmospheres, *Refractional Influences in Astrometry and Geodesy: Proceedings from IAU Symposium No. 89*, eds. E. Tengstrom, G. Teleki, I. Ohlsson, D. Reidel, Dordrecht, p. 73.

Sanders R.H., McGaugh S.S. (2002) Modified newtonian dynamics as an alternative to dark matter, *Annual Review Astronomy Astrophysics* **40**, pp. 263–417.

Savage B.D., Sembach K.R. (1996) Interstellar abundances from absorption-line observations with the Hubble space telescope, *Annual Review of Astronomy and Astrophysics* **34**, pp. 279–329, doi: 10.1146/annurev.astro.34.1.279.

Schatzman E., Praderie F. (1990) *Les Etoiles*, InterEditions et Editions du CNRS, Paris.

Shirley J.H., Fairbridge R.W. (1997) *Encyclopedia of Planetary Sciences*, Kluwer Academic Publisher, Dordrecht.

Smart W.M. (1953) *Celestial Mechanics*, Longmans and Green, London.

Smart W.M. (1965) *Text Book on Spherical Astronomy*, Cambridge University Press, Cambridge.

Sobel D. (1996) *Longitudes*, Penguin Books, Baltimore, MD.

Steves B.A. (1998) The cycles of Selene, *Vistas in Astronomy* **41**, pp. 541–571.

Stumpff P. (1986) Relativistic and Perspective Effects in Proper Motions and Radial Velocities of Stars, in Relativity in Celestial Mechanics and Astronomy, *IAU Symposium 114*, Leningrad, Russia, USSR, May 29–31, 1985, eds. J. Kovalevsky, V.A. Brumberg, Reidel Publishing Company, Dordrecht, ISBN 90-277-2189-0 (HB), 90-277-2190-4 (PB).

Thompson M.J., Christensen-Dalsgaard J., Miesch M.S., Toomre J. (2003) The internal rotation of the Sun, *Annual Review of Astronomy and Astrophysics* **41**, pp. 599–643.

Trumpler R.J., Weaver H. (1953) *Statistical Astronomy*, Dover, New York.

Tyson R.K., Frazier B.W. (2004) *Field Guide to Adaptive Optics, SPIE*. vol. FG03.

Urban S.E., Seidelmann P.K. eds. (2012) *Explanatory Supplement to the Astronomical Almanac*, University Science Books, https://aa.usno.navy.mil/publications/docs/exp_supp.php

Valsecchi G.B. (2001) On the Orbit of the Moon, in Earth, Moon, and Planets vol. 85/86, p. 443, Springer, ISSN: 0167-9295 (Print) 1573-0794 (Online).

Vernet E., Ragazzoni R., Esposito S., Hubin N. (2002) Beyond Conventional Adaptive Optics: A Conference Devoted to the Development of Adaptive Optics for Extremely Large Telescopes, *ESO Conference and Workshop Proceedings 58*, Garching, Germany, ISBN 3923524617.

Wallace P. (1994) *The SLALIB Library, Astronomical Data Analysis Software and Systems III*, ASP Conference Series 61, pp. 481–484, D.R. Crabtree, R.J. Hanisch, J. Barnes, eds., ISBN 0-937707-80-5.

Walter H.G., Sovers O.J. (2000) *Astrometry of Fundamental Catalogues*, Springer, Berlin.

Wegener A. (1966) *Original book Die Entstehung der Kontinente und Oceane, 1929, English translation The Origin of Continents and Oceans*, Dover Edition, New York, ISBN 10: 0486617084.

Weigelt G. (1991), Triple-correlation imaging in Astronomy, *Progress in Optics* **29**, pp. 298–319, http://www.mpifr-bonn.mpg.de/div/ir-interferometry/.

White H.E. (1944) *Atomic Spectra*, Dover, New York.

Will C.M. (1993) *Theory and Experiment in Gravitational Physics*, Cambridge University Press, Cambridge.

Woolard E.W., Clemence G.M. (1966) *Spherical Astronomy*, Academic Press, New York.

Zagar F. (1948) *Astronomia Sferica e Teorica*, Zanichelli, Bologna, reprinted 1988.

Zirin H. (1988) *Astrophysics of the Sun*, Cambridge University Press, Cambridge.

# REFERENCES

Anderson J.D., Laing P.A., Lau E.L., Liu A.S., Nieto M.M., et al. (2002) Study of the anomalous acceleration of Pioneer 10 and 11, *Physics Review D* **65**, Issue 8 id. 082004, American Physical Society, doi: 10.1103/PhysRevD.65.082004.

Archinal B.A., Acton C.H., A'Hearn M.F., Conrad A., Consolmagno G. J., et al. (2018) Report of the IAU working group on cartographic coordinates and rotational elements: 2015, *Celestial Mechanics and Dynamical Astronomy* **130**, Issue 3, article id. 22, 46 pp., doi: 10.1007/s10569-017-9805-5.

Ardila D.R., Van Dyk S.D., Makowiecki W., Stauffer J., Fajardo-Acosta S., et al. (2010) *Astrophysical Journal Supplement* **192**, see also *VizieR On-line Data Catalog:* Spitzer Atlas of Stellar Spectra (SASS), http://simbad.u-strasbg.fr/simbad/.

Aoki S., Guinot B., Kaplan G.H., Kinoshita H.H., McCarthy D.D., et al. (1982) The new definition of universal time, *Astronomy and Astrophysics* **105**, p. 359.

Astronomical Almanac, yearly publication by the United States Naval Observatory and UK Nautical Almanac Office.

Auer L.H., Standish M.E. (2000) Astronomical refraction: Computational method for all zenith angles, *The Astronomical Journal* **119**, pp. 2472–2484.

Babcock H.W. (1953) The possibility of compensating astronomical seeing, *Publications of the Astronomical Society of Pacific* **65** Issue 386, p. 229.

Bailer-Jones C.A.L., Farnocchia D., Meech K.J., Brasser R., Micheli M., et al. (2018) Plausible home stars of the interstellar object Oumuamua found in Gaia DR2, *The Astronomical Journal* **156**, 11 pp.

Basri G., Mohanty S., Allard F., Hauschildt P.H., Delfosse X., et al. (2000) An effective temperature scale for late-M and L dwarfs, from resonance absorption lines of Cs I and Rb I, *The Astrophysical Journal* **538**, pp. 363–385.

Baumgardner J., Wroten J., Mendillo M., Martinis C., Barbieri C., et al. (2013) Imaging space weather over Europe, *Space Weather* **11**, pp. 1–10, doi: 10.1002/swe.20027,2015.

Benn C.R., Ellison S.L. (1998) ING Technical Note 115: *La Palma night-sky brightness*, Isaac Newton Tecnical Group, La Plama, Canary Islands.

Bertotti B., Iess L., Tortora P. (2003) A test of general relativity using radio links with the Cassini spacecraft, *Nature* **425**, p. 374.

Bessel F.W. (1838) *On the parallax of 61 Cygni*, Monthly Notices of the Royal Astronomical Society, **4**, pp. 152–161.

Biermann L. (1951). Kometenschweife und solare Korpuskularstrahlung, *Zeitschrift für Astrophysik* **29**, p. 274. Bibcode:1951ZA.....29..274B.

Black G.J., Nicholson P.D., Thomas P.C. (1995) Hyperion: Rotational dynamics, *Icarus* **117**, pp. 149–171.

Bovy J. (2017) Galactic rotation in Gaia DR1, *Monthly Notices of the Royal Astronomical Society: Letters* **468**, Issue 1, pp. L63–L67, doi: 10.1093/mnrasl/slx027.

Bowen I.S. (1927) The origin of the chief nebular lines, *Publications of the Astronomical Society of the Pacific* **39**, Issue 231, p. 295.

Bowen I.S. (1947) Limiting visual magnitude, *Publications of the Astronomical Society of the Pacific* **59**, p. 253.

Breger M. (1976) A catalog of spectrophotometric scans of stars, *The Astrophysical Journal Supplement Series* **32**, pp. 7–87, doi: 10.1086/190392.

Brosche P.U., Seiler U., Suendermann J., Wuensch J. (1989) Periodic changes in the Earth's rotation due to oceanic tides, *Astronomy and Astrophysics* **220**, pp. 318–320.

Brown A.G.A., Vallenari A., Prusti T., de Bruijne J.H.J., Babusiax C., et al. (2018) Gaia collaboration Gaia DR2- Summary of the contents and survey properties, *Astronomy.and Astrophysics* **616**, p. A1.

Burstein D., Heiles C. (1984) Reddening estimates for galaxies, *Astrophysics Journal Supplement* **54**, p. 33.

Bus S.J., Binzel R.P. (2002) Phase II of the small main-belt asteroid spectroscopic survey. A feature-based taxonomy, *Icarus* **158**, p. 146.

Campins H., Rieke G.H., Lebofsky M.J. (1985) Absolute calibration of photometry at 1 through 5 microns, *The Astronomical Journal* **90**, pp. 896–899.

Capitaine N., Wallace P.T., Chapront J. (2003a) Expressions to implement the IAU 2000 definition of UT1, *The Astronomy and Astrophysics* **406**, pp. 1135–1149, doi: 10.1051/0004-6361:20030817.

Capitaine N., Wallace P.T., Chapront J. (2003b) Expressions for IAU 2000 precession quantities, *Astronomy and Astrophysics* **412**, pp. 567–586, doi: 10.1051/0004-6361:20031539.

Cardelli J.A., Clayton G.C., Mathis J.S. (1989) The relationship between infrared, optical and ultraviolet extinction, *The Astrophysical Journal* **345**, p. 245.

Carpenter K.G., Nielsen K. E., Kober G.V., Ayres T.R., Wahlgren G.M., et al. (2018) The Treasury Advanced Spectral Library (ASTRAL): Reference spectra for evolved M stars, *The Astrophysical Journal* **869**, Issue 2, article id. 157, 17 pp.

Cavazzani S., Ortolani S., Barbieri C. (2011) Fluctuations of photon arrival times in free atmosphere, *Monthly Notices of the Royal Astronomical Society* **411**, p. 271.

Chapront-Touzé M., Chapront J. (1983) Ephémérides Lunaires Parisiennes ELP 2000, *Astronomy and Astrophysics* **124**, pp. 50–62.

Chapront-Touzé, M., Chapront, J. (1988) ELP 2000–85. A semianalytical lunar ephemeris adequate for historical times, *Astronomy Astrophysics* **190**, pp. 342–352.

Charlot P., Sovers O.J., Williams J.G., Newhall X.X. (1995) Precession and nutation from joint analysis of radio interferometric and lunar laser ranging observations, *The Astronomical Journal* **109**, pp. 418–427.

Ciufolini I. (2000) The 1995–1999 measurements of the Lense–Thirring effect using laser-ranged satellites, *Classical Quantum Gravity* **17**, Issue 21 June, pp. 2369–2380.

Colombo G. (1966), Cassini's second and third law, *The Astronomical Journal* **71**, pp. 891–896.

Cooray A. (2003) Kuiper belt object sizes and distances from occultation observations, *The Astrophysical Journal* **589**, pp. L97–100.

Cova S., Ghioni M., Lotito A., Rech I., Zappa F. (2004) Evolution and prospects for single-photon avalanche diodes and quenching circuits, *Journal of Modern Optics* **51**, pp. 1267–1288.

Covino S., Wiersema K., Fan Y.Z., Toma K., Higgins A.B., et al. (2017) The unpolarized macronova associated with the gravitational wave event GW 170817, *Nature Astronomy* **1**, pp. 791–794, doi: 10.1038/s41550-017-0285-z.

de Bruijne H.J., Hoogerwerf R., de Zeeuw P.T. (2001) A Hipparcos study of the Hyades open cluster. Improved colour-absolute magnitude and Hertzsprung-Russell diagrams, *Astronomy and Astrophysics* **367**, nr. 1, pp. 111–147, doi: 10.1051/0004-6361:20000410.

de Sitter W., Brouwer D. (1938) On the system of astronomical constants, *Bulletin of the Astronomical Institutes of the Netherlands* **8**, pp. 213–231.

de Vaucouleurs G., Peters W.L. (1984) The dependence on distance and redshift of the velocity vectors of the sun, the galaxy and the local group with respect to different extragalactic frames of reference, *The Astrophysical Journal* **287**, pp. 1–16.

di Serego A.S., Cimatti A., Fosbury R.A.E., Perez-Fournon I. (1996) Spectropolarimetry of 3C 265, a mis-aligned radio galaxy, *Monthly Notices of the Royal Astronomical Society* **279**, p. L57.

Dicke R.H., Goldberg M. (1974) The oblateness of the Sun, *The Astrophysical Journal Supplement Series* **27**, pp. 131–182.

Dickey J.O., Marcus S.L., de Viron O., Fukumori I. (2002) Recent Earth oblateness variations, *Science* **208**, p. 1975.

Dishon G., Weber T.A. (1977) Redshifts and superluminal velocities of expansion, *The Astrophysical Journal* **212**, p. 31.

Dravins D., Lindegren L., Mezey E., Young T. (1997a) Atmospheric intensity scintillation of stars, I. Statistical distributions and temporal properties, *Publications of the Astronomical Society of the Pacific* **109**, pp. 173–207.

Dravins D., Lindegren L., Mezey E. (1997b) Atmospheric intensity scintillation of stars, II, Dependence on optical wavelength, *Publications of the Astronomical Society of the Pacific* **109**, pp. 725–737.

Dravins D., Lindegren L., Mezey E. (1997c) Atmospheric intensity scintillation of stars, III, Effects for different telescope apertures, *Publications of the Astronomical Society of the Pacific* **110**, pp. 610–633.

Dravins D., Lindegren L., Madsen S. (1999), Astrometric radial velocities. I. Non-spectroscopic methods for measuring stellar radial velocity, *Astronomy and Astrophysics* **348**, p. 1048.

Dyson F.W., Eddington A., Davidson C. (1920) A determination of the deflection of the light by the Sun's gravitational field, from observations made at the total solar eclipse of May 29, 1919, *Philosophical Transactions of the Royal Society of London A* **220**, p. 291.

Edlén B. (1943) Die Deutung der Emissionslinien im Spektrum der Sonnenkorona. Mit 6 Abbildungen. *Zeitschrift fr Astrophysik* **22**, p. 30.

Eichhorn H. (1981) On the computation of acceleration parallaxes, *The Astronomical Journal* **86**, pp. 915–917 and *Astronomy and Astrophysics* **102**, p. 35.

Elliot J.L., French R.J., Frogel J.A., Elias J.H., Mink D.J., et al. (1981) Orbits of nine Uranian rings, *The Astronomical Journal* **86**, p. 444.

Encke J.F. (1822) *Die Entfernung der Sonne von der Erde aus dem Venusdurchgange von 1761*, In der Beckerschen Buchhandlung, Gotha.

Encke J.F. (1824) *Der Venusdurchgang von 1769*, In der Beckerschen Buchhandlung, Gotha.

Evans D. (1971) Photoelectric measurements of lunar occultations, *The Astronomy Journal* **76**, p. 1107.

Evans R.W., Stapelfeldt K.R., Peters D.P., Trauger, J.T., Padgett D.L., et al. (1998) Asteroid trails in hubble space telescope WFPC2 images: First results, *Icarus* **131**, Issue 2, pp. 261–282, doi: 10.1006/icar.1997.5873.

Falcon R.E., Winget D.E., Montgomery M.H., Williams K.A. (2010) A gravitational redshift determination of the mean mass of white dwarfs. DA stars, *The Astrophysical Journal* **712**, Issue 1, pp. 585–595.

Fiala A.D., Dunham D.W., Sofia S. (1994) Variation of the solar diameter from solar eclipse observations, 1715–1991, *Solar Physics* **152**, Issue 1, pp. 97–104.

Fiorucci M., Munari U. (2003) The Asiago Database on Photometric Systems (ADPS). II. Band and reddening parameters, *Astronomy and Astrophysics* **401**, p. 781.

Folkner W.M., Charlot P., Finger M.H., Williams J.C., Sovers O.J., et al. (1994) Determination of the extragalactic-planetary frame tie from joint analysis of radio interferometers and lunar ranging measurements, *Astronomy and Astrophysics* **287**, p. 279.

Fomalont E.B., Shramek S.A. (1975) A confirmation of Einstein general theory of relativity by measuring the bending of microwave radiation in the gravitational field of the Sun, *The Astrophysical Journal* **199**, p. 749, doi: 10.1086/153747.

Fornasier S., Dotto E., Barucci A.M., Barbieri C. (2004) Water ice on the surface of the large TNO 2004 DW, *Astronomy and Astrophysics* **422**, pp. L43–46.

Franz M. (2002) Heliospheric Coordinate Systems, updated version: https://www2.mps.mpg.de/homes/fraenz/systems/

Fricke W. (1971) A rediscussion of Newcomb's determination of precession, *Astronomy and Astrophysics* **13**, p. 298.

Fricke W. (1977) Basic material for the determination of precession and galactic rotation, and a review of methods and results, *Veröffentlichungen Astronomisches* **28**, Rechen-Institut, Heidelberg.

Fricke W., Schwan H., Corbin T.E., Bastian U., Bien R., et al. (1988) Fifth Fundamental Catalogue (FK5), Veroff. Astron. Rechen Inst. Heidelberg nr. 32, Verlag G Braun, Karlsruhe. The FK5 Extension was published in 1991, ibidem nr. 33.

Fukushima T. (2003) A new precession formula, *The Astronomical Journal* **126**, pp. 494–534, doi: 10.1086/375641.

GAIA Collaboration. (2018) GAIA DR2: Observational Hertzsprung-Russel diagrams, *Astronomy and Astrophysics Special Issue*, doi: 10.1051/0004-6361/201832843.

Garstang R.H. (1989) Night sky brightness and observatories and sites, *Publications of the Astronomical Society of the Pacific* **101**, p. 306.

Garstang H.R. (1991) Dust and light pollution, *Publications of the Astronomical Society of the Pacific* **103**, p. 1109.

Garstang R.H. (1995) Radiative hyperfine transitions, *The Astrophysical Journal* **447**, p. 962.

Geballe T.R., Knapp G.R., Leggett S.K., Fan X., Golimowski D.A., et al. (2002) Toward spectral classification of L and T dwarfs: Infrared and optical spectroscopy and analysis, *The Astrophysical Journal* **564**, p. 466.

Germanà C., Zampieri L., Barbieri C., Naletto G., Čadež A., et al. (2012) Aqueye optical observations of the Crab Nebula pulsar, *Astronomy and Astrophysics* **548**, article id. A47, 7 pp.

Gies D.R. (2004) Time- resolved H-alpha spectroscopy of the Be star PLEIONE during a lunar occultation, *The Astronomical Journal* **100**, pp. 1601–1609.

Gil-Hutton R. (2007) Polarimetry of M-type asteroids, *Astronomy and Astrophysics* **464**, Issue 3, pp. 1127–1132, doi: 10.1051/0004–6361:20066348.

Gillon M., Triaud A.H.M.J., Damory B.-O., Jehin E., Agol E., et al. (2017) Seven temperate terrestrial planets around the nearby ultracool dwarf star TRAPPIST-1, *Nature* **542**, pp. 456–460, doi: 10.1038/nature21360.

Gliese W. (1969) *Catalogue of Nearby Stars*, Braun, Karlsruhe.

Gliese W., Jahreiss H. (1979) Nearby star data published 1969–1978, *Astronomy and Astrophysics Supplement* **38**, p. 423.

Graner F., and Dubrulle B. (1994a) Titius-Bode law in the solar system. 1: Scale invariance explains everything, *Astronomy and Astrophysics* **282**, Issue 1, pp. 262–268.

Graner F., and Dubrulle B. (1994b) Titius-Bode law in the solar system. 2: Build your own law from disk models, *Astronomy and Astrophysics* **282**, Issue 1, pp. 269–276.

GRAVITY Collaboration. (2018a) Detection of the gravitational redshift in the orbit of the star S2 near the Galactic centre massive black hole, *Astronomy and Astrophysics* **615**, p. L15, doi: 10.1051/0004-6361/201833.

GRAVITY Collaboration.(2018b) A geometric distance measurement to the Galactic center black hole with 0.3% uncertainty, *Astronomy and Astrophysics* **625**, p. L10, https://www.aanda.org/articles/aa/pdf/2019/05/aa35656-19.pdf.

Hanson R.B. (1975) A study of the motion, membership, and distance of the Hyades cluster, *The Astronomical Journal* **80**, pp. 379–401.

Hapgood M. (1992) Space physics coordinate transformations: A user guide, *Planetary and Space Science* **40**, pp. 711 717 and *Corrigendum Planetary and Space Science* **45**, Issue 8, p. 1047.

Hapgood M. (1995) Space physics coordinate transformations: The role of precession, *Annals of Geophysics* **13**, Issue 7, pp. 713–716.

Harada W., Fukushima T. (2004) A new determination of planetary precession, *The Astronomical Journal* **127**, pp. 531–538.

Hayes D.S., Latham D.W. (1975) A rediscussion of the atmospheric extinction and the absolute spectral distribution of Vega, *The Astrophysics Journal* **197**, p. 593.

Hayes D.S., Oke J.B., Schild R.E. (1979) A comparison of the Heidelberg and Nbs-Palomar spectrophotometric calibrations, *The Astrophysical Journal* **162**, p. 361.

Hazard C., Sutton J., Argue A.N., Kenworthy C.M., Morrison L.V., et al. (1971) 3C 273B - Coincidence of radio and optical positions, *Nature* **233**, p. 89.

Hearnshaw J.B. (1992) Origins of the stellar magnitude scale, *Sky and Telescope* **84**, p. 494.

Hellings R.W. (1986) Relativistic effects in astronomical time measurements, *The Astronomical Journal* **191**, p. 650.

Hewish A., Bell S.J., Pilkington J.D.H., Scott P.F., Collins R.A. (1968) Observation of a rapidly pulsating radio source, *Nature* **217**, Issue 5130, pp. 709–713.

Hiltner W.A. (1952) Photometric, polarization and spectrographic observations of O and B stars, *The Astrophysical Journal* **106**, p. 231.

Hilton J.L., Capitaine N., Chapront J., Ferrandiz J.M., Fienga A., et al. (2006) Report of the International Astronomical Union Division I Working Group on Precession and the Ecliptic, *Celestial Mechanics and Dynamical Astronomy* **94**, Issue 3, pp. 351–367, doi: 10.1007/s10569-006-0001-2.

Holmberg J., Flynn C., Portinari L. (2006) The colours of the Sun, *Monthly Notice of the Royal Astronomical Society* **367**, pp. 449–453.

Huang C.X., Bakos G.Á. (2014), Testing the Titius-Bode law predictions for Kepler multiplanet systems, *Monthly Notices of the Royal Astronomical Society* **442**, Issue 1, pp. 674–681.

Irwin J.B. (1959) Standard light-time curves, *The Astronomical Journal* **64**, p. 149.

Ives H.E., Stilwell G.R. (1938) An experimental study of the rate of a moving atomic clock, *Journal of the Optical Society of America* **28**, Issue 7, p. 215, doi: 10.1364/JOSA.28.000215.

Ives H.E., Stilwell G.R. (1941) An experimental study of the rate of a moving atomic clock. II, *Journal of the Optical Society of America* **31**, Issue 5, p. 369, doi: 10.1364/JOSA.31.000369.

Jacoby G.H., Hunter D.A., Christian C.A. (1984) A library of stellar spectra, *The Astrophysical Journal Supplement Series* **56**, p. 257.

Jehin E., Gillon M., Queloz D., Magain P., Manfroid J., et al. (2011) TRAPPIST: TRAnsiting Planets and PlanetesImals Small Telescope, *The Messenger* **145**, pp. 2–6.

Jewitt D. (2015) Color systematics of comets and related bodies, *The Astronomical Journal* **150**, Issue 6, article id. 201. p. 18. doi: 10.1088/0004-6256/150/6/201.

Johnson H.L., Morgan W.W. (1951) On the color-magnitude diagram of the Pleiades, *The Astrophysical Journal* **114**, p. 522.

Johnson H.L., Morgan W.W. (1953) Fundamental stellar photometry for standards of spectral type on the revised system of the Yerkes spectral atlas, *The Astrophysical Journal* **117**, p. 313.

Johnson H.L. (1966) Astronomical measurements in the infrared, *Annual Review of Astronomy and Astrophysics* **4**, p. 193.

Jones B.F. (1976) Gravitational deflection of light: Solar eclipse of 30 June 1973, *The Astronomical Journal* **81**, p. 455.

Kaplan G.H. (1981) *The 1976 IAU Resolution on Astronomical Constants*, US Naval Observatory Circular No. 163.

Kasten F., Young A.T. (1989) Revised optical air mass tables and approximation formula, *Applied Optics* **28**, pp. 4735–4738.

Keller H.U., Barbieri C., Lamy P., Rickman H., Rodrigo R., et al. (2007) OSIRIS the scientific camera system onboard rosetta, *Space Science Reviews* **128**, Issue 1–4, pp. 433–506, doi: 10.1007/s11214-006-9128-4.

Kinoshita H. (1977) Theory of the rotation of the rigid earth, *Celestial Mechanics* **15**, p. 277.

Knapp G.R., Kerr F.J. (1974) The galactic dust-to-gas ratio from observations of 81 globular clusters, *Astronomy and Astrophysics* **35**, p. 361.

Kovalevsky J., Lindegren L., Perryman M.A.C., Hemenway P.D., Johnston K.J., et al. (1997) The HIPPARCOS catalogue as a realisation of the extragalactic reference system, *Astronomy and Astrophysics* **323**, pp. 620–633.

Kristensen L.K. (1998) Astronomical refraction and airmass, *Astronomische Nachrichten* **319**, pp. 193–198.

Laskar J. (1985) Accurate methods in general planetary theory, *Astronomy and Astrophysics* **144**, Issue 1, pp. 133–146.

Laskar J. (1986) Secular terms of classical planetary theories using the results of general theory, *Astronomy and Astrophysics* **157**, Issue 1, pp. 59–70 and Erratu um *Ibidem* **164**, Issue 2, p. 437.

Laskar J., Joutel F., Robutel P. (1993) Stabilization of the Earth's obliquity by the Moon, *Nature* **361**, Issue 6413, pp. 615–617.

Laskar J., Robutel P., Joutel F., Gastineau M., Correia A.C.M., et al. (2004a) A long-term numerical solution for the insolation quantities of the Earth, *Astronomy and Astrophysics* **428**, pp. 261–285, doi: 10.1051/0004–6361:20041335.

Laskar J., Correia A.C.M., Gastineau M., Joutel F., Levrard B., et al. (2004b) Long- term evolution and chaotic diffusion of the insolation quantities of Mars, *Icarus* **170**, Issue 2, pp. 343–364.

Laskar J., Fienga A., Gastineau M., Manche H. (2011) La2010: A new orbital solution for the long-term motion of the Earth, *Astronomy and Astrophysics* **532**, p. A89, doi: 10.1051/0004–6361/201116836.

Lasker B., Sturch C.R., McLean B.J., Russell J.L., Jenkner H., et al. (1990) The guide star catalog. I-Astronomical foundations and image processing, *The Astronomical Journal* **99**, pp. 2019–2058.

Law N.M., Mackay C.D., Baldwin J.-E. (2006) Lucky imaging: High angular resolution imaging in the visible from the ground, *Astronomy and Astrophysics* **446**, pp. 739–745.

Lazzarin M., Marchi S., Barucci M.A., di Martino M., Barbieri C. (2004) Visible and near-infrared spectroscopic investigation of near-Earth objects at ESO: First results, *Icarus* **169**, pp. 373–384, doi: 10.1016/j.icarus.2003.12.023.

Leão I.C., Pasquini L., Ludwig H.G., de Medeiros J.R. (2019) Spectroscopic and astrometric radial velocities: Hyades as a benchmark, *Monthly Notices of the RAS* **483**, p. 5026.

Leblanc F., Doressoundiram A., Schneider N., Mangano V., López Ariste A., et al. (2008), High latitude peaks in Mercury's sodium exosphere: Spectral signature using THEMIS solar telescope, *Geophysical Research Letters* **35**, Issue 18. ID L18204. doi: 10.1029/2008GL035322.

Leeuwen van F. (2005) Rights and wrongs of the Hipparcos data, *Astronomy and Astrophysics* **439**, p. 805.

Leeuwen van F., Fantino E. (2005) A new reduction of the raw Hipparcos data, *Astronomy and Astrophysics* **439**, p. 791.

Leeuwen van F. (2007) Validation of the new Hipparcos reduction, *Astronomy and Astrophysics* **474**, pp. 653–664, doi: 10.1051/0004–6361:20078357.

Liebscher D.-E., Brosche P. (1998) Aberration and relativity, *Astronomische Nachrichten* **5**, pp. 309–318.

Lieske J.H., Lederle T., Fricke W., Morando B. (1977), Expression for the precession quantities based upon the IAU (1976) system of astronomical constants, *Astronomy and Astrophysics* **58**, pp. 1–16.

Lindegren L., Madsen S., Dravins D. (2000) Astrometric radial velocities. II. Maximum-likelihood estimation of radial velocities in moving clusters, *Astronomy and Astrophysics* **356**, p. 1119.

Mackay C., Rebolo R., Crass J., King D.L., Labadie L., et al. (2014) High-resolution imaging in the visible on large ground-based telescopes, *SPIE 9147–64*, Montreal, June 2014, arxiv: 1408.0117, doi: 10.1117/12.2055907.

MacMillan D., Fey A., Gipson G., Gordon D., Jacobs C., et al. (2018) Galactic Aberration in VLBI analysis, *International VLBI Service for Geodesy and Astrometry* 2018 *General Meeting Proceedings*, NASA/CP-2019-219039, pp. 163–168, Longyearbyen, Svalbard; Norway, Bibcode: 2019ivs..conf..163M.

Madden J.H., Kaltenegger L. (2018) A catalog of spectra, albedos, and colors of solar system bodies for exoplanet comparison, *Astrobiology* **18**, Issue 12, pp. 1559–1573.

Madsen S., Dravins D., Lindegren L. (2002) Astrometric radial velocities. III. Hipparcos measurements of nearby star clusters and associations, *Astronomy and Astrophysics* **281**, p. 446.

Makarov V.V. (2002) Computing the parallax of the Pleiades from the Hipparcos intermediate astrometry data: An alternative approach, *The Astronomical Journal* **124**, p. 3299 and *Ibidem* **126**, p. 2048.

Mangano V., Leblanc F., Barbieri C., Massetti S., Milillo A., et al. (2009) Detection of a southern peak in Mercury's sodium exosphere with the TNG in 2005, *Icarus* **201**, Issue 2, pp. 424–431.

Marini J. W., Murray C. W. (1973) Correction of Laser Range Tracking Data for Atmospheric Refraction at Elevations Above 10 Degrees, NASA GSFC X-591-73-351.

Masoumzadeh N., Boehnhardt H., Li J.-Y., Vincent J.B. (2015) Photometric properties and variations across the surface of asteroid (21) Lutetia, *Icarus* **257**, pp. 239–250, doi: 10.1016/j.icarus.2015.05.013.

Matthews A.T., Sandage A.R. (1963) Optical identification of 3C 48, 3C 196 and 3C 286 with stellar objects, *The Astrophysical Journal* **138**, p. 30.

Mayor M., Queloz D. (1995) A jupiter-mass companion to a solar-type star, *Nature* **378**, Issue 6555, pp. 355–359, doi: 10.1038/378355a0.

Mayor M., Pepe F., Queloz D., Bouchy F., Rupprecht G., et al. (2003) Setting new standards with HARPS, *The Messenger* **114**, pp. 20–24, Bib Code 2003Msngr.114...20M (ISSN0722-6691).

Mendillo M., Baumgardner J., Wroten J., Barbieri C., Umbriaco G., et al. (2012) A stable auroral red arc over Europe, *Astronomy & Geophysics* **53**, Issue 1, pp. 1.16–1.18.

Mignard F., Klioner S.A., Lindegren L., Hernandez J., Bastian U., et al. (2018) Gaia DR2. The celestial reference frame (Gaia-CRF2), *Astronomy & Astrophysics* **616**, article id. A14, 15 pp., doi: 10.1051/0004-6361/201832916.

Milgrom M. (1983) A modification of the newtonian dynamics: implications for galaxy systems, *The Strophysical Journal* **270**, pp. 371–384 and *Ibidem* pp. 384–389.

Minato A., Sugimoto N., Sasano Y. (1992) Optical design of cube-corner retroreflectors having curved mirror surfaces, *Applied Optics* **31**, pp. 6015–6020.

Moniez M. (2003) Does transparent hidden matter generate optical scintillation? *Astronomy and Astrophysics* **412**, p. 105.

Moro D., Munari U. (2000) The Asiago Database on Photometric Systems (ADPS). I. Census parameters for 167 photometric systems, *Astronomy and Astrophysics Supplement* **147**, p. 361.

Morrison L.V., Stephenson F.R. (1998) The sands of time and the Earth's rotation, *Astronomy & Geophysics* **39**, pp. 5–8.

Mróz P., Udalski A., Skowron D.M., Skowron J., Soszyński I., et al. (2019) Rotation curve of the milky way from classical cepheids, *The Astrophysical Journal Letters* **870**, Issue 1, article id. L10, 5 pp., doi: 10.3847/2041–8213/aaf73f.

Nesci R., Mandalari M., Gaudenzi S. (2007) Optical variability of the strong-lined and X-Ray-bright source 1WGA J0447.9−0322, *The Astronomical Journal* **133**, Issue 3, pp. 965–970.

Nesvotný D., Jenniskens P., Levison H., Bottke W.F., Vokrouhlický D., et al. (2010) Cometary origin of the zodiacal cloud and carbonaceous micrometeorites. implications for hot debris disks, *The Astrophysical Journal* **713**, Issue 2, pp. 816–836, doi: 10.1088/0004-637X/713/2/816.

Nobili A.M., Will C.M. (1986) The real value of Mercury's perihelion advance, *Nature* **320**, pp. 39–41.

Nobili A.M., Milani A., Carpino M. (1989) Fundamental frequencies and small divisors in the orbits of the outer planets, *Astronomy and Astrophysics* **210**, pp. 313–336.

Noorwood J., Moses J., Fletcher L., Orton G., Irwin P.G.J., et al. (2015) Giant planet observations with the James Webb space telescope, *Publications of the Astronomical Society of the Pacific* **128**, Issue 959, doi: 10.1088/1538-3873/128/959/018005.

Nota A., Pasquali A., Drissen L., Leitherer C., Robert C., et al. (1996) O stars in transition. I. Optical spectroscopy of Ofpe/WN9 and related stars, *Astrophysical Journal Supplement* **102**, p. 383 and paper II, (1997) *The Astrophysical Journal* **478**, p. 340.

Olnon F.M., Raimond E., Neugebauer G., van Duinen R.J., Habing H.J., et al. (1986) IRAS catalogues and atlases — Atlas of low-resolution spectra, *Astronomy and Astrophysics Supplement Series* **65**, Issue 4, pp. 607–1065, http://irsa.ipac.caltech.edu/IRASdocs/iras.html

Omizzolo A., Barbieri C., Rossi C. (2005) 3C 345: The historical light curve (1967–1990) from the digitized plates of the Asiago observatory, *Monthly Notices of the Royal Astronomical Society* **356**, Issue 1, pp. 336–342.

Owens J.C. (1967) Optical refractive index of air: Dependence on pressure, temperature, and composition, *Applied Optics* **6**, p. 51.

Panagia N., Gilmozzi R., Macchetto D., Adorf H.M., Kirschner R.P. (1991) Properties of the SN 1987A ring and the distance to the LMC, *The Astrophysical Journal* **380**, p. L23.

Paolicchi P., Burns J.A., Weidenschilling S.J. (2002) *Side Effects of Collisions: Spin Rate Changes, Tumbling Rotation States, and Binary Asteroids, Asteroids III*, Part IV, pp. 517–526, W.F. Jr. Bottke, A. Cellino, P. Paolicchi, R.P. Binzel, eds., University of Arizona Press, Tucson, AZ.

Park R.S., Folkner W.M., Konopliv A.S., Williams J.G., Smith D.E., et al. (2017) Precession of mercury's perihelion from ranging to the MESSENGER spacecraft, *The Astronomical Journal* **153**, p. 121 (7 pp.).

Pearce, J.A. (1955) The moving cluster in Taurus, *Publications of the Astronomical Society of the Pacific* **67**, Issue 394, p. 23.

Perryman M.A.C., Brown A.G.A., Lebreton Y., Gomez A., Turon C., et al. (1998) The Hyades: distance, structure, dynamics, and age, *Astronomy and Astrophysics* **331**, pp. 81–120.

Pian P., D'Avanzo P., Benetti S., Branchesi M., Brocato E., et al. (2017) Spectroscopic identification of r-process nucleosynthesis in a double neutron star merger, *Nature* **551**, Issue 7678, pp. 67–70.

Pogson N. (1856) Magnitudes of the 36 Minor Planets for Each Day of Each Month of the Year 1857 Monthly Notices of the Royal Astronomical Society, Vol. XVII. XVII, 12–15. Bibcode:1856MNRAS..17...12P. doi: 10.1093/mnras/17.1.12.

Ragazzoni R. (1996) Pupil plane wavefront sensing with an oscillating prism, *Journal of Modern Optics* **43**, Issue 2, pp. 289–293, doi: 10.1080/09500349608232742.

Ragazzoni R. (1997) On the existence of transverse relativistic aberrations in moving mirror, *Experimental Astronomy* **7**, Issue 3, pp. 209–219.

Ragazzoni R., Marchetti E., Valente G. (2000) Adaptive-optics corrections available for the whole sky, *Nature* **403**, Issue 6765, pp. 54–56, doi: 10.1038/47425.

Ray R.D., Steinberg D.J., Chao B.F., Cartwright D.E. (1994) Diurnal and semidiurnal variations in the Earth's rotation rate induced by oceanic tides, *Science* **264**, pp. 830–832.

Reino S., de Bruijne J., Zari E., d'Antona F., Ventura P. (2018) A Gaia study of the Hyades open cluster, *Monthly Notices of the Royal Astronomical Society* **477**, p. 3197.

Ribas I., Tuomi M., Reiners A., Butler R.P., Morales J.C., et al. (2018) A candidate super-Earth planet orbiting near the snow line of Barnard's star, *Nature* **563**, Issue 7731, pp. 365–368, doi: 10.1038/s41586-018-0677-y.

Richichi A., Calamai G., Stecklum B. (2002) New binary stars discovered by lunar occultations VI, *Astronomy and Astrophysics* **382**, p. 178.

Scarpa R., Marconi G., Gilmozzi R. (2003) Using globular clusters to test gravity in the weak acceleration regime, *Astronomy and Astrophysics* **405**, pp. L15–L18.

Schlesinger F. (1917) On the secular changes in the proper-motions and other elements of certain stars. *The Astronomical Journal* **30**, pp. 137–138.

Schmidt B.P., Kirschner R.P., Eastman, R.G. (1992) Expanding photospheres of type II SNs and the extragalactic distance scale, *The Astrophysical Journal* **395**, p. 366.

Schwarzschild K. (1916) Über das Gravitationsfeld eines Massenpunktes nach der Einsteinschen Theorie, *Sitzungsberichte der Königlich Preußischen Akademie der Wissenschaften (Berlin)* pp. 189–196.

Seeliger H. (1900) Bemerkung über veränderliche Eigenbewegungen, Astronomische Nachrichten **154**, Issue 3, p. 65.

Simon J.L., Bretagnon P., Chapront J., Chapront-Touze M., Francou G., et al. (1994) Numerical expressions for precession formulae and mean elements for the Moon and the planets, *Astronomy and Astrophysics* **282**, p. 663.

Skowron D.M., Skowron J., Mróz P., Udalski A., Pietrukowicz P., et al. (2019) A three-dimensional map of the Milky Way using classical Cepheid variable stars, *Science* **365**, Issue 6452, pp. 478–482, doi: 10.1126/science.aau3181.

Smith S.M., Stober G., Jacobi C., Chau J.L., Gerding M., et al. (2017) Characterization of a double mesospheric bore over Europe, *Journal of Geophysical Research: Space Physics* **122**, pp. 9738–9750, doi: 10.1002/2017JA024225.

Soma M., Aoki S. (1990) Transformation from FK4 system to FK5 system, *Astronomy and Astrophysics* **240**, p. 150.

Standish E. (1981) Two different definitions of the dynamical equator and the mean obliquity, *Astronomy and Astrophysics* **101**, pp. L17–19.

Standish E. (1982) Orientation of the JPL ephemerides, DE 200/LE 200, to the dynamical equinox of J 2000, *Astronomy and Astrophysics* **114**, pp. 297–302.

Standish E. (1990) The observational basis for JPL's DE 200, the planetary ephemerides for the Astronomical Almanac, *Astronomy and Astrophysics* **233**, pp. 252–271.

Standish E. (1998a) JPL, Planetary and Lunar Ephemerides, DE405/LE405. Interoffice memorandum IOM312.F, JPL, Los Angeles.

Standish E. (1998b) Time scales in the JPL and CfA ephemerides, *Astronomy and Astrophysics* **336**, Issue 1, pp. 381–384.

Stephenson F.R., Morrison L.V. (1984) Long term changes in the rotation of the earth: 700 B.C. to A.D. 1980, *Philosophical Transactions of the Royal Society of London, Series A* **313**, p. 47.

Stern A., Weaver H.A., Spencer J.R., Olkin C.B., Gladstone G.R., et al. (2019) Initial results from the New Horizon exploration of 2014 MU68, a small Kuiper Belt object, *Science* **364**, eaaw9771, doi: 10.1126/science.aaw9771.

Stumpff P. (1979) Rigorous Treatment of Stellar Aberration, and Doppler shift, and the barycentric motion of the Earth, *Astronomy and Astrophysics* **78**, Issue 2, pp. 229–238.

Stumpff P. (1985) Rigorous treatment of the heliocentric motion of stars, *Astronomy and Astrophysics* **144**, p. 232.

Swings P. (1965) Cometary spectra, *Quarterly Journal of the Royal Astronomical Society* **6**, p. 28.

Tedesco E.F., Veeder G.J., Fowler J.W., Chillemi J.R. (1992) The IRAS Minor Planet Survey, PL-TR-92-2049.

Torr M.R., Torr G.D., Stencel R. (1979) Zodiacal light surface brightness measurements by Atmospheric Explorer-C, *Icarus* **40**, p. 49.

Tsiganis K., Gomes R., Morbidelli A., Levison H.F. (2005) Origin of the orbital architecture of the giant planets of the Solar System, *Nature* **435**, pp. 459–461, doi: 10.1038/nature03539.

Tubiana C., Guettler C., Kovacs C. Bertini I., Bodewitz D., et al. (2015) Scientific assessment of the quality of OSIRIS images, *Astronomy & Astrophysics* **583**, article id. A46, 9 pp., doi: 10.1051/0004-6361/201525985.

Udalski A., Szymański M.K., Szymański G. (2015) OGLE-IV: Fourth phase of the optical gravitational lensing experiment, *Acta Astronomica* **65**, Issue 1, pp. 1–38, arXiv:1504.05966v1 [astro-ph.SR].

Urban, S.E., Corbin T.E., Wycoff G.L., Martin J.C., Jackson E.S., et al. (1998) The AC 2000: The asrographic catalogue on the system defined by the Hipparcos and Tycho Catalogue, *The Astronomical Journal* **115**, pp. 1212–1223.

Verani S., Barbieri C., Benn C., Cremonese G., Mendillo M. (2001) The 1999 quadrantids and the lunar Na atmosphere, *Monthly Notices of the Royal Astronomical Society* **327**, pp. 244–248.

Vernet E., Ragazzoni R., Esposito S., Hubin N. (2002) Beyond Conventional Adaptive Optics: A Conference Devoted to the Development of Adaptive Optics for Extremely Large Telescopes, *European Southern Observatory Conference and Workshop Proceedings 58*, Garching, Germany, ISBN 3923524617.

Vidal C.R., Cooper J., Smith E.W. (1973) Hydrogen stark-broadening tables, *The Astrophysical Journal Supplement Series* **25**, p. 37.

Whitford A.E. (1951) An extension of the interstellar absorption curve, *The Astrophysical Journal* **107**, p. 102.

Will C.M. (2003) Propagation speed of gravity and the relativistic time delay, *The Astrophysical Journal* **590**, Issue 2, pp. 683–690, doi: 10.1086/375164.

Williams J.G. (1994) Contributions to the earth's obliquity rate, precession and nutation, *The Astronomical Journal* **108**, pp. 711–724.

Wing W.H. (2003) On the aberration of light from a moving retroreflector, *Optic Communications* **220**, pp. 1–6.

Womack M., Sarid G., Wirzchos K. (2017) CO in distantly active comets, *Publications of the Astronomical Society of the Pacific* **129**, Issue 973, p. 031001, doi: 10.1088/1538-3873/129/973/031001.

Zacharias N. (2018) Astrometric Surveys in the Gaia Era, Astrometry and Astrophysics in the Gaia Sky, *Proceedings of the IAU, IAU Symposium*, Volume 330, pp. 49–58, Nice, France.

Zampieri L., Čadež A., Barbieri C., Naletto, G., Calvani M., et al. (2014) Optical phase coherent timing of the Crab nebula pulsar with Iqueye at the ESO new technology telescope, *Monthly Notices of the Royal Astronomical Society* **439**, Issue 3, pp. 2813–2821.

Zampieri L., Richichi A., Naletto G., Barbieri C., Burtovoi A., et al. (2019) Lunar occultations with Aqueye+ and Iqueye, *The Astronomical Journal* **158**, Issue 5, article id. 176, 7 pp., doi: 10.3847/1538-3881/ab3979.

Zanna P., Sigismondi C. (2004) Dúngal, Letterato e Astronomo https://arxiv.org/ftp/arxiv/papers/1211/1211.3687.pdf

Zellner B., Tholen D.J., Tedesco E.F. (1985) The eight-color asteroid survey- Results for 589 minor planets, *Icarus* **61**, pp. 355–416.

## WEB SITES

This Section contains a list, by no means exhaustive, of useful websites, several of them cited in the text and others of general interest.

For the **definition of fundamental constants and units**, see https://www.bipm.org/en/measurement-units/ and https://physics.nist.gov/cuu/Units/index.html

Sites containing **IAU** and **IERS** resolutions, graphs, tables of useful and recent data, products such as accurate subroutines for astrometric applications are the following:

https://www.iau.org/science/scientific_bodies/divisions/A/ for Fundamental Astronomy; http://www.usno.navy.mil/astronomy; http://www.iers.org/.

The site: http://iers.obspm.fr/icrs-pc/newwww/icrf/ provides data of successive realizations of the International Celestial Reference Frame (ICRF) by VLBI. The third release, ICRF3, is the new fundamental celestial reference frame adopted by the International Astronomical Union at its XXXth General Assembly (20-31 August 2018) as a replacement of ICRF2 as of 1 January 2019.

The *Astronomical Almanac Online!* is mirrored in two sites:

http://asa.hmnao.com/ in the UK and
http://asa.usno.navy.mil/ in the USA.

The two main astronomical **software packages** (USNO NOVAS and IAU SOFA respectively) are found in: http://aa.usno.navy.mil/software/novas/novas.info.php and http://www.iausofa.org/.

Notice that the data calculated by the US Naval Observatory utilizes the NOVAS package, while the data prepared in UK rely on the IAU Standard of Fundamentals Astronomy (SOFA) software package. Results generally agree to better than few microarcseconds.

The **Multiyear Interactive Computer Almanac (MICA)**, based on the US Naval Observatory calculations, is available from Willman-Bell, Richmond, also on CD ROM

For information on **timescales** and on historical and modern **clocks** see. http://tycho.usno.navy.mil/ and the IERS site (https://www.iers.org/).

For instance, the **duration of the day** since 1623 is given by: https://www.iers.org/IERS/EN/Science/EarthRotation/LODsince1623.html.

The uncertainties still affecting theories of **Earth rotation, precession and nutation** can be appreciated by reading the 2019 report of the IAU/IAG Joint Working Group on Theory of Earth Rotation and Validation, by Ferrándiz and collaborators: https://syrte.obspm.fr/astro/journees2019/journees_pdf/SessionIV_2/FERRANDIZ_JSR19_JWG_TERV.pdf.

The report contains many useful references to recent work.

The site http://star-www.rl.ac.uk/star/docs/sun67.htx/sun67.html contains a library of **SLALIB** routines, intended to make accurate and reliable positional-astronomy applications. A number of telescope control systems around the world make use of such library.

The updated version of Franz's **Heliographic coordinate systems** is in: https://www2.mps.mpg.de/homes/fraenz/systems/

For the **Earth Gravity Models EGM)** in their successive releases by the National Geo-Spatial Intelligence Service (NGA) see: https://www.nga.mil/ProductsServices/GeodesyandGeophysics/Pages/EarthGravityModel.aspx

Information on the **Sun-Earth interactions and space weather** in general (aurorae, meteorites, sunspot number, cosmic ray fluxes etc.) see: https://spaceweather.com.

For a description of the **Lageos** program see:
https://ilrs.cddis.eosdis.nasa.gov/missions/satellite_missions/current_missions/lag1_general.html

**Ephemerides of planets, comets, and asteroids**

Several sites provide ephemerides of planets, comets, and asteroids for each location on Earth, with a variety of possible choices. A widely used program is **Horizon** by the Jet Propulsion Laboratory http://ssd.jpl.nasa.gov/horizons_doc.html, which has the option of siting the observer on several spacecraft.

Another site is the one of the Paris Astronomical Observatory **MCCE**, https://www.imcce.fr/.

The **transits of Mercury and Venus** in front of the solar disk are described in detail in:

https://eclipse.gsfc.nasa.gov/transit/catalog/MercuryCatalog.html,
https://eclipse.gsfc.nasa.gov/transit/catalog/VenusCatalog.htm, respectively.

The latter web site discusses the **SAROS**, https://eclipse.gsfc.nasa.gov/SEsaros/SEsaros.html.

The Nov.11, 2019 transit of Mercury was observed for the first time in extreme UV light from space by NASA's Solar Dynamics Observatory **SDO**: sdo.gsfc.nasa.gov/gallery/potw/item/713#:~:text=On%20May%209%2C%202016%2C%20Mercury,      a%20transit%20of%20the%20Sun.&text=NASA's%20SDO%20studies%20the%20Sun, %3A%20Solar%20Dynamics%20Observatory%2C%20NASA.

Another website for the calculations of all **solar eclipses and transits** for several millennia, is **CalSky**, https://www.calsky.com/.

**For lunar eclipses, the circumstances of recent and upcoming lunar eclipses for any location worldwide can be obtained from** http://aa.usno.navy.mil/data/docs/LunarEclipse.html.

**Another useful web site for eclipses is NASA's** https://www.nasa.gov/eclipse, reporting calculations by Fred Espenak, with Google maps of solar eclipse paths on Earth surface

For the **Moon**
Useful animations regarding the **lunar aspect and orbit** are given by NASA's site: https://svs.gsfc.nasa.gov/4442.

**Web sites specific for lunar occultations are:**

http://www.lunar-occultations.com/entersite.htm,
http://tdc-www.harvard.edu/occultations/occultations.html.

The latter **site gives information about stellar occultations in general and separately for each planet, with external links to other useful sites**.

For **lunar occultations of X-ray sources** see the EXOSATE and **ROSAT** satellites sites:
https://www.cosmos.esa.int/web/exosat and http://wave.xray.mpe.mpg.de/rosat.

**Specifically for Sco-X1 see:** http://heasarc.gsfc.nasa.gov/docs/objects/heapow/archive/compact_objects/rosat_scox1_occult.html.

For the lunar **GRAIL** mission see: https://www.nasa.gov/mission_pages/grail/main/index.html.

The Project **Apollo** to determine high precision distances to the Moon by **laser ranging** with the 3.5 m Apache Point Telescope is described in https://tmurphy.physics.ucsd.edu/apollo/apollo.html.

There are other projects devoted to the same task, for instance in France at Grasse and Wettzel in Germany.

For **Mercury**;
For Mariner 10: https://solarsystem.nasa.gov/missions/mariner-10/in-depth/
For NASA Mercury Messenger: http://messenger.jhuapl.edu/
For ESA BepiColombo: http://sci.esa.int/bepicolombo/

For **asteroids**: http://asteroid.lowell.edu provides observing services, including finding charts with respect to USNO stars.

The site https://newton.spacedys.com/neodys/ provides information and services for **Near Earth Asteroids**.

See also the **Near-Earth Object Wide-field Infrared Survey Explorer** on the NASA/ IPAC Infrared Science archive **at** https://irsa.ipac.caltech.edu/frontpage/.

For the details of the **NEAR mission to asteroid 433 Eros** see http://near. jhuapl.edu/.

For an overview of NASA's spacecraft DAWN which visited **Vesta and Ceres** see https://solar-system.nasa.gov/missions/dawn/overview/.

For an overview of the ESA's spacecraft Rosetta, which visited he third largest asteroid **Lutetia**, the small asteroid **Steins** and the comet **67P/C-G** of the Jupiter family see: https://www.esa.int/Our_Activities/Space_Science/Rosetta

For the Solar Heliospheric Observatory **SOHO**: https://sohowww.nascom.nasa.gov/.

For **Ulysses**: http://sci.esa.int/ulysses/.

Most data on the **planets and moons of the outer solar system** came from the **Voyagers**. Today, the spacecraft provide data on the so-called **heliopause** beyond 140 AU. See: https://voyager.jpl.nasa.gov/.

**Pluto and a Kuiper Belt Object** dubbed ultima Thule were visited by the NASA spacecraft **New Horizons**, http://Pluto.jhuapl.edu/.

Several software packages allow the determination of the **orbit** from the observations.

See for instance for the programs OrbFit and Find_Orb. The site; http://www.projectpluto.com/find_orb.htm, with Italian and French versions, describes Find_Orb, a given in the MPC (Minor Planet Center) for the NEODyS or AstDyS formats, and finds the corresponding orbit.

For **exoplanets:**
A list of exoplanets and their properties can be found in the *Extrasolar Planets Encyclopaedia* at: http://exoplanet.eu/catalog/.

See also the NASA Exoplanet Science Institute at https://nexsci.caltech.edu/.

For individual dedicated satellites:
NASA Kepler: https://www.nasa.gov/mission_pages/kepler/main/index.html
NASA TESS: https://heasarc.gsfc.nasa.gov/docs/tess/https://tess.gsfc.nasa.gov/
CNES COROT: http://sci.esa.int/corot/
ESA CHEOPS: http://sci.esa.int/cheops/
ESA PLATO: i http://sci.esa.int/plato/
ESA ARIEL: http://sci.esa.int/ariel

The **difficulties of imaging** a very faint planet next to a bright star are really challenging. See for instance the coronagraphic systems SPHERE at the ESO VLT: https://www.eso.org/sci/facilities/paranal/instruments/sphere.html.

The system suppresses the light of the star and permits to obtain images of planets typically a fraction of arcsec away.

A comprehensive archive of data from **Infrared Surveys**, from ground and Space, is provided by the NASA/ IPAC Infrared Science archive at https://irsa.ipac.caltech.edu/frontpage/

**For satellites of cosmological interest:**

- NASA WMAP https://map.gsfc.nasa.gov/
- ESA Planck: https://www.cosmos.esa.int/web/planck

## Atmospheric optics

Among the useful sites on the atmosphere we quote: https://www.atoptics.co.uk/; in particular for the ionosphere: http://www.ngdc.noaa.gov.

### Active and Adaptive Optics

All major observatories describe their **Active Optics** systems for the correction of structural flexures and optical misalignments, and the **Adaptive Optics** systems employed to optimize the image quality by correction of seeing effects. These systems are often aided by lasers to exploit the neutral Na layer at 100 km. See for instance https://www.eso.org/public/search/?q=adaptive+optics.

### Interferometers in the extended visible band

**CHARA**, Mt. Wilson California, http://www.chara.gsu.edu/;

ESO **VLTI**, Cerro Paranal Chile: https://www.eso.org/sci/facilities/paranal/telescopes/vlti.html;

Large Binocular Telescope **LBT**, sited on Mt Graham in Arizona: http://www.lbto.org/.

### Millimeter/Submillimeter telescopes and interferometers:

For the Atacama Large Millimeter/submillimeter Array (*ALMA*) *see:* https://www.eso.org/public/italy/teles-instr/alma/

IRAM: https://www.iram-institute.org/EN/content-page-8-1-8-0-0-0.html

*For the **radio Very Long Baseline Interferometry (VLBI)** see for instance*: http://people.rses.anu.edu.au/lambeck_k/pdf/82.pdf.

For the European VLBI Network (**EVN**) see: https://www.evlbi.org/

## Catalogues

### The Hipparcos and Tycho Catalogues

ESA SP-1200, 17 Volumes, (https://www.cosmos.esa.int/web/hipparcos).

A new reduction of the data has generated **Hipparcos-2**: https://www.cosmos.esa.int/web/hipparcos/hipparcos-2.

The site has several useful subsites, e.g. for searching a particular star see: https://www.cosmos.esa.int/web/hipparcos/search-facility.

The data is available from the Centre des Donnés Stellaires (CDS, http://cdsweb.u-strasbg.fr) through the VizieR browser: https://vizier.u-strasbg.fr/viz-bin/VizieR.

**The Hipparcos-2 Catalogue is also available from ESASky (sky.esa.int) and from the Gaia Archive (https://gea.esac.esa.int/archive/).**

The original Tycho-1 Catalog has been replaced by the **Tycho-2** Catalog, namely an astrometric and photometric reference catalogue of the 2.5 million brightest stars on the entire sky, accessible through VizieR. See also Erik Høg's homepage: http://www.astro.ku.dk/%7Eerik/Tycho-2/, last updated 6 January 2010, with many useful references.

The **USNO-A2.0 Catalog** (http://tdc-www.harvard.edu/catalogs/ua2.html) provides astrometric information and the blue and red magnitude of 526,280,881 stars, based on a re-reduction of the Precision Measuring Machine (PMM) scans that were the basis for the USNO-A1.0 catalog. The major difference between A2.0 and A1.0 is that A1.0 used the Guide Star Catalog (see Lasker et al., 1990, in the References) as its reference frame, whereas A2.0 uses the **ICRF** as realized by the USNO ACT catalog (Urban et al., 1998, in the References section).

**UCAC5**, namely the fifth version of the U.S. Naval Observatory CCD **Astrograph Catalog** (see e.g. Zacharias N. 2018 in the References section) with stellar positions on the Gaia coordinate system, is available also from VizieR.

Let us recall that the photographic **Sky Surveys** is a set of all-sky 6.5x6.5 sq. degree plates in E, V, J, R, and N photometric bands conducted with the Palomar and UK Schmidt telescopes.

The **digitized** versions of the **Sky Surveys** (DSS), covering the entire celestial sphere, can be accessed through the web site of the HST (http://www.stci.edu). The tool to recover a field is:

http://archive.stsci.edu/cgi-bin/dss_plate_finder/.

This site provides also the Guide Star Catalog and the Guide Star Photometric Catalog.

The **Canadian Astronomy Data Centre**, http://cadcwww.hia.nrc.ca, contains a wealth of telescope data products, including the Canadian Version of the **DSS**.

The **ESO Skycat** tool http://www.eso.org/sci/observing/tools/skycat.html combines visualization of images and access to catalogs and archive data for astronomy, including access to the DSS and HST Guide Star Catalog.

As further examples of digitized photographic archives, see the Archives of Photographic PLates for Astronomical USE (**APPLAUSE**), https://www.plate-archive.org/applause/

and the **Digital Plate Archives of Hamburger Sternwarte**, https://plate-archive.hs.uni-hamburg.de/index.php/en/.

The VizieR browser links to other digital surveys, e.g. Byurakan's, http://cdsarc.u-strasbg.fr/viz-bin/VizieR?-meta.foot&-source=VI/116

The already quoted **Strasbourg Astronomical Data Center** (CDS, http://cdsweb.u-strasbg.fr) collects and distributes astronomical data catalogs, related to observations of stars and galaxies, and other galactic and extragalactic objects. Catalogs about the solar system bodies and atomic data are also included. The catalogs and tables managed by CDS can be summarized as follows:

5620 Catalogs, of which 4581 are available online (as full ASCII or FITS files), and 4264 are also available through the VizieR browser.

The NASA **Extragalactic Database** NED (http://nedwww.ipac.caltech.edu/) gives coordinates, magnitudes, redshifts, cross references, finding charts, etc., for 7,600,000 extragalactic objects, as well as *thesis abstracts* of doctoral dissertations on extragalactic topics.

*SkyView* (http://skyview.gsfc. nasa.gov/) is a Virtual Observatory generating images of any part of the sky at wavelengths in all regimes from Radio to Gamma-Ray.

The **Virtual Observatory** (VO, http://ivoa.net/) is an international alliance to coordinate access to astronomical datasets and other resources, defining the needed technical standards.

References to a wealth of **photometric systems** are given by: Mermilliod J.-C., Hauck B., Mermilliod M. *The General Catalogue of Photometric Data,* University of Lausanne, Switzerland, http://obswww.unige.ch/gcpd/gcpd.html, and http://cdsarc.u-strasbg.fr/viz-bin/cat/II/168

**Spectroscopic Catalogues**

The ESA **International Ultraviolet Observatory** (IUE) and **Infrared Space Observatory** (ISO websites are http://sci.esa.int/iue/31297-archive/ and https://www.cosmos.esa.int/web/iso

The **ELODIE** library is in http://atlas.obs-hp.fr/elodie/

The **HST** site http://www.stsci.edu/institute/ can be searched for UV and optical spectra. For instance, CALSPEC contains the composite stellar spectra that are flux standards on the HST system.

Current and future research programs can be viewed in: https://www.stsci.edu/stsci-research/research-topics-and-programs/.

Searching the HST archive, http://archive.stsci.edu/, links to the Barbara A. Mikulski archive for Space Telescopes (**MAST**), https://mast.stsci.edu/portal/Mashup/Clients/Mast/Portal.html. MAST can be used to search multiple collections of astronomical datasets to find astronomical data, publications, and images.

For the Infrared Astronomical Satellite (**IRAS**) spectra see: http://irsa.ipac.caltech.edu/IRASdocs/iras.html

For the infrared Space Observatory (**ISO**) see: https://www.cosmos.esa.int/web/iso

For the **Herschel** Space Observatory (formerly called FIRST): https://sci.esa.int/web/herschel

For the **Spitzer** Space Telescope see: https://www.nasa.gov/mission_pages/spitzer/main/index.html

In general, for photometric and spectral catalogues, consult the already cited *VizieR On-line Data Catalog.*

# Index

α Boo (Arcturus) 125, 242, 247
α Car (Canopus) 242
α CMa *see Sirius*
α Cen (Rigil Kent) 117, 198, 242
α Lyr (Vega) 33, 121, 137, 242, 253–258, 270, 274, 275, 286, 292
α UMi (Polaris) 24, 61, 76, 242
α Vir (Spica) 60, 74

Aberration of light **93**
    annual **96, 98–100**, 106, 118, 145, 146
    diurnal **102**, 119, 146
    elliptic 28, 95, 100–102
    E-terms 100
    planetary 73, 93, **103**
    relativistic 99
    secular or galactic 93, 106
    solar **94–96**, 100
    stellar 93, 101, 103, 106
Abetti Giorgio 292
Absorption, extinction, *see also Atmosphere (Earth's), Spectroscopy*
    interstellar 31, 257, **263**, 264, 269, 271
    galactic 264, 265
Adams John C. 212, 215, 216
Adams Walter S. 199, 286
Airy George B. 20, 92, 98
Albedo
    geometric 266, 296
    Bond 266, 271
Alt–Azimuth *see Coordinate systems, Telescopes*
Analemma 56
Andromeda constellation 33, 127
Angstrom Anders J. 273
    unit Å 241, 245, 277, 279
Anomaly 193
    eccentric 152, 189, 190, 201
    equation of 176
    mean 50, 56, 181, 190, 207, 210, 213
    of the gravity 19, 24, 214
    true 47, 50, 178, 180, 189
Apex of
    Earth motion 97–102
    solar motion 120, 135, 137, 139
    stellar motions **135**, 136
Apollolunar laser ranging 29, 132, 150, 212
ApolloNASA lunar missions 116, 132
Apses, line of 95, 149, 176, 178, 189, 207, 227, 228
Aquarius constellation 30, 60
Arecibo Radio Observatory 222
ARIEL ESA satellite 238, 316
Aries constellation 26, 30, 60
Aristarchus of Samos 121
Armellini Giuseppe 196, 204
486958 Arrokoth (Ultima Thule, KBO) 75, 224, 297, 313

Asiago Observatory 10, 24, 56, 159, 235, 251, 269, 274–278, 293
ASIAGO imager 159
Asterisms 33
Asteroids 1, 10, 13, 17, 25, 62, 75, 93, 110, 114, 127, 166, 182, 183, 192, 196, 204, 205, 222, 225, 231
    angular sizes 115
    ephemerides 314
    Main Belt 196, 215, 234
    Near Earth (NEAs) 315
    photometry 241, 266, 267, 270, 316
    polarimetry 267, 276, 316
    spectra 273, 296
    taxonomy 270
    trails 121
    tumbling 223
Asteroseismology 238, 281
Atacama Large mm Array (ALMA) 300, 317
Atmosphere (Earth's) 98, 99, 113, **157**, 159, 316
absorption,extinction 160, 229, 241, **249**, 250, 269
adiabatic lapse 167
airmass 250, 265, 269, 276
    chemical composition 157–159
    chromatic dispersion **165**, 166
    gas law **166**
    geocorona 159, 185
    hydrogen 167
    mass 59, 157, 167
    molecules, velocities, weight 166–168
    optical depth 249, 260
    refraction **160–164**
    scale height 167
    scintillation, seeing, turbulence **168**
    vertical structure 157
Auwers Arthur von 28

Baade Walter 262
Baily Francis 225
    beads 225, 232
Balmer limit, lines, series *see Hydrogen*
Barnard's star 127, 130, 134, 236
BepiColombo ESA mission 223, 315
BesselFriedrich W. 28, 62; *see also Eclipses, Year*
    daily numbers, star constants 71, 102
    Sirius B proper motion 199
    61 Cyg parallax 121
Binary stars 94, 173, 204
    angular resolution 41, 168
    Campbell elements 201, 203
    masses 178, 179, 218
    orbital elements **189, 198**, 203
    parallaxes 120
    Thiele–Innes elements 203
    Zwier method 200
Binet's formula 176, 180

Binomial series expansion 4, 92, 209
Black Body **252**
    color indices 255–257
    Planck formula 252, 254
    Rayleigh – Jeans approximation 253
    Russel's formulae 255
    Stefan–Boltzmann law 199, 253, 254, 260–262, 288
    Wien approximation 253
    Wien law 253
Black Holes 128
    massive black holes (MBH) 128
    Schwarzschild radius 104, 128, 212
Blazars 277, 278
Bohr radii 283
Boltzmann
    constant 185, 252, 253
    excitation equation 281–285
    isothermal atmospheric model 167
Bouguer Pierre 173
    line 250
Bradley James 27
    aberration 27, 94, 96
    nutation 61, 62, 68, 69, 77
Brahe Tycho 62, 115, 157, 183, 211
Brigg's formulae 4, 6
Brown Ernest W. 210
Bunsen Robert W. 273

Calendar 10, 30, 48
    Gregorian 53–57, 146
    Julian 53–**55**
    lunar 55, 212
    proleptic 53
Callandreau Octave P. 92
Campbell's elements *see Binary stars*
Cardinal points 18, 19, 74
Carrington Richard C. 33
Cartesian space 1, 5; *see also Coordinate systems*
Cassini Jean Dominique 62, 157; *see also Moon, Saturn*
    Galilean moons 93
    solar parallax 115
Cassini–Huygens NASA-ESA-ASI spacecraft 106, 208, 298
Cauchy's atmospheric formula 165
Celestial Intermediate Origin (CIO) 153, 183
Celestial Mechanics 1, 12, 62, 91, 107, 203, 208, 210,
    216, 221
Ceres 75, 316
    apparent magnitude, photometry 242, 265, 266
    classification 62
    dimensions 265
    discovery 196
Cerro Paranal Observatory 168, 169, 316
Cesium 152
o Cet (Mira) 294
ChandlerSeth C. 89
    wobble 89
CHARA optical interferometer 171, 316
Charge-Coupled Devices (CCDs) *see Detectors*
Charlier Carl 138
Charon moon of Pluto 75, 183, 224, 316
CHEOPS ESA satellite 238, 316
Clairaut Alexis C. 77, 91, 92, 205, 207, 221
Clark Alvan 199
Clavius Christof S.J. 55

Clocks 24, 27, 28, 56, 62, 74, 90, 96, 145–155, 314
    moving, relativity 126
Color indices *see Black Body, Photometry*
Coma cluster of galaxies 31
Comets 25, 32, 35, 74, 93
comae 275, 296, 299, 300
    1P/Halley 195, 275, 276
    1I/2017 U1 (Oumuamua) 179
    2I/Borisov 179
    67P/Churyumov-Gerasimenko (C-G) 75, 219, 248,
        300, 316
    ISON C/2012 S1, 298, 200
    Jupiter family 219, 269, 316
    molecules 251, 273, 299, 300
    Near Earth (NECs) 296
    Oort's cloud 219
    orbits 103, 110, 114, 275
    polarization 276
    Shoemaker-Levy 219
    spectra 275, 276, 294, 298–300
    tails 296
    1864 II Tempel 273
Cones *see Eye*
Constellations 17, 19, 30, 32, 33, 60, 61, 73, 241
Coordinate systems 10, 12, **17**, 29, 30, 32
    Alt-Azimuth **18**, 29
    apparent (*see Atmosphere (Earth's), Gravitational
        Deflection of Light*)
    barycentric 73, 75, 103, 104, 119–125, 129, 154,
        173–180, 206, 216
    cartesian 1, 5–11, 26, 30, 35, 70, 73, 75, 80, 81, 90, 113,
        122, 129, 135, 180, 189, 192, 196, 230
    declination 19, **25**, 26–29, 37, 38, 45
    ecliptic 23, **29, 63**
    equatorial **26**, 65, 80, 81, 135, 136
    equinoxes 26, 27–31, 48–56
    galactic **31**, 32, 140
    galacto-centric 32
    geocentric 10, 19, 23, 24, 110, 114, 154, 193, 196,
        198, 230
    geodetic 19–24, 60, 89, 110, 111, 154
    heliocentric 35, 71, 73, 111, 116–118, 119, 122–128,
        135–141, 192–198, 206, 207
    heliographic 314
    horizon
    hour angle **25**, 37, 38
    latitude,longitude (*see Ecliptic, Galactic, Terrestrial*)
    mean (*see Precession, Nutation*)
    polar, spherical **7**, 11
    reduced 192, 202
    right ascension **26**–29, 42–51
    terrestrial 8, **9**, 10, 18, 39
    topocentric 63, 73, 103–115, 122, 164, 193, 198
    transformations 11, **35**, 39, 45, 73, 104, 110, 138, 193
    true (*see Precession, Nutation*)
Coordinated Universal Time (UTC) *see Times*
Copernicus Nicolaus 1, 50, 60–62, 115
    1.8m Asiago telescope 24, 159, 235
COROT CNES mission 237, 316
Coudèfocus 42
Crab nebula (M1), pulsars 155, 171
Curve of growth *see Spectroscopy*
Cygnus constellation 138
61 Cyg 121, 293

d'Alembert Jean le Rond 70, 77, 191, 205
date
    Julian 55
    century 53
    line 10
    Modified (MJD) 55
Day
    Julian **55**, 57, 153
    short-term fluctuations 60, 150
    secular variations 60, 150
    side real 48, 49, 52, 84, 87, 88, **145**, 153, 155
    solar 48, 49–53, **146–153**, 183, 222, 223
Decibel (dB) 242
Deimos moon of Mars 219, 223
Delaunay Charles-Eugene 207, 212, 214, 220
De la Rue Warren 225
Detailed balance condition (*see Spectroscopy*)
Detectors
    CCD 224, 248, 249, 270, 274, 275, 277, 292, 293,
        299, 300
    eye 8, 24, 32, 33, 168, 241, 242, 246, **247**, 249, 254,
        293, 298
    photoelectric photometers 248, 276
    photographic emulsion 243, 245, 247, 248, 275, 276
    photon counting, SPADs 171, 235, 237, 247, 248, 269
Deuterium 289, 290
Donati Giovanni B. 273
Doppler Christian 125
    broadening (*see Spectroscopy*)
    effect, shift 120–134, 141–143, 279, 280
Dorado constellation 27
Draco constellation 26
Draconic (draconitic) *see Month, Year*
γ Dra (Eltanin) 33, 61, 62, 94, 96
ω Dra 27
Dust
    atmospheric 251
    cometary 208, 269, 275, 276, 289, 290, 300
    interplanetary 271, 296
    interstellar 140, 242, 263–265
    meteoric 229, 289
Dwarf *see Planets, Stars*
Dynamics, *see also General Relativity, Celestial
        Mechanics*
    chaotic 173
    Modified Newtonian (MOND) 207
    Newtonian 94, 128, 173, 207

Earth, *see also Atmosphere (Earth's), Precession,
        Nutation, Free Rotation*
    angular velocity, rotation 27, **47**, 86
    escape velocity 167, 184
    figure, dimensions **9**, 14, 19–23, *see also WGS84*
    flattening 20, 78, 222
    Gravitational Model (EGM, EGM96) 20, 314
    magnetic field 32, 152, 159, 160, 221
    mass, $GM_\oplus$, 79, 83, 178
    moments of inertia 64, 78, 83, 84
    orbit **47**, 59, 63, 94–101, 117, 119. 129, 134, 211, 212
    rotation angle ERA 145, 147, 149, 151, **153**, 155
    surface gravity 78, 92
Earth–Moon system
    barycenter 96, 98, 101, 116, 119, 129
    distance 150

dynamics, perturbations 173, 183, 186, 210, 211
    Lagrangian points 218
Easter date 57
ECAS, photometric and taxonomic system 267, 270
Eclipses **225**, 315
    ancient 149, 225
    Besselian elements **230**, 231, 239
    exeligmos 232
    fundamental plane 230
    lunar 229
    magnitudes 230
    Medicean moons 93
    Metonic cycles 232
    number and repetitions 231
    Saros cycles 232, 233, 315
    solar 227
Ecliptic **17**, *see also Coordinate systems*
    definition 62
    movements (*see Precession, Nutation*)
    obliquity 17, 26, 62–70, 79–86, 102, 222
    poles 26, 27
Eddington Arthur 105, 225
Einstein Albert 98, 104, 205, 282
Electron pressure *see Spectroscopy*
Elongations *see Planets*
Encke Johan Franz 115, 116, 182
Energy, *see also Radiation*
    atomic and molecular levels 152, 248, 273–299
    constant, kinetic, potential, total 173–187, 194, 206,
        219, 280
    dark 289
    precessional 84
    rotational 85, 87, 149, 150, 212
Ephemerides 24, 29, 93, 104, 115, 146–152, **189**, 204–211,
        234, 239, 265, 314
    from orbital elements **192**
Ephemeris
    meridian 151, 153
    time (ET) (*see Times*)
Equation of the center (EC) 50, 192, 211
Equation of time 48–56, 95
Equator *see Coordinate Systems*
Equatorial
    coordinates (see *Coordinate systems*)
    telescopes 29, 170
Equinoxes *see Coordinate systems, Sun, Precession,
        Nutation, Times*
Eris 224
Eros asteroid 115, 116, 315
Euler Leonhard 52, 77, 205, *see also Earth's Free Rotation*
    nutation 19, 59, 60, 77, 89
    rotation angles 86, 88
Evection 211
Excitation *see Spectroscopy*
Exoplanets 218, 222, 223, 225, 235–238, 316
EXOSAT ESA satellite 225, 315
Extinction *see Absorption*
Eye *see Detectors*

Filters 243–248z
Fizeau Hyppolite 94, 171
Flammarion Camille 214
Flamsteed John 28
Fornax galaxy 262

Foucault Leon 94, 273
Fourier
    series 171, 191
    spectroscopy 274
Fowler Alfred 285
Fragmentation *see Planets*
Fraunhofer Joseph von
    A and B bands 249, 251, 273
    optical instruments 121
Fresnel fringes 234
Fried's parameter *see Atmosphere (Earth's)*
Full Width Half Maximum (FWHM) *see Atmosphere
            (Earth's) Filters, Spectroscopy*

GAIA ESA satellite 8, 13, 29, 60, 94, 106, 107, 119, 121,
            134–141, 218, 235, 236, 244, 262, 270, 317
Galaxies 1, 8, 30–32, 41, 98, 119, 127, 173, 178, 289,
            296, 317
    active 273, 276
    counts, interstellar absorption 263–265, 269
    Markarian 276, 278
    photometry 241, 242, 257, 262, 271
    radial velocities 127, 142
    rotation curves 207
    spectra 275
Galaxy *see Milky Way*
Galileo Galilei 1, 24, 56, 93, 213, 241
Galileo Global Navigation system (GNSS) 152
Galileo NASA mission 219
Galileo 1.2m telescope Asiago 235, 274
Galileo 3.5m telescope (TNG) 24, 30, 236
Galle Johan G. 216
    Γ relativistic parameter 104, 106
Garstang formula *see Atmosphere (Earth's)*
Gauss Carl Friedrick 5, 56, 196, 205
    constant 183, 184
    function 138, 168, 280
    groups 6, 14
    magnetic strength unit 265, 281, 296
Geocorona *see Atmosphere (Earth's)*
General relativity *see Relativity*
Geodesic, *see also Precession*
    orbit 223
    path 9
Geodetic *see Coordinate systems*
Gladstone-Dale law 166
Global Positioning System (GPS) 24, 152
GLONASS navigation system 152
GRAIL NASA lunar mission 214, 315
Grating, grisms 273, 274
Gravitation, gravitational
    assist 218, 219
    constant *G* 128, 177, 216
    field, force 47, 66, 74–79, 91, 92, 106, 128, 135,
            **173**–177, 184, 186, 195, 205, 206
    deflection of light 35, 71, 96, **104**–107, 133
    Earth's model EGM 20
    *vs.* inertial mass 211
    lensing 238
    perturbations 135, 149, 186, 222
    potential, energy 152, 167, 179, 219
    redshift 32, 128, 134
    waves 32, 208, 291
Gravity waves 159

Greenwich, *see also Times*
    ephemeris meridian 151
    meridian 9, 10, 14, 15, 27, 88, 154
    Observatory 9
Gregorius XIII Pope 53, 55
Guldberg-Wage mass function *see Molecules*

Halley Edmond
    comet (*see Comets*)
    Moon mean motion 211
    proper motions 125
    solar corona 225
Hanbury Brown Robert 171
Hansen Peter A. 149, 210
Harvard (HD) *see Starsspectral classification*
Haute-Provence Observatory 301
Heisenberg's uncertainty principle 279, 280
Helioseismology 33, 281
Henderson Thomas 121
Hermes, Hermean *see Mercury*
Herschel William
    (and John) stellar gauges 31
    proper motions 137
    Uranus 207
Herschel ESA IR telescope 238, 301, 318
Hertzsprung Einar 261
Hertzsprung–Russell (H–R) diagram *see Photometry*
Hill George W. 210
    sphere 218, 219
Hipparchus of Nicea 5, 27, 60, 61, 74, 116, 125, 211, 229,
            232, 241
Hipparcos ESA satellite 8, 13, 28, 29, 60, 94, 119, 121,
            130, 134, 136, 139, 236, 262, 317
Horizon *see Coordinate systems*
Hour angle *see Coordinate systems*
Hubble Edwin 127, 265
Hubble–Lemaître constant 127
Hubble Space Telescope (HST) 23, 74, 103, 114, 121, 129,
            159, 218, 242–244, 301, 317, 318
Huggins William Sir 273
Huygens Christiaan 155, 160, *see also Cassini, Cassin–
            Huygens spacecraft*
Hyades open cluster 136, 137, 262
Hydrogen 286, 288, 289, *see also Atmosphere (Earth's)*
    Balmer discontinuity, limit, lines 235, 256, 275, 277,
            280, 281, 285, 288, 292–296
    mass 185
    molecular 158, 185, 297
    21-cm (1420 MHz) line, galactic system 31, 31, 264
Hyperion moon of Saturn 223

Inequalities *see Moon*
Inglis-Teller formula 281
Integrals of motion 206
Interferometry
    Intensity (HBTII) 171, 172
    optical arrays 171, 203
    radio, VLBI 8, 28, 90, 91, 106, 134, 141, 154, 164,
            314, 317
    speckle 171
International Astronomical Union (IAU) 10, 179, 291, 314
    aberration theory 101
    catalogues, reference systems 28, 29, 32, 48, 63,
            74, 154

dwarf planets definition 195, 224
galactic coordinates 31, 40, 41, 134
nutation, precession, times 54, 56, 59, 67, 74, 75, 81,
        83, 90, 91, 146, 147, 151, 152, 154
    software (SOFA) 314
    solar parallax 116
    system of constants 28, 29, 60, 67, 74, 106, 184
International Celestial Reference Frame (ICRF) 28, 29,
        134, 141, 154, 314, 317
International Celestial Reference System (ICRS) 28, 29,
        48, 63, 154, 314
International Earth Rotation Service (IERS) 20, 28, 90,
        150–155, 314
Interstellar medium 165, 171, 263, 287
Io Jupiter moon 93, 183, 223, 242, 266
Ionization see Spectroscopy
IRAM mm telescope 300, 317

Jacobi constant, integral 217, 218
James Webb Space Telescope (JWST) 237, 244, 298
Jet Propulsion Laboratory (JPL) 29, 205, 221, 314–316
Johnson and Morgan photometric system see Photometry
Joy Alfred H. 286
Julius Caesar 53, 55
Jupiter planet, see also Comets
    apparent magnitude, phases 195, 242, 266
    atmosphere 298
    flattening, rotation 222
    magnetic field 32, 222
    mass, physical data 32, 79, 104, 175, 183, 241, 272, 289
    orbital data 122, 123, 184, 194–196, 215
    perturbations 59, 62, 63, 79, 106, 119, 175, 178,
        202–209, 215–219
    spectra 241, 298
    trojans 218

Kant Immanuel 149
Kapteyn Jacobus C. 138, 139
Kepler Johannes 149, 183, 215
    equation 189, 190, 202
    laws 43, 49, 59, 84, 94, 115, 116, 120, 149, 150, 175,
        178, 181, 183, 192, 198, 200, 221, 239
    orbit, rotation curve 120, 143, 194, 206–210, 215, 221
Kepler NASA mission 204, 225, 236, 237, 316
Kirchhoff first law 282
Kohlschütter Arnold 286
Kolmogorov Andrej N. 168
Kuiper Belt Objects (KBO) 75, 196, 234, 238, 270, 298,
        316
Kuiper NASA Observatory 238
Küstner Karl F. 89

Lacaille Nicolas-Luis de 115, 116
Lacaille star 293
LAGEOS geodetic satellite 90, 223, 314
Lagrange Joseph M. 21, 205, 216
    formalism 87
    function 174
    Lagrangian points 217, 218, 229, 238
    planetary equations 206, 210, 220, 221
Lalande Joseph Jérôme de 116, 212
Laplace Pierre-Simon 157, 191, 196, 205
    Earth's potential 83
    invariable plane 206

lunar acceleration 212
planet–moon plane 223
resonances 215, 223
Large Binocular Telescope (LBT) 171, 317
Large Magellanic Cloud (LMC) 24
Laser, see also Adaptive Optics
    ranging 29, 90, 107, 116, 134, 150, 212, 315
    transitions 282
Latitude see Coordinate systems
Lemaître Georges 127
Lense–Thirring effect 223
Leonardo da Vinci 229
Le Verrier Urbain 62, 116, 207, 208, 215, 216
Libra constellation 26, 30, 33
Librations see Moon
Light, see also Aberration, Atmosphere (Earth's),
                Gravitational deflection
    velocity 93, 94, 96, 184
Light-year 118, 236
Lilius Luigi 55
Limb darkening 195, 225, 234, 267
Lindblad Bertil 134, 263, 292
Line profiles see Spectroscopy
Local Standard of Rest (LSR) 134, 137, 138
Local Thermodynamic Equilibrium (LTE)
                see spectroscopy
Longitude see Clocks, Coordinate systems
Lorentz
    function 168, 280
    transformations 98, 99, 126
Luminosity see Photometry, Radiation Quantities,
                Spectroscopy
Luna 3 Soviet lunar mission 214
Lunar Orbiter NASA missions 214
21 Lutetia asteroid 75, 267, 316
Lyra constellation 120

M3 globular cluster 262
M31 galaxy 127
M57 (Ring Nebula) planetary nebula 276
Magnitude see Eclipses, Photometry
Main Belt (MB) see Asteroids
Markarian see Galaxies
Mariner 10 NASA mission 10, 315
Mars planet 10, 115, 116, 265
    apparent magnitude, phases 195, 242
    flattening, rotation 195, 222
    mass, density 79, 288
    moons 62, 219, 221, 223
    orbital data 75, 183, 184, 194–196, 208
    perturbations 215, 219, 221, 223
    spectra 296, 297
    trojans 218
Mascons 214
Maskelyne Nevil 28
Mass spectroscopy 273, 290
Matrix rotation, transformations see Coordinate systems
Mauna Kea Observatory 250, 269, 294
Maxwell's law 167, 185, 280
Mechanics see Dynamics
Medicean moons of Jupiter 24, 93, 116
Mercury planet 10, 115, 222, 269
    advance of perihelion 104, 173, 205–208, 223
    apparent magnitude 242

Mercury planet (*cont.*)
    atmosphere 185, 251
    flattening, rotation 222, 223
    gravitational deflection of light 105, 106
    mass, physical data 183, 184, 288
    orbital data, elongations 29, 184, 194, 195, 196, 222
    perturbations 222
    seasons 223
    spectra 296, 297
    transits 104, 194, 204, 225, 315
Meridians *see Coordinate systems*
Messenger NASA mission 10, 315
Meteorites 289, 290
Metonic cycle *see Eclipses*
Mie scattering of light 263
Mile 10, 15
Milky Way, Galaxy 1, 119, 138, 142, 159, 236, 301
    absorption, extinction 257, 263, 264 (*see also*
        *galaxies counts*)
    center 31, 32, 128, 138, 141, 182
    force field, tide 31, 59, 133, 135, 139, 186
    magnetic field, polarization 265
    rotation 140, 141 (*see also Oort's constants*)
    velocity field 126, 127, 138 (*see also Local
        Standard of Rest*)
Milne Edward A. 285
Mirror 29, 42, 94, 171, 203, 270, 274
    reflectivity 243–245
Modified Newtonian dynamics (MOND) 207
Molecules, *see also Atmosphere (Earth's), Comets*
    dissociation 251, 287, 299
    equilibrium distance 287
    ionization 287, 299
    sublimation 299
    velocity distribution 185
Month
    calendar 53, 55
    date 54–57
    lunar 17, 61, 81, **211–213**, 231
Moon, *see also Earth–Moon, Eclipses, Month,
        Precession, Nutation, Laser ranging, Tides,
        Times*
    albedo, apparent magnitude 226, 242, 257, 266
    angular size 8, 211
    ashen light, cinerea lux 229
    atmosphere 185
    cartography 32
    Cassini's laws, states 213, 219, 223
    dimension, distance, parallax 10, 13, 17, 43, 44, 79,
        111–116, 121, 150, 184, 211, 212
    illusion 170
    inequalities 211
    librations 205, 214, 232
    mass, density 79, 116, 184, 288
    occultations 104, 213, 225, 233–238
    orbit, perturbations 24, 25, 30, 47, 94, 116, 150,
        205–215, 239
    phases, syzygies 116, 195, 211, 212, 225, 226
    refraction, rise and set 43, 44, 114, 170, 171
    rotation 222, 223
    spectra 296
Morgan–Keenan–Kellman (MKK, MK), *see
        Starsspectral classification*
Mullard Radio Telescope 155

NEAR NASA mission 115, 315
Near Earth Asteroids, Objects (NEAs, NEOs) 296, 315
Nebulium 274, 300
Neptune planet
    discovery 215, 216
    flattening, obliquity, rotation 222
    mass, physical data 184
    orbital data 184, 195, 196, 216, 223, 298
    spectra 298
    trojans 218
Newcomb Simon 60
    ecliptic 62
    ephemeris time 149
    Eulerian free rotation 89
    precessional angles 62, 71, 76
    solar oblateness 208
    solar parallax 116
    solar time 49, 52, 146, 147, 151
    tropical year 52
New Horizons NASA mission 75, 121, 224, 297, 316
New Technology Telescope (NTT) ESO 171
Newton Isaac 56
    dynamics 77
    lunar theory 210
    luni-solar precession 61
    spectroscopy 273
NICE model of the Solar System 215
Night
    astronomical,civil 42, 43
    sky 247, 251, 269, 276, 300
    vision 247
Non-gravitational forces, perturbations 186, 208, 221
Nutation 27–29, 59, 62, **68–74**, 133, 145, 314, *see also
        Aberration, Precession, Times*
    Bradley's discovery 27, 61, 62, 96, 98
    dynamics **77–91**
    Eulerian or free 19, 59, 60, 89, 90
    IAU theory 29, 59, 74, 75, 81, 91
    planetary 62

Obliquity
    of ecliptic 17, 26, 29, 31, 37, 61–70, 79–81, 86, 102
    of Mars 62
    of planets and Pluto 222
Occultations 149, 171, 195, 204, 211, **225**, 315
    KBO 238
    lunar 213, 235, 238, 247
    stellar 233, 235
Octans constellation 24
σ Octanctis 24, 61
Oort Jan H. 140, 141, *see also Comets*
    constants 134, **140–143**
    galactic extinction 263
    galactic force field 186
Ophiuchus constellation 30
Oppolzer Theodor von 62, 88, 89, 238
Opposition effect 266
Optical depth **249**
Optical Gravitational Lensing Experiment (OGLE)
    141, 238
Optics
    active 39, 316
    adaptive 39, 159, 169–171, 203, 316
OSIRIS imaging system on Rosetta 248, 249, 267

Parallax
  annual **116**–118, 121, 130, 230
  diurnal 10, 15, **110**–114, 122
  dynamical **120**, 203
  from radial velocity and angular expansion 120, 141
  geocentric, topocentric 111, 115
  group parallaxes **135**
  horizontal equatorial 110–114, 229
  interplanetary 121
  lunar 10, 43, 44, 111, **115**, 116, 121, 211, 228–230
  parallactic factors 114, 115
  parallactic inequality 116, 211
  secular **120**, 137–139
  solar 43, 93, 96, **111–118**, 121, 211, 228–230, 239
  spectroscopic 286
  statistical 139
  stellar 71, 96, 106, 110, 118–121, **131–136**, 203, 235, 239, 260
  trigonometric **109**, 125, 130
Parallels *see Coordinate systems*
Parsec 118
Partition functions, ions *see Spectra*
Pauli exclusion principle 282
Penumbra *see eclipses*
Perigee **47**, 48–54, 95, 100, 129, 149, 155, 210
  dates of **155**
  lunar 209–214, 228
  motion 149
  satellite 221
Perihelion 54, 55, 80, 94, 95, 100, 104, 115, 128, 178, 182, 189, 192, 193, 300
Period, synodic, sidereal **195**
Perturbation theories 205, *see also Earth rotation, Earth-Moon, Ecliptic Planets*
Phase angle, function 226, 266, 267, *see also Moon*
Phobos moon of Mars 219, 221, 223
Photometry 157, 238, **241**, 247, 251, 257, 267, 270, 271, 294, 317, 318
  absolute magnitudes, luminosity **258**–260
  apparent magnitudes, luminosity 241–243
  black-body color indices 256
  bolometric magnitudes, luminosity 243
  calibration in physical units **257**
  color excess 263
  color indices 255
  distance indicators 120
  H–R diagrams 136, **261–264**, 270
  photometric diameters **258**, 261
  photometric sequences 241
  photometric systems 244, 246, 260, 262, 266, 268, 270, 318
  reddening-free parameter 264
  relationship with spectral classification **294**
  surface brightness 242
  two–color diagrams **255–257**, 262, 264, 294
Piazzi Giuseppe 121, 196
Picard Jean 115
Pickering Edward C. 291, 292
Pisces constellation 30, 60
Plana Giovanni A. 210, 212
Planck's law, *see Black Body*
Planets 17
  cartography 10, 32
  configurations **194**, 195

catalogues, reference systems 28, 29, 32, 48, 63, 74, 154
  ephemerides 29, 104, 146, 150, **189**, 192, 204, 205, 314, 316
  flattening, obliquity rotation 221, **222**
  masses, physical data 79, 184
  orbital data 184, 195
  perturbations **205**
PLATO ESA satellite 238, 316
Pluto dwarf planet, *see also New Horizons*
  classification 195, 216
  discovery 62, 216, 224
  flattening, obliquity, rotation 221, 222
  mass, physical data 184
  orbital data 29, 184, 195, 223, 297
Poincaré Henry 205, 207, 215
Poinsot Louis 88
Polarization 29, **263**, 265, 267, 276, 281
Poles *see Coordinate Systems, Earth free rotation, Nutation, Precession*
Precession 19, 27, 53, **59–75**, *see also Dynamics of Earth's rotation, Nutation, Proper Motions and Radial Velocities*
  Bessel's daily numbers 71
  constants, theory 28, 29, 48, 52, 59, 67, 74, 75, 133 (*see also IAU*)
  Copernicus 60, 61
  equinox 60–70, 74–79, 86, 94, 96, 104, 119, 122, 123, 133, 145–149, 153, 154, 213
  general 27, 62, 63, 66, 70, 91, 132, 213
  geodesic 71, 133, 134
  independent day numbers 71, 102
  Newcomb angles 62, 71, 76
  Newton's explanation 61
  planetary 62, 66
  position angles 73
  Solar System bodies orbital elements 73, 74
Prisms, objective prisms 165, 274
Proper motions 93, 118, 120, 125, **129**, 130, 133, 141
  convergence 136, 137
  equatorial components 125, 126, 131
  exoplanets 236
  interplay with parallaxes 120, 130, 139
  interplay with precession 132–134
  perspective acceleration 132
  position angles 131
  Sirius A and B 199
Proxima Centauri 117, 121
Proxima-b, exoplanet 121, 236
Ptolemaeus (Ptolemy) Claudius 5, 50, 212, 241
Pulsars 106, 155, 171, 235

Quantum efficiency QE 247–249, 270
Quantum numbers 281
Quasi Stellar Objects (QSOs, quasars) 28, 98, 127, 134, 142, 154, 166, 257, 265, 275, 276

Radar 106, 113, 115, 116, 150, 196, 212, 222, 223, 300
Radiation
  energy 160, 168, 169, 241, 243, 245, 252, 254, 255, 258, 260, 271
  pressure 221, 186
  quantities **268**, 269
Radioactivity 272

Radioastronomy, signals, telescopes 8, 31, 90, 104, 140,
            152, 158, 164, 171, 172, 218, 221, 235, 251, 253,
            265, 300, 318, *see also VLBI*
        sources 28, 31, 32, 106, 127, 128, 152, 155, 165,
            172, 225
Rayleigh scattering of light 250, 251, 263
Reference systems *see Coordinate systems*
Reflectivity *see Asteroids, Mirrors*
Refraction *see Atmosphere (Earth's)*
Relativity
    Galilean 99, 178
    general 71, 94, 104–107, 120, 127, 128, 133, 145, 152,
            173, 178, 205, 208, 212, 225
    special or restricted 98, 99, 126, 127
Resonances 215, 222, 223
Retro-reflectors 107, 116
Roche's curve 217, 218
Rods *see Eye*
RömerÖleg 93, 96
Roque de los Muchachos Observatory 158, 165, 251
Rosetta ESA mission 75, 219, 248, 249, 266, 267, 300, 316
Rosette movement 176, 177
Rotation, *see also Earth, Energy, Planets, Milky Way*
    matrix 11, 35, 36
    synchronous 213, 223, 238
Rowland Henry A. 273
Russell Henry N. 255, 285, *see also H–R diagram*
    formulae 255, 261

Sagittarius constellation 30
Sgr A*, 128
Saha equation **284**, 285, 287
Santini Giovanni 28
Satellites
    artificial 83, 109, 111, **185**, 208, 219–224
    natural 223
Saturn planet, *see also Cassini, Cassini–Huygens
                spacecraft*
    apparent magnitude 242
    flattening, obliquity, rotation 222, 223
    mass, density physical data 79, 184, 289
    orbital data 79, 119, 122, 123, 184, 195, 298
    perturbations 104, 205, 206, 215, 219
    rings 274
    seasons 298
    spectra 274, 298
    trojans 218
Scattering
    dynamical 219
    of light 229, 250, 251, 263, 265, 266, 273, 276
Schalen Carl 292
Schlesinger Frank 121, 134, 141
Schwarzschild Karl
    black holes, radius 104, 128, 212
    stellar currents, motions 138
Scintillation *see Atmosphere (Earth's)*
Scorpio constellation 30
Seasons 43, 53, **54**, 149, 165
Secchi Angelo S.J. 225, 273
Seeing *see Atmosphere (Earth's)*
Seeliger Hugo von 134
Sensitivity *see Detectors*
Series developments, expansions 2, 4, 13, 21, 82, 127,
            133, 141

Shapiro delay 106, 208
Sidereal *see Month, Time, Year*
Signal-to-Noise ratio (S/N) 271
Sirius (α CMa) 128, 198, 199, 203, 242, 247, 253, 254
Sirius B white dwarf 128, 198, 199, 289
Snell's law 160
Solar Dynamic Observatory (SDO) 204
Solar Heliospheric Observatory (SOHO) 218, 316
Solar System 1, 8, 317, 318, *see also* Aberration, Deflection
            of light, JPL. LSR, Titius-Bode law
    angular momentum 281
    barycenter,ephemerides, reference systems 28–31, 59,
            73, 119, 122, 129
    chemical composition 289
    geocentric model 1
    heliocentric model 1
    invariable (Laplace) plane 206, 223
    position in the Galaxy 31
scale, structure, age 31, 78, 89, 115, 178, 195, 214, 215,
            219, 223
    secular motion, apex 120
Sosigenes 53
Spectroscopy 300
    absorption bands 275, 288, 297, 298
    absorption lines 256, 263, 273, 275–277, 285, 288, 292
    collisional damping 280, 281
    curve of growth 279, 280
    detailed balance condition **281**, 282
    dispersion,reciprocal dispersion, resolution 246,
            273–277, 285, 288, 291
    Doppler broadening 279–281
    electron pressure 280, 285, 286
    emission bands 273, 275, 298
    emission lines 235, 251, 257, 275–277, 291–296, 300
    equivalent area, width 277–279
    gas pressure 286
    ionization 273, 277, 283–287, 299
    line strength 282
    Local Thermodynamic Equilibrium (LTE) 282, 284
    oscillator strength 279, 280, 282
    partition function 283, 284, 287
    spectrographs 168, 274, 275, 301
    spectrophotometry 166, 251, 276, 300
    spectropolarimetry 251
Stark broadening 280, 281, 285
    Zeeman effect 280, 281, 296
Spencer Jones Harold 94, 116
Spherical astronomy **1**
    trigonometry, triangles **5**
    excess 6
SS433 variable star 127
Stars, *see also Photometry, Spectroscopy*
    atmospheres 279, 281, 286, 288
    binaries (*see Binary stars*)
    chemical composition 262, 273, 279, 285, 286, 288
    circumpolar 24, 25, 28, 39, 45
    diameters 234, 235, 258–261
    magnetic fields 288
    magnitudes, color indices (*see Photometry*)
    masses 218, 238, 285
    motions, currents **135**, 136
    names 30, 32, 33
    solar analogs 258
    spectral classifications **288**, 291

spectral atlases and libraries 301, 318
Stefan–Boltzmann law 199, 253, 254, 260–262, 288
2867 Steins asteroid 75
Strehl ratio *see Atmosphere (Earth's)*
Strömgren Bengt G.D. 138
Struve Friedrich G.W. von 62, 123
Struve Otto L. 300
Subaru telescope 294
Sun 1, *see also Coordinate systems, Nutation, Precession,
        LSR, Times*
    aberrated 93, **94**, 96
    absolute and apparent magnitude, luminosity 260, 261
    angular diameter 8, 47, 114, 260
    chromosphere 185, 225, 235, 275
    color indices, bolometric correction **258**, 259
    corona 165, 185, 225, 273, 275, 284, 285, 300
    culmination, transit 43
    day, night 25, 42, 43
    diameter, radius, solid angle 184, 260
    distance, parallax 111, **116**
    ecliptic, orbit 5, 8, 17, 25, 28, 42, 47–51, 56, 63
    equinoxes, solstices, seasons 25, 26, 54, 55, 60
    escape velocity 184
    fictitious, mean 49–54
    flares 281
    galactic position 32, 106, 140
    gravity, mass, $GM\odot$, 78, 79, 84, 104, 115, 128, 178, 184
    heliographic coordinates 32, 33
    peculiar motion **120**, **137**
    photosphere, temperature 252, 260, 275, 284
    precession 19, **60**, 61, 63, **77**–80, 84–86, 91
    rotation 32, 33, 208
    sidereal time 42, 48, 56
    solar time, calendars 48, 49, 52, 53, 56
    spectrum, classification 249, 273, 277, 291, 292
    wind 159, 185, 299, 300
Swan bands 273

TAI *see Times*
Tatarski V.I. 168
Taylor series expansion 3, 141
Telescopes mechanical mounts 27, **29**, 38, 39, 42, 170
Telescopio Nazionale Galileo (TNG) 24, 29, 30, 236, 237,
        244, 297
Terminator 195
Three-body problem 205, 206, 216, 218, 219
Tides 60, 79, 90, 149, 150, 186, 205, 212
Times
    astronomical 47, 48, **145**, 147
    atomic (TAI) 47, 145, 146, **152**
    of the barycenter (TDB, TCB) 153, 155
    coordinate, proper 126, 145, 152, 155
    ephemeris (ET) 146–153
    equation of (E) **48**, 49, 51, 56, 95
    Greenwich apparent sidereal time (GAST) 147,
        153, 154
    Greenwich mean sidereal time (GMST) 43, 146, 147,
        153, 154
    sidereal (*ST*) 27, 37–44, **48**–59, 103, 122, 145, 147, 151
    solar (T⊙) 44, **48**–52, 146, **147**, 151
    terrestrial (TDT, TT) 147, 150, 152, 153
    universal (UT, UT0, UT1, UT2, UTC) 27, 29, 43,
        52–57, 146, 147, 150–152
    zones 110

Timocharis 60
Tisserand Félix 214
    criterion 218–220
Titius–Bode law **194**–**196**, 204, 219
Tombaugh Clyde 216
Transformations
    coordinates 7, 11, 23, 32–44, 64, 73, 101, 145
    errors 45
    Galilean 97
    Lorentz 98, 99, 126, 154
    projective 200
Transiting Exoplanet Survey Satellite (TESS) NASA
        mission 225, 236, 237, 316
Transmissivity
    atmosphere 243
    filters, optics 243, 245, 270
Trans-Neptunian Objects (TNO) 196, 224, 225, 234, 297
TRAPPIST telescope 225, 236
Trappist-1 exoplanet 236
Triangles, Trigonometry *see Spherical Astronomy*
Triton moon of Neptune 195
Trojans 270, *see also Jupiter, Mars, Neptune,
        Saturn,Uranus*
Trumpler Robert J. 31, 263
Two-body problem **173**, 178, 186, 206
Two-color diagrams *see Photometry*

Ulysses spacecraft 219, 220, 316
Ultima Thule *see 486958 Arrokoth*
Umbra *see Eclipses*
Universe 120, 264
    chemical composition, evolution 289
    expansion, age 127, 128, 242, 262, 265, 269, 289
Uranus planet
    discovery 195
    flattening, obliquity, rotation 221, 222
    mass, radius, physical elements 184, 195, 196
    perturbations 207, 215, 216
    rings 225
    spectrum 298
    trojans 218

Vector calculus 11
Vega *see α Lyr*
Velocity
    angular 12, 48
    areal 48, 175, 178
    astrometric **134**
    circular 140, 184, 186
    escape 167, 184, 185
    indicative, cosmological 128
Venus planet
    advance of perihelion 208
    apparent magnitude, albedo 242, 260, 282
    atmosphere 297
    catalogues, reference systems 28, 29, 32, 48, 63, 74, 154
    flattening, obliquity, rotation 195, 222
    mass, physical data 79, 183, 184
    orbital data 79, 184, 194–196
    parallax 111
    perturbations, resonances 59, 62, 63, 104, 223
    transits 115, 204, 225, 315
Very long baseline radio-interferometers (VLBI) 8, 28, 29,
        90, 91, 106, 134, 141, 154, 164, 314, 317

Vesta asteroid 62, 75, 196, 316
Virgo constellation 30, 32, 127
Virtual Observatory (VO) 318
Vis viva integral 180, 181
Voyagers NASA missions 106, 223, 225, 316
Vulcan, vulcanoids 208

Weaver Harold F. 31
WGS84 20, 24, 31, 153, 154
White dwarfs 128, 138, 199, 256, 262, 276, 281, 289,
        291, 296
Wurm Johann F. 195

Year 48, **147**, *see also Calendar*
    anomalistic **149**, 155

Besselian 32, **53**, 54, **148**
civil 47
draconic, draconitic, eclipse **149**
Gaussian **149**, 184
Julian 53, 54, 67
leap, (Latin, bi-sextus) 53, 54, 55
platonic 60
sidereal 89, 94, **148**, 183
tropical 17, 28, 49, **52–54**, 61–64, 80, 147, 148, 151

Zeeman effect *see Spectroscopy*
Zeipel Edward H. von 210
Zodiac 30
Zodiacal light 269, 271, 296
Zwier Hendrikus J. 200

Printed in the United States
by Baker & Taylor Publisher Services

Printed in the United States
by Baker & Taylor Publisher Services